A Pocket Book of Marine Engineering Rules and Tables

SEATON & ROUNTHWAITE'S

MARINE ENGINEERING POCKET-BOOK.

A POCKET BOOK OF
MARINE ENGINEERING
RULES AND TABLES.

FOR THE USE OF

MARINE ENGINEERS, NAVAL ARCHITECTS,
DESIGNERS, DRAUGHTSMEN,
SUPERINTENDENTS,

AND ALL ENGAGED IN THE DESIGN AND CONSTRUCTION OF

MARINE MACHINERY, NAVAL & MERCANTILE.

BY

A. E. SEATON, M.INST.C.E., M.INST.MECH.E.,
M. OF COUNCIL INST.N.A.,

AND

H. M. ROUNTHWAITE, M.INST.MECH.E., M.INST.N.A.

THIRD EDITION, REVISED AND ENLARGED.

WITH DIAGRAMS.

LONDON: CHARLES GRIFFIN & COMPANY, LIMITED;
EXETER STREET, STRAND.
NEW YORK: D. VAN NOSTRAND COMPANY,
23 MURRAY STREET AND 27 WARREN STREET.
1895.

PREFACE.

——:o:——

A special Pocket-book of Memoranda, Tables, &c., has long been a desideratum with Marine Engineers. In the existing pocket-books, Marine Engineering matters are only dealt with generally, and such information as is given is in some cases very restricted, in others obsolete, and in all too scattered to be useful. We, ourselves, have experienced this want, and have heard on all hands the desire expressed for a Pocket-book in which Marine Engineering questions are dealt with thoroughly, are easy to find, and not "mixed up" with general information in such a way as to render the seeking of them difficult and tedious.

We therefore trust that in presenting this book to the public we have not only fulfilled the task we set ourselves, but have supplied this long felt want in a manner that will prove satisfactory to all engaged in Marine Engineering affairs. While we have been careful to make the book of special value to Marine Engineers, we have omitted nothing, so far as we know, that would be of use and importance to others having to do with Ships and their machinery; at the same time, we have avoided the introduction of extraneous matter of only general interest, which would make the volume so bulky, and the arrangement of it so complex, as to very materially detract from its usefulness. Hence, we have, while not altogether neglecting past experience, but omitting information now almost only historic, devoted our attention generally to the most modern and approved practice.

We have dealt with steel as the material in general use, and not, as heretofore, an exceptional thing to be found only in high-class structures; the Tables of Weights, &c., are, therefore, given fully for this material.   *a*

Inasmuch as the practice in a considerable part of the Mercantile Marine is now more nearly approaching that followed in Naval ships, as to speed and economy of weight, than was formerly the case, the information and formulæ pertaining to light fast-running machinery have been elaborated and based on the most recent practice of the leading firms of Manufacturing Engineers.

In conclusion, we trust that the book may be received favourably, and found of use by practical men, and that any shortcomings may be overlooked on the score that it is the production of the spare moments of busy men, rather than of those having ample leisure.

A. E. S.
H. M. R.

November, 1893.

GENERAL TABLE OF CONTENTS.

LIST OF TABLES.

MARINE ENGINEERING RULES
AND TABLES.

HORSE-POWERS.

Nominal Horse-power, as understood by Watt, was both a measure of the commercial value of an engine, and of the power it might be expected to develop in ordinary work. Gradually, however, as it became possible to construct boilers to supply steam at higher than atmospheric pressure, the powers developed by engines exceeded the nominal horse-powers, until it was necessary to alter Watt's rule, or to devise another that would give values more in accordance with facts. Alterations of this kind were made from time to time, by various private persons and public bodies, in different parts of the country; but they rarely obtained general acceptance, and, being without any proper scientific basis, were soon rendered useless by the rapid advances made in engineering construction.

Thus, for many years, and until about ten years ago, a marine engine was expected to indicate about five times its nominal horse-power. The rule then in use was as follows:—

$$N.H.P. = \frac{\text{Sum of squares of piston diameters}}{30 \text{ to } 33 \text{ (according to district)}}$$

boiler pressure, piston speed, &c., being left entirely out of consideration. It is, of course, hardly necessary to say that such a rule was quite useless for any scientific purpose; whilst, even for commercial purposes, it gave only a very imperfect idea of the relative values of different engines. Its use, however, still survives in some districts, the divisor being 30, the normal stroke ·618 of diameter of L.P. cylinder, and the normal heating surface 16 square feet per N.H.P.

The matter was still in this chaotic condition when, in 1883, Mr Seaton, in his "Manual of Marine Engineering," suggested the use of E.H.P., or **Estimated Horse-power**, and gave the following formula for calculating it:—

Rule 1.
$$E.H.P. = \frac{D^2 \times \sqrt{P} \times R \times S}{8500}$$

where D is diameter of L.P. cylinder; P, absolute boiler pressure; R, revolutions per minute; and S, stroke in feet.

1

This formula, which had already been tested by ten years' use, was at once adopted by several firms,—some employing E.H.P. in lieu of N.H.P., and others using $\dfrac{\text{E.H.P.}}{5}$ as being more easily comparable with the previously used N.H.P.

About 1887, the then recently-formed North-East Coast Institution of Engineers and Shipbuilders turned attention to the matter, and, towards the end of 1888, issued the report of their committee on the subject, which,—further developing the E.H.P. idea,—proposed the following very complete formulæ as a solution of the question :—

Rule 2. $\text{N.I.H.P.} = \dfrac{(D^2\sqrt[3]{S} + 3H)\sqrt[3]{P}}{100}$

Where N.I.H.P. = Maximum normal indicated horse power, on loaded trial trip, of surface-condensing screw engines, working at any pressure between 50 and 250 lbs., under "normal" conditions.

     D = Diameter of L.P. cylinder, in inches (if more than one D^2 must equal sum of squares).

     S = Stroke, in inches.

     P = Working pressure, in lbs., above atmosphere.

     H = Heating surface of boilers in sq. feet.

     P_m = Mean pressure, in lbs., referred to L.P. cylinder.

The conditions assumed as "normal" are as follows :—That

(1) Steam of all pressures is expanded down to the same terminal pressure ;

(2) Expansion is effected with same degree of efficiency for all pressures ;

(3) Piston speeds are proportional to cube roots of strokes, and, further, actual loaded trial-trip piston speed may be taken as $144\sqrt[3]{S}$;

(4) In all cases where relative proportions of engine and boiler prevent (1) being fulfilled without violating (3), the coal consumption will not be affected, but will be constant for the same boiler pressure ;

(5) Boilers are of usual proportions and construction, and the horse-power proportional to heating surface (H), and to cube root of pressure ($\sqrt[3]{P}$); and further, actual loaded trial-trip horse-power may be taken as $\dfrac{H\sqrt[3]{P}}{16}$;

(6) Efficiency of engine mechanism is constant, and the propeller such that engines will utilize boiler power, referred to in (5), in manner prescribed in (3) and (4).

As a result of (1) and (2), it follows that mean pressure referred to L.P. cylinder (P_m) may be assumed as proportional to cube root of boiler pressure ($\sqrt[3]{P}$), and further, that its actual loaded trial-trip value may be taken, without sensible error, as $5.6\sqrt[3]{P}$.

The normal relation between engines and boilers is expressed by the equation.

$$H = \frac{D^2 \sqrt[3]{S}}{3 \cdot 25}.$$

It is remarked that the results obtained by the above formula for N.I.H.P., if divided by 6, give quantities comparing very nearly with those given by the old nominal horse-power rule.

It is also claimed that,—for machinery of same type and design, proportions and arrangement, built of similar materials, under similar circumstances, to similar factors of safety, and not differing very widely in size,—the weights and costs will vary almost exactly as N.I.H.P.

For paddle engines the same rule may be used, with the co-efficients altered to suit the piston speeds usual for these engines. Assuming that, under (4), piston speed of paddle engines may be taken at $90\sqrt[3]{S}$, the rule may, without sensible error, be written,—

Rule 3. $\quad$ N.I.H.P. $= \dfrac{(D^2 \sqrt[3]{S} + 5H) \sqrt[3]{P}}{160}$

and the normal relation between engines and boilers will be expressed by,—

$$H = \frac{D^2 \sqrt[3]{S}}{5 \cdot 2}.$$

It is, of course, evident that for any rule to be general in its application, or to be possible at all, all engines must be assumed to conform to certain "normal" conditions ; and it also seems clear that relative values of engines and boilers can be most conveniently and aptly expressed in terms of horse-power that would be indicated under such conditions ; and it would therefore appear that the above rules satisfactorily comply with these preliminary considerations, and, having otherwise a thoroughly sound scientific basis, they should come into general use, and be of great value, both to professional and commercial men.

Another rule, now very generally accepted in the North East Coast district is,—

Rule 4. $\quad$ N.H.P. $= \dfrac{D_1{}^2 + D_2{}^2 + D_3{}^2}{92} \times \sqrt[3]{S}$

where $D_1 D_2 D_3$ are the diameters of the cylinders in inches and S,—the stroke,—also in inches.

The "Standard practice" of that district, for triple engines, working at 160 lbs. pressure, is based upon this rule, and the proportions adopted are shown by the following Table :—

Table I.—Sizes of cylinders, &c., and corresponding N.H.P.

N.H.P.	Diars. of cyls. in inches.			Stroke in ins.	N.H.P.	Diars. of cyls. in inches.			Stroke in ins.
	H.P.	M.P.	L.P.			H.P.	M.P.	L.P.	
20	8	13	22	18	140	19	31½	51½	36
30	9½	15½	26	21	150	20	32½	53	36
40	11	18	30	21	160	20	33	54	39
50	12½	20	33	24	170	20½	34	56	39
60	13½	22	36	24	180	21	35	57	39
70	14	23	38	27	190	21½	36	59	39
80	15½	25	40	27	200	22	36½	60	42
90	16	26	43	30	225	23	38	63	42
100	16½	27	45	30	250	24½	40	66	45
110	17	28	46	33	275	25	41½	69	45
120	18	29	48	33	300	26	43	71	48
130	18½	30	50	33	...	...	...	...	...

For registration purposes the Board of Trade still use the old rule for nominal horse-power above referred to, viz.:—

Rule 5. $$\text{N.H.P.} = \frac{S}{30}$$

where S is the sum of the squares of the diameters, in inches, of all the cylinders.

The rule now used by Lloyds, for determining amount of survey fees, &c., was brought into use about three years ago, and is as follows:—

Rule 6.   N.H.P. (of Triple engines) $= \frac{1}{2}\left(\frac{D^2 \times \sqrt{S}}{100} + \frac{H}{15}\right)$

where D = diameter of L.P. cylinder in inches.
       S = stroke in inches.
       H = heating surface in square feet.

Indicated Horse-power.—The indicated horse-power of an engine may be defined as the measure of work done in the steam cylinder, as shown by the indicator diagrams, and is equal to (area of piston in square inches × mean pressure in lbs. per square inch × number of feet travelled through by piston, per minute) ÷ 33,000; or,—

Rule 7. $$\text{I.H.P.} = \frac{A \times P \times S}{33,000}$$

Piston speed, S, is equal to stroke in feet × 2 × number of revolutions per minute.

In the case of engines having more than one steam cylinder, the I.H.P. of each cylinder is determined separately, and the sum of these is the I.H.P. of the engine.

Where accuracy is required, the sectional area of any piston rod, or rods, should be deducted in calculating the areas of pistons.

The mean pressure is determined from the indicator diagrams as follows :—

Let fig. 1 represent a pair of diagrams from the L.P. cylinder of a compound engine. Draw two perpendiculars to the atmospheric line AB, one at each end of, and touching the diagrams. Then divide the space between these perpendiculars into ten equal parts, placing the division marks so that there shall be half a space at each end, and draw a vertical ordinate through each mark. Then take the scale corresponding to the spring used in the indicator, measure off the breadths of the diagrams at each ordinate, and figure them on ends of ordinates as shown, keeping the figures referring to each diagram in a separate column.

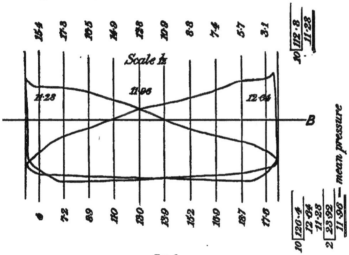

Fig. 1.

If these columns of figures are now added up, and the sum of each divided by ten, two mean pressures are obtained,—one of which refers to each side of the piston,—and the mean of these two is the mean pressure required.

In noting down the breadths of diagrams it is convenient and accurate enough for all ordinary purposes, to use one decimal place as shown.

Planimeter.—Where there are many diagrams to be calculated or compared with one another, it is quicker to use a planimeter in place of the method given above. The method is as follows :—Measure the

area of the space ... means of the instrument, and divide it by the
length ... when the quotient will be the mean breadth in inches;
... multiplied by the 'scale' of the spring used (the number
of pounds required to compress it one inch, will give the mean
pressure required. For purposes of comparison it is, of course, sufficient
to use the areas of the figures.

The Index averaging instrument is a form of planimeter which has
been specially designed for dealing with indicator diagrams. It leaves
at the arms two needle points, the distance between which is the mean
breadth of the figure,—and thus performs mechanically the process of
dividing area of figure by length.

The following equivalents may be useful in calculations connected
with the above :—

> Pounds per sq. inch × ... = Kilogrammes per sq. centimetre.
> Kilogrammes per sq. centimetre × 14·22 = Pounds per sq. inch.

> Foot-pounds × ... = Kilogrammetres.
> Kilogrammetres × ... = Foot-pounds.

> Horse-power × ... = Chevaux.
> Chevaux × ... = Horse-power.

See also "Tables of Pounds per square inch and Kilogrammes per
square centimetre," pages 353 and 354.

The Continental "Cheval" is equal to 4500 kilogrammetres, or
32,546 foot-pounds per minute, as against 33,000 foot-pounds per
minute,—the value of the English "horse-power."

The following Table will considerably facilitate the computation of
indicated horse-power :—

Table II.—Constant Multipliers for I.H.P.

Constant = $\dfrac{\text{area of cylinder}}{33,000}$; and I.H.P = constant × mean press. × piston speed.

Diameter of Cylinder.	Constant.	Diameter of Cylinder.	Constant.	Diameter of Cylinder.	Constant.	Diameter of Cylinder.	Constant.
6	·00085	16½	·00648	34	·02751	60	·08569
¼	·00092	¾	·00668	½	·02833	61	·08856
½	·00100	17	·00688	35	·02916	62	·09149
¾	·00108	¼	·00708	½	·02999	63	·09447
7	·00116	½	·00728	36	·03084	64	·09748
¼	·00125	¾	·00749	½	·03171	65	·10055
½	·00134	18	·00771	37	·03258	66	·10368
¾	·00143	¼	·00792	½	·03347	67	·10684
8	·00152	½	·00814	38	·03437	68	·11005
¼	·00162	¾	·00836	½	·03528	69	·11332
½	·00172	19	·00859	39	·03620	70	·11663
¾	·00182	¼	·00882	½	·03713	71	·11998
9	·00192	½	·00905	40	·03808	72	·12339
¼	·00203	¾	·00928	½	·03904	73	·12683
½	·00214	20	·00952	41	·04001	74	·13033
¾	·00226	½	·01000	½	·04099	75	·13388
10	·00238	21	·01049	42	·04198	76	·13748
¼	·00250	½	·01100	½	·04299	77	·14112
½	·00262	22	·01152	43	·04401	78	·14481
¾	·00275	½	·01205	½	·04504	79	·14854
11	·00288	23	·01259	44	·04608	80	·15232
¼	·00301	½	·01314	½	·04713	81	·15615
½	·00314	24	·01371	45	·04820	82	·16003
¾	·00328	½	·01428	½	·04927	83	·16398
12	·00342	25	·01487	46	·05036	84	·16795
¼	·00357	½	·01547	½	·05146	85	·17198
½	·00372	26	·01609	47	·05257	86	·17604
¾	·00387	½	·01671	½	·05370	87	·18016
13	·00402	27	·01735	48	·05483	88	·18432
¼	·00417	½	·01800	½	·05598	89	·18853
½	·00433	28	·01866	49	·05714	90	·19280
¾	·00449	½	·01933	½	·05832	91	·19710
14	·00466	29	·02001	50	·05950	92	·20146
¼	·00483	½	·02071	51	·06191	93	·20587
½	·00500	30	·02142	52	·06436	94	·21030
¾	·00517	½	·02214	53	·06685	95	·21480
15	·00535	31	·02287	54	·06940	96	·21937
¼	·00553	½	·02362	55	·07200	97	·22394
½	·00571	32	·02437	56	·07464	98	·22859
¾	·00590	½	·02513	57	·07733	99	·23328
16	·00609	33	·02592	58	·08006	100	·23799
¼	·00628	½	·02671	59	·08285	101	·24280

area of the figure by means of the instrument, and divide it by the length AB, when the quotient will be the mean breadth in inches ; and this, multiplied by the "scale" of the spring used (the number of pounds required to compress it one inch) will give the mean pressure required. For purposes of comparison it is, of course, sufficient to note the areas of the figures.

The Coffin averaging instrument is a form of planimeter which has been specially designed for dealing with indicator diagrams. It leaves on the card two needle pricks, the distance between which is the mean breadth of the figure,—and thus performs mechanically the process of dividing area of figure by length.

The following equivalents may be useful in calculations connected with the above :—

$$\begin{cases} \text{Pounds per sq. inch} \times \cdot 07 = \text{Kilogrammes per sq. centimetre.} \\ \text{Kilogrammes per sq. centimetre} \times 14 \cdot 22 = \text{Pounds per sq. inch.} \end{cases}$$

$$\begin{cases} \text{Foot-pounds} \times 7 \cdot 233 = \text{Kilogrammetres.} \\ \text{Kilogrammetres} \times \cdot 138 = \text{Foot-pounds.} \end{cases}$$

$$\begin{cases} \text{Horse-power} \times 1 \cdot 0139 = \text{Chevaux.} \\ \text{Chevaux} \times \cdot 9863 = \text{Horse-power.} \end{cases}$$

See also "Tables of Pounds per square inch and Kilogrammes per square centimetre," pages 353 and 354.

The Continental "Cheval" is equal to 4500 kilogrammetres, or 32,549 foot-pounds per minute, as against 33,000 foot-pounds per minute,—the value of the English "horse-power."

The following Table will considerably facilitate the computation of indicated horse-power :—

Table II.—Constant Multipliers for I.H.P.

$$\text{Constant} = \frac{\text{area of cylinder}}{33,000} \; ; \text{ and I.H.P} = \text{constant} \times \text{mean press.} \times \text{piston speed.}$$

Diameter of Cylinder.	Constant.	Diameter of Cylinder.	Constant.	Diameter of Cylinder.	Constant.	Diameter of Cylinder.	Constant.
6	·00085	16½	·00648	34	·02751	60	·08569
¼	·00092	¾	·00668	½	·02833	61	·08856
½	·00100	17	·00688	35	·02916	62	·09149
¾	·00108	¼	·00708	½	·02999	63	·09447
7	·00116	½	·00728	36	·03084	64	·09748
¼	·00125	¾	·00749	½	·03171	65	·10055
½	·00134	18	·00771	37	·03258	66	·10368
¾	·00143	¼	·00792	½	·03347	67	·10684
8	·00152	½	·00814	38	·03437	68	·11005
¼	·00162	¾	·00836	½	·03528	69	·11332
½	·00172	19	·00859	39	·03620	70	·11663
¾	·00182	¼	·00882	½	·03713	71	·11998
9	·00192	½	·00905	40	·03808	72	·12339
¼	·00203	¾	·00928	½	·03904	73	·12683
½	·00214	20	·00952	41	·04001	74	·13033
¾	·00226	½	·01000	½	·04099	75	·13388
10	·00238	21	·01049	42	·04198	76	·13748
¼	·00250	½	·01100	½	·04299	77	·14112
½	·00262	22	·01152	43	·04401	78	·14481
¾	·00275	½	·01205	½	·04504	79	·14854
11	·00288	23	·01259	44	·04608	80	·15232
¼	·00301	½	·01314	½	·04713	81	·15615
½	·00314	24	·01371	45	·04820	82	·16003
¾	·00328	½	·01428	½	·04927	83	·16398
12	·00342	25	·01487	46	·05036	84	·16795
¼	·00357	½	·01547	½	·05146	85	·17198
½	·00372	26	·01609	47	·05257	86	·17604
¾	·00387	½	·01671	½	·05370	87	·18016
13	·00402	27	·01735	48	·05483	88	·18432
¼	·00417	½	·01800	½	·05598	89	·18853
½	·00433	28	·01866	49	·05714	90	·19280
¾	·00449	½	·01933	½	·05832	91	·19710
14	·00466	29	·02001	50	·05950	92	·20146
¼	·00483	½	·02071	51	·06191	93	·20587
½	·00500	30	·02142	52	·06436	94	·21030
¾	·00517	½	·02214	53	·06685	95	·21480
15	·00535	31	·02287	54	·06940	96	·21937
¼	·00553	½	·02362	55	·07200	97	·22394
½	·00571	32	·02437	56	·07464	98	·22859
¾	·00590	½	·02513	57	·07733	99	·23328
16	·00609	33	·02592	58	·08006	100	·23799
¼	·00628	½	·02671	59	·08285	101	·24280

When the effective or resultant pressure on the piston at each point
in the stroke is required,—as for instance, when it is desired to
calculate the twisting moment on the crankshaft,—diagrams should
be constructed from the indicator diagrams, as follows :—

First,—draw line of no pressure, CD, at such a distance below the
atmospheric line, AB, that AC=BD=14·7 lbs., to the same scale as
the diagrams. Then, dealing with one stroke at a time, the curve
EFG represents the varying pressures on one side of the piston,
whilst the opposing pressures are represented by the curve JKB,
which forms a part of the diagram from the other side of the piston.

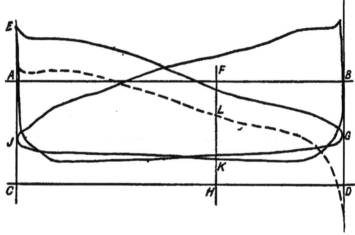

FIG. 2.

Draw any ordinate FH, and set off HL=HF - HK ; then L is a
point in the required resultant diagram in which any number of other
points may be found in a similar way ; and at the point H, in the
stroke CD, the effective pressure is HL. When the quantity corresponding to HF - HK is minus, it must be set off below the line of no
pressure.

It is very important that the pipes leading from the ends of the
cylinder to the indicator shall be large, short, of equal length, and as
free from bends as possible ; as, otherwise, there will be loss of area
in the diagram, and the apparent I.H.P. will be less than that really
exerted in the cylinder.

These pipes vary, in common practice, from ¾-inch to 1-inch
diameter, according to their length and to the piston speed of the
engine, but are still, beyond doubt, the cause of very perceptible loss
of area in the diagrams.

To make accurate tests of engines, with a view to determining water consumption per I.H.P., &c., all indicator pipes should be removed, and a separate indicator fitted direct to each end of each cylinder. This practice is becoming more general now, and should always be adopted where anything like accuracy is required.

It should not be forgotten that the indicator shows only *differences* between the pressures of the steam and of the atmosphere, and not absolute pressures.

To ascertain the weight of steam accounted for by any diagram, take a point A in the expansion curve of the diagram (Fig. 3), just

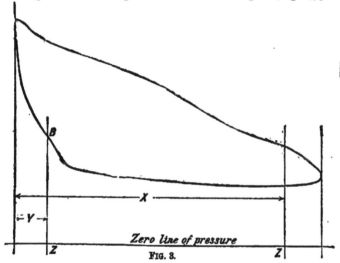

FIG. 3.

above the point of release, and measure the absolute pressure at this point; then take another point B in the compression curve, just above the point at which the exhaust closes, and measure the absolute pressure here also; and from Table CX., page 333, ascertain the weight of a cubic foot of steam at each of the pressures AZ and BZ.

Now calculate the volume, in cubic feet, swept by the piston while travelling through the distance X, and multiply it by the weight per cubic foot at pressure AZ; also calculate the volume corresponding to the travel Y, and multiply it by the weight per cubic foot at pressure BZ; then subtract the second product from the first, and the remainder will be the number of pounds of steam accounted for by the diagram during the stroke.

A similar calculation from the other diagram of the pair will give the amount of steam accounted for during the return stroke, and the sum of the two—multiplied by the number of revolutions—will give the amount accounted for by the diagrams per minute or per hour.

The clearances need not be considered in these calculations if the points A and B be taken at the same distance above the zero line of pressure.

This method is mainly useful in determining condensation and re-evaporation that occur during passage of steam through a series of cylinders. No conclusion as to economy can be derived from diagrams only. They tell nothing about water that may be present in cylinders. Feed water must be measured to determine real economy of engine.

EFFICIENCY OF ENGINES.

The indicated horse-power, calculated from diagrams as above described, gives only a measure of the work done in the cylinder, and the amount delivered at the propeller is, of course, less than this by the amount expended in working the pumps and in overcoming the friction of the various parts of the mechanism.

The mechanical efficiency of an engine is usually expressed by the decimal fraction whose value is,—

$$\text{Efficiency of engine} = \frac{\text{Net, or Brake, horse-power.}}{\text{Indicated horse-power.}}$$

The expense and difficulty of making brake tests of large marine engines are so great that, up to the present, none have been made, and their efficiency can therefore only be estimated from a comparison of the results of experiments made on small engines of various types, and on engines employed in driving dynamos, with regard to which (and more especially to the latter class) a large amount of accurate information has been recorded.

The efficiency of the best class of small compound non-condensing engines varies from about ·84 to about ·92, and a fair average value is ·88 ; and for compound condensing dynamo engines the average efficiency may be taken at ·90 to ·94 according to size, &c.

But before any comparison can be made between engines of these classes and marine engines, it is necessary to subtract from the gross I.H.P. of the marine engine, the power absorbed in working the pumps—thus obtaining what may be called the corrected I.H.P. Assuming then that the corrected I.H.P. is 95 per cent. of the gross— in cases where air, circulating, feed, and bilge pumps are all driven from the main engines—and taking into consideration the various other circumstances tending to affect efficiency, there is ground for believing that the efficiency of vertical triple-expansion marine engines, of good modern design, is between ·82 and ·88, and that an average specimen delivers at the propeller fully 85 per cent. of the gross power developed in the cylinders.*

When the efficiency of any engine (not driving its own pumps) is tested at various powers, and the curves of indicated and brake horse-power are laid down on the same diagram to equal horizontal and vertical scales, the one curve is seen to be sensibly parallel to the other : that is to say—the power absorbed in overcoming the friction

* This, of course, assumes that the thrust block is maintained in a perfectly efficient condition.

of the engine is practically a constant quantity, and is therefore the smallest percentage of the gross I.H.P. when the latter is a maximum.

As there is considerable difficulty in experimentally determining the power absorbed in overcoming the friction of an engine, the following graphic method (suggested by the late Dr Froude) is of interest, and may sometimes prove useful :—

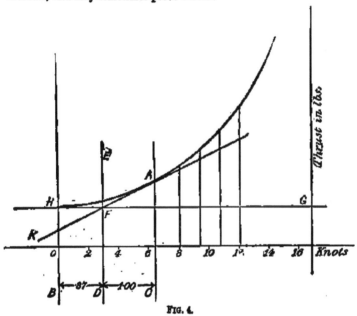

FIG. 4.

A series of progressive trials having been carried out, and the results carefully recorded—calculate the indicated thrusts (from the formula—

$$\text{Indicated thrust} = \frac{\text{I.H.P.} \times 33,000}{P \times R}$$ where P is pitch of propeller in feet,

and R revolutions per minute) for each speed, and set them up as ordinates from a base line on which the speeds are set off—as shown in fig. 4. Then, supposing A to be the lowest known point on the curve, draw the tangent KA ; divide BC at D so that BC = 1·87 DC, and through D draw the vertical line DE, cutting the tangent in F ; and through F draw HG parallel to the base line. The height OH then represents the constant friction of the engine, and the point H is the vertex of the thrust curve which may now be completed.

The lengths of the ordinates intercepted between HG and the thrust curve represent power expended in overcoming ship's net resistance, augment of resistance due to propeller, and friction of propeller blades, and are proportional to the ship's true resistance.

Efficiency as affected by Jacketing.

The most recent investigations show that all types of steam engines are rendered more efficient by the addition of steam jackets, and that the more completely the hot surfaces of cylinders, receivers, &c., are jacketed, the greater is the saving effected.

This amounts to saying, in other words, that for every pound of steam condensed in the jackets, some greater quantity is saved in the cylinders. The ratio of steam saved in the cylinders to steam expended in the jackets, varies from a little under 2 to 1 in some types of engine, to over 5 to 1 in other types.

The gain that may be expected to result from jacketing an engine depends on such a multitude of considerations—relating not only to the design of the engine and of the boiler, but also to the management of the machinery under steam—that it can only be very generally stated as lying between 5 and 25 per cent. of the total feed-water evaporated ; but in the case of modern marine machinery, of good construction, it is not probable that the gain is over 10 per cent. of the total feed-water evaporated.

The limit of usefulness of jacketing is reached when the exhaust is perfectly free from particles of water in suspension.

Small cylinders are more benefited by jacketing than large ones, in consequence of the ratio of area of hot surface to cubic contents being greater than in large ones.

It is important that there should be a thorough circulation of steam in the jackets, but the plan of passing the steam through the jackets on its way to the H.P. cylinder should be avoided.

The transmission of heat to the steam in the cylinder varies inversely as the thickness of the cylinder wall.

THE RESISTANCE AND PROPULSION OF SHIPS.

In dealing with cubes, or with parallelepipeds of similar forms, immersed in water until the uppermost face is just flush with the surface, it is found, on making the necessary calculations, that the wetted surface is exactly proportional to the $\frac{2}{3}$rd power (or the square of the cube root) of the displacement. Taking the case of cubes,—

Let L = length of edge,
D = displacement,
W = wetted surface ;
then $D = L^3$, or $L = \sqrt[3]{D}$,
and $W = 5 \times L^2 = 5 \times (\sqrt[3]{D})^2$.
That is, W varies as $D^{\frac{2}{3}}$.

It may also be noted that $L^2 = (\sqrt[3]{D})^2$,—that is, L^2 (which corresponds to mid-ship section) also varies as $D^{\frac{2}{3}}$; and therefore W varies as L^3, or, in other words, wetted surface varies as area of mid-ship section.

These results are not quite accurate for parallelepipeds which are not similar in form (that is, whose lengths, breadths, and depths are not of the same relative proportions), but the inaccuracy is only slight within considerable limits; so that, if ships of ordinary proportions are substituted for cubes or parallelepipeds, it is practically correct to say that wetted surface varies as area of mid-ship section, and also as $\frac{2}{3}$rd power of displacement.

For similar vessels displacement is also a measure of the fineness of the lines, since, when length and mid-ship section are the same, it varies directly as the prismatic co-efficient of fineness.

Now the resistances of ships depend almost entirely upon these two elements,—wetted surface, and form, or fineness of lines, and may be classed under the three heads,—

(1) Resistance due to skin friction ;
(2) Resistance due to eddy making ;
(3) Resistance due to wave making.

The first of these depends on the extent and nature of the wetted surface, and the depth of immersion; the third, on the lines of the ship, and on her degree of fineness; and the second, on all of these combined.

It is therefore evident, from the above considerations, that the old speed and power formulæ rest on a sound basis, and are capable, in experienced hands, of giving fairly accurate results. These formulæ are,—

Rule 8.
$$\text{I.H.P.} = \frac{D^{\frac{2}{3}} \times S^3}{C}$$

$$\text{and I.H.P.} = \frac{\text{area of immersed mid-ship section} \times S^3}{K}$$

where D, is displacement in tons; S, speed in knots; and C and K, co-efficients.

The peculiar value of these formulæ lies in the fact that they can be applied at a very early stage of the work, before such data as angles of obliquity of stream lines can be obtained with any degree of accuracy : and thus the power, approximate weight, and an outline drawing of the machinery can be got out simultaneously with the design of the vessel,—a great advantage where time is limited, as is usually the case in preparing tenders.

It is important to notice, in connection with the above formulæ that—although the resistance of a ship, moving uniformly at any speed, varies as the *square* of that speed,—the power required to overcome the resistance, and propel it at any speed, varies as the *cube* of that speed. For,—

Let S = speed in feet per minute ;
R = resistance in pounds at that speed ;
Then R = $S^2 \times C$,—where C is a co-efficient ;

and, multiplying both sides by S

$$R \times S = S^3 \times C.$$

But R × S is the work done, in foot-pounds per minute, in overcoming the resistance R through the space S, and, divided by 33,000, is equal to the horse-power required to drive the ship. This law, of course, only holds for similar ships, driven at corresponding speeds (speeds proportional to the square roots of the linear dimensions) since it is evident that the resistance due to wave making can only be proportional under these conditions.

The second formula is useful as a check, or corrective to the first, where the vessels under consideration are not absolutely similar, but have same ratio of length to breadth and draught, with a variation in the rise of floor.

The following Table (given by Sir W. H. White), shows the values of C for some typical ships of very different classes, at various speeds, and, whilst indicating generally the range of the variations that occur, serves also to show the difficulties that the naval architect has to encounter in obtaining high speeds in vessels of small dimensions. (For method of using curves of values of C, *see* page 23.)

The figures for horse-powers are "round." The "Medusa's" figures for 20 knots are those obtained from Stokes Bay trials, and strikingly illustrate the retarding effect of shallow water on the speed of a vessel ; the figures for the other ships at that speed are estimated for deep water.

It is, perhaps, scarcely necessary to add that these co-efficients of performance represent the *combined* efficiency of ships and machinery, and that they are therefore just as liable to be affected by an unsuitable propeller as by a foul bottom or unsuitable lines ; and also that for every model there is, as a rule, only one speed of maximum efficiency, though it is evident from the type of curve usually obtained, that there may be two speeds (one above and one below that of maximum efficiency) at which the efficiency is equal and slightly below the maximum.

It should also be borne in mind that accurately determined co-efficients of performance, &c., generally apply to more or less smooth water conditions, and that a form of vessel which gives highest speed with least power, on such trials may be far from the best for an ocean-going steamer. As a rule length assists speed in a sea-way, and, within very wide limits, the deeper the draught the higher the speed.

Table III.—Relation of Powers and Speeds.

Description of Vessel	Length in feet.	Breadth in feet.	Draught (mean). ft. In.	Displacement in tons.	I.H.P's. and Admiralty Co-efficients							
					10 Knots.		14 Knots.		18 Knots.		20 Knots.	
					I.H.P.	Co-eff.	I.H P.	Co-eff.	I.H.P.	Co-eff.	I.H.P.	Co-eff.
Torpedo Boat, Torpedo Gunboat,	135	14	5 1	103	110	200	260	232	870	147	1,130	156
"Sharpshooter" Class, 3rd Class Cruiser,	230	27	8 3	735	450	181	1100	203	2,500	190	3,500	186
"Medusa," 2nd Class Cruiser,	265	41	16 6	2,800	700	284	2100	259	6,400	181	10,000	159
"Terpsichore," 1st Class Cruiser,	300	43	16 2	3,330	800	279	2400	255	6,000	217	9,000	198
"Edgar," 1st Class Cruiser,	360	60	23 9	7,390	1000	380	3000	347	7,500	2 95	11,000	276
"Blenheim," 1st Class Cruiser,	375	65	25 9	9,100	1500	290	4000	298	9,000	282	12,500	278
Atlantic Passenger Steamer,	525	63	21 3	11,550	2000	255	4600	304	10,000	297	14,500	281

Names.	A.	B.	C.	D.	E.	F.	G.	H.	J.	K.	L.	M.
Length, perpendiculars,	450'-0"	388'-0"	840'-0"	370'-0"	313'-6"	300'-0"	300'-0"	311'-6"	270'-0"	280'-0"	250'-0"	280'-0"
Breadth, extreme,	45'-2"	43'-0"	41'-0"	41'-0"	35'-6"	38'-0"	34'-6"	36'-0"	32'-6"	31'-0"	33'-0"	26'-5"
Mean draught water,	23'-7"	17'-7½"	17'-3"	18'-11"	14'-11½"	18'-0"	18'-9"	11'-6"	11'-1"	18'-6"	12'-0"	11'-4"
Displacement (tons),	8500	4315	4125	4635	2480	2500	2200	1790	1270	1940	1370	1050
Area Imm. mid section,	925	615	680	656	422	453	425	392	343	350	320	280
Wetted skin,	82,578	22,330	20,340	22,633	15,200	14,594	14,043	12,028	10,094	12,532	10,028	8494
Length, fore body,	129'-0"	142'-0"	105'-0"	123'-0"	107'-6"	106'-10"	119'-0"	152'-6"	140'-0"	72'-9"	100'-0"	89'-3"
Angle of entrance,	17°-16'	14°-0'	19°-36'	16°-4'	14°-56'	18°-36'	15°-16'	12°-30'	12°-20'	18°-0'	15°-10'	14°-30'
$\frac{\text{Displacement} \times 35}{\text{Length} \times \text{I. mid area}}$	0·714	0·632	0·685	0·668	0·658	0·64	0·604	0·51	0·48	0·603	0·599	0·614
Speed (knots),	15·05	16·82	14·0	13·8	15·94	14·6	18·2	18·7	19·7	15·08	14·9	14·9
Indicated horse-power,	4900	4660	4195	2500	2243	2601	4398	4000	4250	2373	2063	1538
I.H.P. per 100 feet of wetted skin,	15·04	20·87	14·9	11·04	14·75	17·8	31·33	31·67	42·11	19	20	17·92
H. P per 100 feet of wetted skin reduced to 10 knots,	4·42	4·4	5·4	4·2	4·09	5·7	5·19	4·84	5·94	5·8	6·35	5·41
$\frac{D^3 \times S^3}{\text{I.H.P.}}$	289	299	235	292	236	220	231·7	240	194	225	194	224
$\frac{\text{I. mid area} \times S^3}{\text{I.H.P.}}$	642	626	564	689	678	542	582	641	570	505	502	559

oming
equal
purse,
speeds
it is
propor-

first,
but
on in

blues
eeds,
that
as
is.

's "
and
eed
ated

s of
ery,
un-
also
ium
ned,
t of
htly

co-
both
peed
ean-
and,
ed.

Kirk's Analysis.

The following very simple and useful method of estimating the power required to drive a vessel at any given speed was devised by the late Dr A. C. Kirk, and is generally known as " Kirk's analysis " :—

A diagram, resembling fig. 5, of what is called the " block model " is first made,—its dimensions bearing certain fixed relations to the dimensions of the vessel under consideration,—and the wetted surface of this " block model " (the whole surface minus that of the upper face) is then, within a very small error,—stated below,—equal to that of the proposed vessel. (Rule 9.) The I.H.P. is then found by assuming that 5 I.H.P. will drive 100 feet of wetted surface at 10 knots, and that the power required varies as the cube of the speed. When the vessel is exceptionally well proportioned, the bottom quite clean, and the efficiency of the machinery high, as low a rate as 4 I.H.P. for 100 feet of wetted surface may be assumed.

The dimensions of the " block model," fig. 5, are determined as follows :—

Length AB — length of ship (from forward side of stem to after side of stern-post, at mean trial draught).

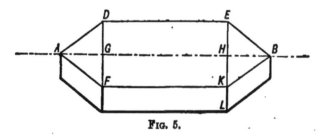

Fig. 5.

Depth KL — depth of ship from mean trial draught to top of keel (if any).

Breadth EK — $\dfrac{\text{Area of immersed mid-ship section.}}{\text{KL}}$

AH — GB — $\dfrac{\text{Displacement in tons} \times 35^*}{\text{Area of immersed mid-ship section.}}$

Length of AD — $\sqrt{AG^2 + GD^2}$.

Then, the wetted surface of the " block model " is,—

$$(EK \times AH) + (2KL \times FK) + (4KL \times AD).$$

* 35 cubic feet of salt water weigh one ton, and are therefore equal to one ton displacement.

The angle of entrance is EBK ; EBH is half that angle ; and tan. EBH $= \frac{EH}{HB}$; or, the tangent of half the angle of entrance is equal to

$\frac{\text{Half breadth of model}}{\text{Length of fore-body}}$, and from this, by means of a table of natural tangents, the angle of entrance may be obtained.

The lengths of fore-body and angles of entrance for different types of vessel are as follows :—

Table IV.   Angles of Entrance, &c.

Description of Vessel.	Angle of Entrance.	Ratio of length of fore-body to total length of block.
Ocean-going merchant steamers whose speed is from 14 knots upwards, . .	18° to 15°	·3 to ·36
Ocean-going steamers whose speed is from 12 to 14 knots, . . .	21° to 18°	·26 to ·3
Cargo steamers whose speed is from 10 to 12 knots, ...	30° to 22°	·22 to ·26

Example :—To find the I.H.P. necessary to drive a ship at 15 knots, the wetted skin of "block-model" being 16,200 square feet.

The I.H.P. per 100 square feet $= \left(\frac{15}{10}\right)^3 \times 5 = 16·875$

and I.H.P. required $= 16·875 \times 162 = 2744.$

In ordinary practice the wetted surface of the "block model" is found to exceed that of the actual ship by 2 per cent. (in the case of full ships), by 3 to 5 per cent. for ordinary steamers, and as much as 8 per cent. in the case of very fine steamers ; but the error is in the right direction, and for all ordinary purposes it is sufficient to take the surface of the "block model."

The following Table gives, on inspection, the horse-powers required per 100 feet of wetted surface at various speeds and rates, and will facilitate calculations by the above method :—

Table V.—I.H.P. per 100 feet of wetted surface at different speeds.

Speeds in Knots.	I.H.P. per 100 sq. feet of wetted surface at 10 knots.												
	4·0	4·1	4·2	4·3	4·4	4·5	4·6	4·7	4·8	4·9	5·0	5·1	5·2
9	2·92	2·99	3·06	3·14	3·21	3·28	3·35	3·42	3·49	3·57	3·64	3·71	3·78
9·5	3·43	3·51	3·60	3·6?	3·77	3·85	3·94	4·02	4·11	4·19	4·28	4·36	4·45
10	4·00	4·10	4·20	4·30	4·40	4·50	4·60	4·70	4·80	4·90	5·00	5·10	5·20
10·5	4·63	4·75	4·86	4·98	5·09	5·21	5·32	5·44	5·56	5·79	5·79	5·90	6·02
11	5·32	5·46	5·59	5·72	5·86	5·99	6·12	6·25	6·39	6·52	6·65	6·79	6·92
11·5	6·08	6·23	6·39	6·54	6·69	6·84	6·99	7·15	7·30	7·45	7·60	7·76	7·91
12	6·91	7·08	7·26	7·43	7·60	7·78	7·95	8·12	8·29	8·47	8·64	8·81	8·99
12 5	7·81	8·01	8·20	8·40	8·59	8·79	8·98	9·18	9·37	9·57	9·76	9·96	10·15
13	8·79	9·01	9·23	9·45	9·67	9·89	10·10	10·32	10·54	10·76	10·98	11·20	11·42
13·5	9·84	10·09	10·33	10·58	10·82	11·07	11·32	11·56	11·81	12·05	12·30	12·55	12·79
14	10·98	11·25	11·52	11·80	12·07	12·35	12·62	12·90	13·17	13·44	13·72	13·99	14·27
14·5	12·19	12·50	12·80	13·11	13·41	13·72	14·02	14·33	14·63	14·94	15·24	15·55	15·85
15	13·50	13·84	14·17	14·51	14·85	15·18	15·52	15·86	16·19	16·53	16·87	17 21	17·54
15·5	14·39	15·27	15·64	16·01	16·38	16·76	17·13	17·50	17·87	18·25	18·62	18·99	19·36
16	16·38	16·79	17·20	17·61	18·02	18·43	18·84	19·25	19·66	20·07	20·48	20·89	21·30
16·5	17·97	18·42	18·87	19·31	19·76	20·21	20·66	21·11	21·56	22·01	22·46	22·91	23·36
17	19·65	20·14	20·63	21·12	21·62	22·11	22·60	23·09	23·58	24·07	24·56	25·06	25·55
17·5	21·44	21·97	22·51	23·04	23·58	24·12	24·65	25·19	25·72	26·26	26·80	27·33	27·87
18	23·33	23·91	24·49	25·08	25·66	26·24	26·83	27·41	27·99	28·58	29·16	29·74	30·33
18·5	25·33	25·96	26·59	27·22	27·86	28·49	29·12	29·76	30·39	31·02	31·66	32·29	32·92
19	27·44	28·12	28·81	29·49	30·18	30·86	31·55	32·24	32·92	33·61	34·29	34·98	35·67
19·5	29·66	30·40	31·14	31·88	32·62	33·37	34·11	34·85	35·59	36·33	37·07	37·81	38·56

Rankine's Rules.

The following rules relating to the propulsion of vessels were given by the late Professor Rankine ; but it must be understood that they are only applicable to ships whose lines are of the " wave line " type :—

Rule I.—In order that the resistance may not increase faster than the square of the speed (S^2) the length of the after-body in feet must not be less than $S^2 \times \cdot375$.

To fulfil the same condition, the length of the fore-body must not be less than that of the after-body, as given above, and may with advantage be $1\frac{1}{2}$ times as long.

Rule II.—The greatest speed in knots suited to a given length of after-body (l) is $\sqrt{\frac{8}{3}l}$.

Rule III.—When the speed does not exceed the limit given by Rule II., the probable resistance in pounds (R) will be,—

$$R = G \times L \times (1 + 4M) \times S^2 \times C.$$

Where G = mean immersed girth in feet ;

L = length on water line in feet ;

M = $\begin{cases} \text{mean square of sines of angles of obliquity of} \\ \text{stream-lines + mean of fourth powers of sines ;} \end{cases}$

S = Speed in knots ;

C = $\begin{cases} \cdot01 \text{ for clean painted iron vessels ;} \\ \cdot009 \text{ to } \cdot008 \text{ for clean coppered vessels ;} \\ \cdot011 \text{ and upwards for moderately rough iron-vessels;} \end{cases}$

The product $G \times L \times (1 + 4M)$ is called the " Augmented surface."

Rule IIIa.—An approximate value of R, in well-designed steamers with clean painted bottoms, is given by,—

$$R = S^2 \times D \times K$$

Where D = the displacement in tons;
    K = ·8 to 1·5 for different types of steamers.

Rule IV.—The net, or effective horse-power (E.H.P.) expended in propelling the vessel is,—

$$E.H.P. = \frac{R \times S}{326}.$$

Rule IVa.—The gross, or indicated horse-power (I.H.P.) will be the E.H.P., as found by Rule IV., divided by the combined efficiency of engine and propeller, which is ordinarily between ·6 and ·625, and averages say ·613 ; therefore,—

$$I.H.P. = \frac{R \times S}{200}.$$

Note.—To obtain the mean immersed girth, proceed as follows :—

(*a*) Divide length on water line into an even number of intervals, and number the dividing points (the number of which will be one more than the number of spaces) 1, 2, 3, 4 &c.

(*b*) Measure the lengths of the half-girths at each point.

(*c*) Add together the two end-most lengths ; four times the intermediate even-numbered lengths ; and twice the intermediate odd-numbered lengths ; and divide the sum by 3, and again by half the number of intervals,—and the result will be the mean immersed girth required.

In determining the "co-efficient of augmentation" (1 + 4M), the lines giving the half angles of entrance at the different water lines must be tangents to the curves at the points of contrary flexure.

Mr Mansel's Formulæ.

Mr Robert Mansel of Glasgow, who has devoted much time to the investigation of the relation of power to speed in steam-ships,—in drawing attention to the generally admitted imperfections of the

formula $I.H.P. = \frac{D^{\frac{2}{3}} \times S^3}{C}$, or $I.H.P. = \frac{D^{\frac{2}{3}}}{C} S \times S^2$ as he sometimes

writes it,—emphasizes the fact that the so-called constant is rarely or never constant for two different speeds (but decreases in value as the speed increases), and suggests the following modified form of the expression :—

$$I.H.P. = \frac{D^{\frac{2}{3}}}{C} S \times 10^{aS}$$

$$\text{or } I.H.P. = \frac{D^{\frac{2}{3}}}{C} S \times 10^{aS}.$$

Where, instead of using the square of the speed, ho uses the number whose logarithm is the speed multiplied by a small co-efficient a,—the value of which usually lies between ·06 and ·13. It should be mentioned, however, that in trying the same vessel through any great range of speeds, the value of a is not always the same throughout.

Speed and Power Curves, etc.

The most reliable method of determining the I.H.P. required to drive any proposed vessel at a given speed is to base the calculations upon the results obtained from the trials of "similar" vessels; the basis of which is the fact that "similar" vessels, driven at "corresponding" speeds have the same co-efficient of performance, when the efficiency of the machinery is the same.

"Similar" vessels are those having the same ratio of length to breadth, and to draught, and the same degree of fineness; and "corresponding" speeds are those which are proportional to the square roots of the linear dimensions of the respective vessels (e.g. proportional to the square roots of the lengths).

Froude found that the resistances of such vessels varied almost exactly as wetted surface $\times$ (speed)².

But to render the results of former trials readily accessible for such a purpose it is very desirable to have them plotted down as a series of curves, somewhat in the following manner :—

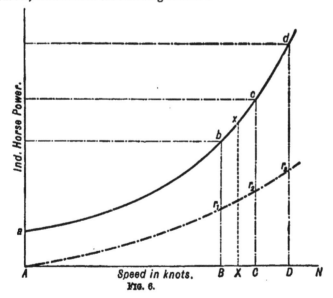

Fig. 6.

Let P_1, P_2, P_3 be the indicated horse-powers developed in obtaining the speeds S_1, S_2, S_3 knots, with R_1, R_2, R_3 revolutions per minute.

Take a line AN (Fig 6.) as a base line, and on it take points B, C, D, so that AB, AC, AD, are proportional to S_1, S_2, S_3; at the points B, C, D erect ordinates Bb, Cc, Dd proportional to P_1, P_2, P_3, and through the points b, c, and d draw the curve $d\ c\ b\ a$, which is called the "curve of power" or "curve of I.H.P." The nature of this curve is then such that if an ordinate be drawn through any other point X in the line AN, Xx will represent the power necessary to obtain the speed AX from the same vessel, or another vessel of the same form and dimensions.

If the curve be accurately drawn, it will be found that it does not pass through A, but at a distance Aa above that point, thus signifying that a certain amount of power is developed even at zero speed; Aa thus represents the power required to overcome the constant friction of the engines. (See also page 11, under "Efficiency of Engines.")

Similarly a curve of revolutions may be constructed by taking points r_1, r_2, r_3 in the ordinates so that Br, Cr, Dr, are proportional to R_1, R_2, R_3.

The slip may also be shown by a curve whose ordinates are proportional to the slips at the speeds S_1, S_2, S_3.

Examination of the curves will show :—

(1) The I.H.P., revolutions, and slip corresponding to any speed intermediate to those observed ;

(2) The constant friction, and therefore general efficiency of the engines ;

(3) The suitability of the lines of the ship for the speeds,—a sudden rise of the curve towards the higher part, showing an undue increase of resistance at the higher speeds ;

(4) The power of the speed with which the I.H.P. increases for the particular type of ship ;

5. The suitability of the propeller to the ship,—any sudden rise in the slip curve showing that the propeller is defective either as regards diameter or surface, or both.

Perhaps the most useful curve, however, for the purpose of determining the power required to propel some "similar" ship, of different size, at any given speed, is one constructed as shown in fig. 7, where the abscissæ represent speeds in knots, and the ordinates numerical co-efficients of performance, derived, either from the Admiralty formula $\left(\text{I.H.P.} = \dfrac{D^{\frac{2}{3}} \times S^3}{C} \right)$ or other similar expression.

Rule 10.—Suppose fig. 7 to represent the curve given by a vessel 250 feet long, and 2400 tons displacement, and that it is required to determine the power necessary to drive a "similar" vessel of 360 feet long, and 7200 tons displacement, at say 13 knots. The question then is,—what co-efficient of performance must be assumed?—and it is answered as follows :—

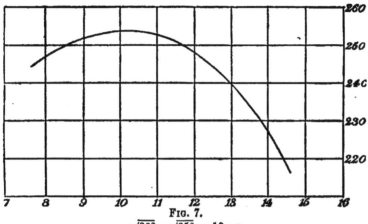

FIG. 7.

$$\sqrt{360} : \sqrt{250} :: 13 : x$$

or $x = \dfrac{\sqrt{250} \times 13}{\sqrt{360}} = 10 \cdot 8 = \begin{cases} \text{``corresponding'' speed of} \\ \text{first, or type ship.} \end{cases}$

and running up the ordinate for 10·8 knots until it cuts the curve, and then along the abscissa from this point of intersection the figure 253 is found, and this, when used in connection with the formula, gives 3250 as the required I.H.P.

In cases where there are no records of exactly "similar" ships, the value of the estimate made will of course depend very largely on the experience of the estimator.

Wetted Surface.—Mr Mumford's Method.

Mr E. R. Mumford has devised the following formula for determining wetted surface :—

$$S = (L \times D \times 1 \cdot 7) + (L \times B \times C)$$

where

S = wetted surface in sq. ft.	D = middle draught in feet.
L = length between perpendiculars in feet.	B = beam in feet.
	C = block co-efficient.

It gives a closer approximation to accuracy than "Kirk's Analysis."

Determination by Model Experiments.

Another method, employed by the Admiralty, and by a few of our leading firms of shipbuilders, for determining the power required to propel any new type of vessel is, to ascertain the resistance of a model of the new vessel in the experimental tank, and to calculate the power from the results obtained ; and, where widely divergent types have to be dealt with, the method is no doubt of great value, but the expense is necessarily great.

The models are usually made of paraffine, and are from 12 feet to 20 feet long, and from ¾-inch to 1¼-inch thick ; they are cast nearly to shape, and then trimmed to the exact form on a special shaping machine. the speeds and pulls on the tow-rope are automatically recorded on paper drums driven by clock-work. The height and position of the waves created,—which are of special importance in the case of paddle vessels,—can also be noted and recorded.

The horse-power required is calculated from the resistance of the model, by the same principle of "corresponding" speeds referred to above, as follows :—

Let l and L = lengths of model and vessel, respectively ;
v and V = corresponding speeds ;
r and R = corresponding resistances ;

$$\text{Then } \frac{V}{v} = \sqrt{\frac{L}{l}}$$
$$\text{and } \frac{R}{r} = \left(\frac{L}{l}\right)^3.$$

Example.—Suppose the E.H.P. (effective horse-power, or gross horse-power less that required to overcome friction of engines and propeller) necessary to drive a ship of 300 feet long at 15 knots is required. Let the length of the model be 12 feet, then the "corresponding" speed for it will be given by,—

$$v = 15 \sqrt{\frac{12}{300}} = 3 \text{ knots.}$$

Assume the resistance of the model at this speed to be 4 lbs., then resistance of ship, at 15 knots will be,—

$$R = 4 \times \left(\frac{300}{12}\right)^3 = 62,500 \text{ lbs.};$$

$$\text{and } \frac{2,500 \times 15 \times 6080}{60 \times 33,000} = 2879 ;$$

or E.H.P. required is approximately 2879,—a slight correction having to be made for skin friction.

Co-efficients of Fineness.

The block co-efficient expresses the ratio borne by the displacement volume to that of the parallelepiped circumscribing the immersed body.

Let V = displacement in cubic feet ;
L = length on water line ;
B = greatest immersed breadth ;
D = draught of water ;
K = displacement co-efficient.

$$\text{Then } K = \frac{V}{L \times B \times D}.$$

The prismatic co-efficient,—which gives a truer measure of fineness of lines than the above,—expresses the ratio borne by the displacement volume to that of the prism swept by moving the immersed mid-ship section through the length at load water-line.

Surface Friction.

The following Table gives a general statement of the results of Froude's experiments on this subject; they were made on boards ³/₁₆-inch thick and 19-inch deep, which were coated with the substances to be experimented on, and towed edgeways through the water. The resistances are given in lbs. per square foot at the standard speed of 600 feet per minute, and, as the power of the speed to which the friction is proportional is also given, the resistance at other speeds is easily calculated.*

Columns A give the power of the speed to which the resistance is approximately proportional; columns B give the mean resistance per square foot of the whole surface of a board of the lengths stated in the table ; columns C give the resistance of a square foot of surface at the distance sternward from the cutwater stated in the heading.

Table VI.—Resistances of Surfaces.

| Nature of Surface. | Length of surface, or distance from cutwater, in feet. | | | | | | | | | | | |
| | 2 Feet. | | | 8 Feet. | | | 20 Feet. | | | 50 Feet. | | |
	A	B	C	A	B	C	A	B	C	A	B	C
Varnish, .	2·00	·41	·390	1·85	·325	·264	1·85	·278	·240	1·83	·250	·226
Paraffine, .	1·95	·38	·370	1·94	·314	·260	1·93	·271	·237	...	...	...
Tinfoil, .	2·16	·30	·295	1·99	·278	·263	1·90	·262	·244	1·83	·246	·232
Calico, . .	1·93	·87	·725	1·92	·626	·504	1·89	·531	·447	1·87	·474	·423
Fine sand,	2·00	·81	·690	2·00	·583	·450	2·00	·480	·384	2·06	·405	·337
Medium ,,	2·00	·90	·730	2·00	·625	·488	2·00	·534	·465	2·00	·488	·456
Coarse ,,	2·00	1·10	·880	2·00	·714	·520	2·00	·588	·490	..	...	...

True Mean Speed.

To determine the true mean speed of a vessel when the runs are taken on the measured mile, half with the tide, and half against :—

* NOTE.—See Rule 11, page 26.

Rule :—Find the means of consecutive speeds continually found until only one remains.

Example.

Runs.	Observed Speeds.	1st Means.	2nd Means.	3rd Means.	4th Means.	Mean of Means.
1st	18·5					
		15·90				
2nd	13·3		15·625			
		15·35				
3rd	17·4		15·525	15·5750		
		15·70			15·52500	
4th	14·0		15·425	15·4750		15·478125
		15·15			15·43125	True mean speed.
5th	16·3		15·350	15·3875		
		15·55				
6th	14·8					

6)94·2

15·70

Ordinary mean speed.

4)61·925

15·48125

Ordinary mean of second means.

The ordinary mean of second means is generally taken,—as unavoidable errors of observation render the third and following decimal places of very doubtful value.

Relation of Speeds and Powers.

Given two speeds of a vessel, and the corresponding horse-powers, to find what power of the speed the horse-power varies as :—

Let s and S = the two speeds.

,, p and P = the corresponding powers.

,, x = power of s and S that p and P vary as

Then $\dfrac{S^x}{s^x} = \dfrac{P}{p}$

or $x\,(\text{Log } S - \text{Log } s) = \text{Log } P - \text{Log } p$

Rule 11.   And $x = \dfrac{\text{Log } P - \text{Log } p}{\text{Log } S - \text{Log } s}$.

Table of Times and Speeds.

Given the time, in minutes and seconds, occupied by the vessel in running the measured mile, the following Table gives, on inspection, the speed in knots :—

TABLE VII.—TIMES AND SPEEDS. 27

TIMES AND SPEEDS.

The number in this Table corresponding to the time in which a vessel passes over the measured mile is her speed in knots.

Secs.	2 min.	3 min.	4 min.	5 min.	6 min.	7 min.	8 min.	9 min.	10 min.	11 min.	12 min.	13 min.	14 min.
0	30·000	20·000	15·000	12·000	10·000	8·571	7·500	6·667	6·000	5·455	5·000	4·615	4·286
1	29·752	19·890	14·938	11·960	9·972	8·561	7·484	6·654	5·990	5·446	4·998	4·609	4·281
2	29·508	19·780	14·876	11·921	9·945	8·531	7·469	6·642	5·980	5·438	4·986	4·604	4·275
3	29·268	19·672	14·815	11·881	9·917	8·511	7·453	6·630	5·970	5·430	4·979	4·598	4·270
4	29·032	19·565	14·754	11·842	9·890	8·491	7·438	6·618	5·960	5·422	4·972	4·592	4·265
5	28·800	19·459	14·694	11·803	9·863	8·471	7·423	6·606	5·950	5·414	4·965	4·586	4·260
6	28·571	19·355	14·634	11·765	9·836	8·451	7·407	6·593	5·941	5·405	4·959	4·580	4·255
7	28·346	19·251	14·575	11·726	9·809	8·431	7·392	6·581	5·931	5·397	4·952	4·574	4·250
8	28·125	19·149	14·516	11·688	9·783	8·411	7·377	6·569	5·921	5·389	4·945	4·568	4·246
9	27·907	19·048	14·458	11·650	9·756	8·392	7·362	6·557	5·911	5·381	4·938	4·563	4·240
10	27·692	18·947	14·400	11·613	9·730	8·372	7·347	6·545	5·902	5·373	4·931	4·557	4·235
11	27·481	18·848	14·343	11·576	9·704	8·353	7·332	6·534	5·892	5·365	4·925	4·551	4·230
12	27·273	18·750	14·286	11·538	9·677	8·333	7·317	6·522	5·882	5·357	4·918	4·545	4·225
13	27·068	18·653	14·229	11·502	9·651	8·314	7·302	6·510	5·873	5·349	4·911	4·540	4·220
14	26·866	18·557	14·173	11·465	9·626	8·295	7·287	6·498	5·863	5·341	4·905	4·534	4·215
15	26·667	18·461	14·118	11·429	9·600	8·276	7·273	6·486	5·854	5·333	4·898	4·528	4·210
16	26·471	18·367	14·062	11·392	9·574	8·257	7·258	6·475	5·844	5·325	4·891	4·523	4·206
17	26·277	18·274	14·008	11·356	9·549	8·238	7·243	6·463	5·835	5·318	4·885	4·517	4·201
18	26·087	18·182	13·953	11·321	9·524	8·219	7·229	6·452	5·825	5·310	4·878	4·511	4·196
19	25·899	18·090	13·900	11·285	9·499	8·200	7·214	6·440	5·816	5·302	4·871	4·506	4·191
Secs.	2 min.	3 min.	4 min.	5 min.	6 min.	7 min.	8 min.	9 min.	10 min.	11 min.	12 min.	13 min.	14 min.

Table VII.—continued.

TIMES AND SPEEDS.

The number in this Table corresponding to the time in which a vessel passes over the measured mile is her speed in knots.

Secs.	2 min.	3 min.	4 min.	5 min.	6 min.	7 min.	8 min.	9 min.	10 min.	11 min.	12 min.	13 min.	14 min.
20	25·714	18·000	13·846	11·250	9·474	8·182	7·200	6·429	5·806	5·294	4·865	4·500	4·186
21	25·532	17·910	13·793	11·215	9·449	8·163	7·186	6·417	5·797	5·286	4·858	4·494	4·181
22	25·352	17·822	13·740	11·180	9·424	8·145	7·171	6·406	5·788	5·279	4·852	4·489	4·176
23	25·175	17·734	13·688	11·146	9·399	8·126	7·157	6·394	5·778	5·271	4·845	4·483	4·171
24	25·000	17·647	13·636	11·111	9·375	8·108	7·143	6·383	5·769	5·263	4·839	4·478	4·167
25	24·828	17·561	13·585	11·077	9·351	8·090	7·129	6·372	5·760	5·255	4·832	4·472	4·162
26	24·658	17·476	13·534	11·043	9·326	8·072	7·115	6·360	5·751	5·248	4·826	4·466	4·157
27	24·490	17·391	13·483	11·009	9·302	8·054	7·101	6·349	5·742	5·240	4·819	4·461	4·152
28	24·324	17·308	13·433	10·976	9·278	8·036	7·087	6·338	5·732	5·233	4·813	4·455	4·147
29	24·161	17·225	13·383	10·942	9·254	8·018	7·073	6·327	5·723	5·225	4·806	4·450	4·143
30	24·000	17·143	13·333	10·909	9·231	8·000	7·059	6·316	5·714	5·217	4·800	4·444	4·138
31	23·841	17·062	13·284	10·876	9·207	7·982	7·045	6·305	5·705	5·210	4·794	4·439	4·133
32	23·684	16·981	13·235	10·843	9·184	7·965	7·031	6·294	5·696	5·202	4·787	4·433	4·128
33	23·529	16·901	13·187	10·811	9·160	7·947	7·018	6·283	5·687	5·195	4·781	4·428	4·124
34	23·377	16·822	13·139	10·778	9·137	7·930	7·004	6·272	5·678	5·187	4·774	4·423	4·119
35	23·226	16·744	13·091	10·746	9·114	7·912	6·990	6·261	5·669	5·180	4·768	4·417	4·114
36	23·077	16·667	13·043	10·714	9·091	7·895	6·977	6·250	5·660	5·172	4·762	4·412	4·110
37	22·930	16·590	12·996	10·682	9·068	7·877	6·963	6·239	5·651	5·165	4·756	4·406	4·105
38	22·785	16·514	12·950	10·651	9·045	7·860	6·950	6·228	5·643	5·158	4·749	4·401	4·100
39	22·642	16·488	12·903	10·619	9·023	7·843	6·936	6·218	5·634	5·150	4·743	4·396	4·096
Secs.	2 min.	3 min.	4 min.	5 min.	6 min.	7 min.	8 min.	9 min.	10 min.	11 min.	12 min.	13 min.	14 min.

TABLE VII.—TIMES AND SPEEDS. 29

Table VII.—*continued.*

TIMES AND SPEEDS.

The number in this Table corresponding to the time in which a vessel passes over the measured mile is her speed in knots.

Secs.	2 min.	3 min.	4 min.	5 min.	6 min.	7 min.	8 min.	9 min.	10 min.	11 min.	12 min.	18 min.	14 min.
40	22·500	16·364	12·857	10·588	9·000	7·826	6·923	6·207	5·625	5·143	4·737	4·390	4·091
41	22·360	16·290	12·811	10·557	8·978	7·809	6·910	6·196	5·616	5·136	4·731	4·385	4·086
42	22·222	16·216	12·766	10·526	8·955	7·792	6·897	6·188	5·607	5·128	4·724	4·379	4·082
43	22·086	16·143	12·721	10·496	8·933	7·775	6·883	6·175	5·599	5·121	4·718	4·374	4·077
44	21·951	16·071	12·676	10·465	8·911	7·759	6·870	6·164	5·590	5·114	4·712	4·369	4·072
45	21·818	16·000	12·632	10·435	8·889	7·742	6·857	6·154	5·581	5·106	4·706	4·364	4·068
46	21·687	15·929	12·587	10·405	8·867	7·725	6·844	6·143	5·573	5·099	4·700	4·358	4·063
47	21·557	15·859	12·544	10·375	8·845	7·709	6·831	6·133	5·564	5·092	4·693	4·353	4·059
48	21·429	15·789	12·500	10·345	8·824	7·692	6·818	6·122	5·556	5·085	4·687	4·348	4·054
49	21·302	15·721	12·457	10·315	8·802	7·676	6·805	6·112	5·547	5·078	4·681	4·343	4·049
50	21·176	15·652	12·414	10·286	8·780	7·660	6·792	6·102	5·538	5·070	4·675	4·337	4·045
51	21·053	15·584	12·371	10·256	8·759	7·643	6·780	6·091	5·530	5·063	4·669	4·332	4·040
52	20·930	15·517	12·329	10·227	8·738	7·627	6·767	6·081	5·521	5·056	4·663	4·327	4·035
53	20·809	15·451	12·287	10·198	8·717	7·611	6·754	6·071	5·513	5·049	4·657	4·322	4·031
54	20·690	15·385	12·245	10·169	8·696	7·595	6·742	6·061	5·505	5·042	4·651	4·316	4·027
55	20·571	15·319	12·203	10·141	8·675	7·579	6·729	6·050	5·496	5·035	4·645	4·311	4·022
56	20·455	15·254	12·162	10·112	8·654	7·563	6·716	6·040	5·488	5·028	4·639	4·306	4·018
57	20·339	15·190	12·121	10·084	8·633	7·547	6·704	6·030	5·479	5·021	4·633	4·301	4·013
58	20·225	15·126	12·081	10·056	8·612	7·531	6·691	6·020	5·471	5·014	4·627	4·296	4·009
59	20·112	15·063	12·040	10·028	8·592	7·516	6·679	6·010	5·463	5·007	4·621	4·291	4·004
Secs.	2 min.	3 min.	4 min.	5 min.	6 min.	7 min.	8 min.	9 min.	10 min.	11 min.	12 min.	18 min.	14 min.

Should the speed be required in kilometres,—multiply the speed in knots, from the above Table, by 1·853.

See also Tables of knots, miles, kilometres, &c., pp. 340 to 342.

The following Table of the ⅔ powers of numbers will be of service in all calculations for which displacement is taken as a basis.

Table VIII.—Two-thirds powers of numbers.

Number.	⅔rd power.	Number.	⅔rd power.	Number.	⅔rd power.	Number.	⅔rd power.
100	21·54	480	61·30	860	90·43	1240	115·42
110	22·96	490	62·15	870	91·13	50	116·04
120	24·33	500	62·99	880	91·83	60	116·66
130	25·66	510	63·83	890	92·52	70	117·27
140	26·96	520	64·66	900	93·22	80	117·89
150	28·23	530	65·49	910	93·91	90	118·50
160	29·47	540	66·31	920	94·59	1300	119·11
170	30·69	550	67·13	930	95·28	10	119·72
180	31·88	560	67·94	940	95·96	20	120·33
190	33·05	570	68·74	950	96·64	30	120·94
200	34·21	580	69·54	960	97·32	40	121·55
210	35·33	590	70·34	970	97·99	50	122·15
220	36·44	600	71·13	980	98·66	60	122·75
230	37·54	610	71·92	990	99·33	70	123·35
240	38·62	620	72·71	1000	100·00	80	123·95
250	39·68	630	73·49	10	100·66	90	124·55
260	40·74	640	74·26	20	101·33	1400	125·14
270	41·78	650	75·03	30	101·99	10	125·74
280	42·80	660	75·80	40	102·65	20	126·33
290	43·81	670	76·57	50	103·30	30	126·92
300	44·81	680	77·33	60	103·96	40	127·51
310	45·80	690	78·08	70	104·61	50	128·10
320	46·78	700	78·84	80	105·26	60	128·69
330	47·75	710	79·59	90	105·91	70	129·28
340	48·71	720	80·33	1100	106·56	80	129·87
350	49·66	730	81·07	10	107·20	90	130·45
360	50·61	740	81·81	20	107·85	1500	131·03
370	51·54	750	82·55	30	108·49	10	131·61
380	52·46	760	83·28	40	109·13	20	132·19
390	53·38	770	84·01	50	109·76	30	132·77
400	54·29	780	84·73	60	110·40	40	133·35
410	55·19	790	85·46	70	111·03	50	133·93
420	56·08	800	86·18	80	111·67	60	134·50
430	56·97	810	86·89	90	112·30	70	135·08
440	57·85	820	87·61	1200	112·92	80	135·65
450	58·72	830	88·32	10	113·55	90	136·23
460	59·59	840	89·03	20	114·17	1600	136·80
470	60·45	850	89·73	30	114·80	10	137·37

TABLE VIII.—TWO-THIRDS POWERS OF NUMBERS. 31

Table VIII.—Two-thirds powers of numbers—*continued.*

Number.	⅔ power.	Number.	⅔ power.	Number.	⅔ power.	Number.	⅔ power.
1620	137·93	2080	162·94	2920	204·28	3760	241·80
30	138·50	2100	163·99	40	205·22	80	242·65
40	139·06	20	165·02	60	206·15	3800	243·51
50	139·63	40	166·05	80	207·08	20	244·36
60	140·19	60	167·09	3000	208·01	40	245·22
70	140·75	80	168·12	20	208·93	60	246·07
80	141·32	2200	169·15	40	209·85	80	246·91
90	141·88	20	170·17	60	210·76	3900	247·76
1700	142·44	40	171·19	80	211·68	20	248·61
10	143·00	60	172·20	3100	212·59	40	249·45
20	143·55	80	173·22	20	213·51	60	250·29
30	144·11	2300	174·24	40	214·42	80	251·14
40	144·66	20	175·24	60	215·33	4000	251·98
50	145·22	40	176·25	80	216·24	20	252·82
60	145·77	60	177·25	3200	217·15	40	253·65
70	146·32	80	178·26	20	218·05	60	254·49
80	146·87	2400	179·26	40	218·95	80	255·33
90	147·42	20	180·25	60	219·85	4100	256·16
1800	147·97	40	181·24	80	220·75	20	257·00
10	148·52	60	182·23	3300	221·65	40	257·83
20	149·06	80	183·22	20	222·54	60	258·67
30	149·61	2500	184·20	40	223·44	80	259·49
40	150·15	20	185·18	60	224·34	4200	260·31
50	150·70	40	186·16	80	225·22	20	261·14
60	151·24	60	187·14	3400	226·11	40	261·96
70	151·78	80	188·11	20	226·99	60	262·78
80	152·32	2600	189·08	40	227·88	80	263·60
90	152·86	20	190·05	60	228·76	4300	264·42
1900	153·40	40	191·02	80	229·64	20	265·24
10	153·94	60	191·98	3500	230·52	40	266·06
20	154·47	80	192·93	20	231·40	60	266·87
30	155·01	2700	193·89	40	232·27	80	267·69
40	155·54	20	194·85	60	233·14	4400	268·51
50	156·08	40	195·80	80	234·02	20	269·32
60	156·61	60	196·75	3600	234·89	40	270·13
70	157·14	80	197·71	20	235·76	60	270·95
80	157·68	2800	198·66	40	236·62	80	271·76
90	158·21	20	199·60	60	237·49	4500	272·56
2000	158·74	40	200·54	80	238·36	20	273·37
20	159·79	60	201·48	3700	239·22	40	274·17
40	160·84	80	202·42	20	240·08	60	274·98
60	161·89	2900	203·35	40	240·94	80	275·78

Table VIII.—Two-thirds powers of numbers—*continued.*

Number.	⅔rd power.	Number.	⅔rd power.	Number.	⅔rd power.	Number.	⅔rd power.
4600	276·58	6150	335·67	8300	409·93	10,900	491·61
20	277·39	6200	337·49	50	411·57	11,000	494·61
40	278·19	50	339·30	8400	413·22	100	497·60
60	278·99	6300	341·11	50	414·85	200	500·58
80	279·78	50	342·91	8500	416·49	300	503·56
4700	280·58	6400	344·71	50	418·12	400	506·53
20	281·38	50	346·50	8600	419·75	500	509·48
40	282·17	6500	348·29	50	421·37	600	512·43
60	282·96	50	350·07	8700	423·00	700	515·38
80	283·76	6600	351·85	50	424·62	800	518·31
4800	284·55	50	353·62	8800	426·24	900	521·23
20	285·33	6700	355·39	50	427·85	12,000	524·15
40	286·11	50	357·16	8900	429·46	100	527·05
60	286·90	6800	358·93	50	431·06	200	529·95
80	287·68	50	360·68	9000	432·67	300	532·83
4900	288·47	6900	362 43	50	434·27	400	535·72
20	289·26	50	364·18	9100	435·86	500	538·60
40	290·05	7000	365·93	50	437·45	600	541·48
60	290·84	50	367·67	9200	439·04	700	544·34
80	291·62	7100	369·41	50	440·64	800	547·20
5000	292·40	50	371·13	9300	442·23	900	550·04
50	294·34	7200	372·86	50	443·82	13,000	552·88
5100	296·27	50	374·58	9400	445·40	100	555·70
50	298·21	7300	376·31	50	446·97	200	558·53
5200	300·15	50	378·02	9500	448·54	300	561·35
50	302·06	7400	379·74	50	450·11	400	564·16
5300	303·98	50	381·44	9600	451·68	500	566·96
50	305·89	7500	383·15	50	453·25	600	569·76
5400	307·80	50	384·85	9700	454·82	700	572·54
50	309·68	7600	386·55	50	456·39	800	575·33
5500	311·58	50	388·24	9800	457·95	900	578·10
50	313·46	7700	389·93	50	459·50	14,000	580·88
5600	315·34	50	391·62	9900	461·06	100	583·63
50	317·21	7800	393·30	50	462·61	200	586·38
5700	319·09	50	394·98	10,000	464·16	300	589·13
50	320·95	7900	396·66	100	467·25	400	591·88
5800	322·81	50	398·33	200	470·33	500	594·61
50	324·66	8000	400·00	300	473·39	600	597·34
5900	326·51	50	401·66	400	476·44	700	600·07
50	328·35	8100	403·32	500	479·49	800	602·80
6000	330·19	50	404·97	600	482·54	900	605·51
50	332·02	8200	406·63	700	485·57	15,000	608·22
6100	333·85	50	408·28	800	488·60		

TRIPLE VERSUS COMPOUND ENGINES.

The great economy resulting from the use of high-pressure steam is due to the fact that the increased pressure is obtained by the expenditure of an amount of fuel quite insignificant compared with the additional amount of energy it renders available. The following tabular statement will make this clear at a glance:—

Table IX.—Pressures and efficiencies.

Working pressure lbs. per sq. in.	Absolute pressure lbs. per sq. in.	Evaporation commences from	Evaporation effected at	Total heat of evaporation in Thermal units per pound.	Specific volume cubic feet per pound.	Total effective external work by one pound of steam in ft. lbs.	Heat used per 1000 ft. lbs. of work done Thermal units.	Saving of fuel at each step per cent.
30	45	100° F.	274° F.	1098	9·26	126 256	8·696	...
75	90	100° F.	320° F.	1112	4·81	174,472	6·373	26·7
125	140	100° F.	353° F.	1122	3·17	206,994	5·420	14·9
175	190	100° F.	377° F.	1129	2·38	231,200	4·833	9·9

Note.—The release pressure is taken to be 10 lbs. per square inch, absolute, in each case, and a back pressure of 4 lbs. per square inch, absolute, is assumed.

That is to say,—an engine using steam of 125 lbs. pressure should, other things being equal, consume nearly 15 per cent. less fuel than one using steam of 75 lbs. pressure; and one using steam of 175 lbs. pressure should effect a further saving of nearly 10 per cent.,—so that, by using 175 lbs. steam in place of 75 lbs., a saving of over 23 per cent. should be effected.

In practice, however, it is found that a compound engine, using steam of 140 lbs. pressure, shows very little economy over a similar engine using 90 lbs. steam; whilst a triple engine using 140 lbs. steam, shows even a greater economy than is theoretically due to the increased pressure,—owing, no doubt, chiefly to the diminished range of temperature in each cylinder.

The actual amount of fuel saved by using a modern triple engine, working at 160 lbs. pressure, in place of a compound engine using, say 75 lbs. steam is nearly 25 per cent.

The following Table shows the comparative performances of some Compound and Triple engines, and confirms the above statements :—

Table X.—Comparison of Triple and Compound Engines.

Name.	Cylinders and Stroke.	Work-ing press-ure.	I.H.P.	Range of temper-ature in cyls.	Consumpt. water per I.H.P. from diagrams.	Percentages of fuel saved.
"Northern,"	$\dfrac{26'' \text{ and } 56''}{60''}$	115 lbs.	1235	H.P. 102° L.P. 100°	15·4 lbs.	"Draco" over "Northern," 8·4 per cent.
"Kovno,"	$\dfrac{25'' \text{ and } 50''}{45''}$	90 lbs.	509	H.P. 93° L.P. 100°	16·6 lbs.	"Draco" over "Kovno," 15 per cent.
"Draco,"	$\dfrac{21'', 32'', \text{ and } 56''}{36''}$	110 lbs.	618	H.P. 64° M.P. 68° L.P. 66°	14·1 lbs.	"Finland" over "Draco," 8·6 per cent.
"Finland,"	$\dfrac{20'', 32'', \text{ and } 56''}{36''}$	150 lbs.	806	H.P. 70° M.P. 70° L.P. 83°	12·88 lbs.	"Finland" over "Northern," 16 per cent. "Finland" over "Kovno," 22·4 per cent.
"Bolama,"	$\dfrac{24'' \text{ and } 45''}{33''}$	80 lbs.	561	H.P. 91° L.P. 92°	16·8 lbs.	"Dynamo" over "Bolama," 22·6 per cent.
"Dynamo,"	$\dfrac{17'', 27'', \text{ and } 48''}{27''}$	150 lbs.	638	H.P. 73° M.P. 70° L.P. 80°	13·0 lbs.	"Finland" over "Bolama," 23·3 per cent.

When the average magnitudes of the strains set up in the two types of engine are compared, the result is again greatly in favour of the triple.

Suppose a three-crank triple engine and an ordinary compound, having same size of L.P. cylinder, same length of stroke, and developing the same power,—the one using steam at 150 lbs., and the other at 75 lbs. ; also let the equivalent mean pressure be 24 lbs. in each case, and let the efficiency of the expansion be the same in the two cases. Then, further, let the L.P. area be represented in each case by the number 14 ; the H.P. area of triple by 2, and the H.P. area of compound by 4, and the M.P. area of triple by 5. Then, if the equivalent, or referred, mean pressure is equally divided between the cylinders in each case, the relative work done will be as follows :—

Triple engine.	Compound engine.
H.P. cylinder, 2×56 or 112.	H.P. cylinder 4×42 or 168.
M.P. ,, $5 \times 22\cdot4$ or 112.	L.P. ,, 14×12 or 168.
L.P. ,, 14×8 or 112.	

That is,—the average strain on the rods, columns, guides, &c., is 50 per cent. more with the compound than with the triple engine. The triple engine is also better balanced than the compound, works with less vibration, gives a more equable turning moment, and consequently a higher efficiency of propeller.

It is not surprising, therefore, that the triple compares so favourably with the compound as regards wear and tear.

RATIOS OF CYLINDERS.

Two-stage or **Compound engines.**—Engines of this type are, for various reasons, still commonly fitted in paddle vessels, and in small screw steamers; they are also used for many auxiliary purposes on board ship, such as, for driving independent air and circulating pumps, centrifugal circulating pumps, feed pumps, dynamos, &c.

When fitted in **Paddle vessels,** the boiler pressure is usually from 70 to 100 lbs. for oscillating engines, and from 90 to 120 lbs. for other types,—the latter pressure having been used in many recent first-class vessels with diagonal engines.

With these boiler pressures, and a cut-off of say ·75 in the H.P. cylinder, the ratios of cylinders should be,—to give the same degree of economy as is usual in mercantile screw engines,—from about 1 : 5·25 for 90 lbs., up to about 1 : 6·75 for 120 lbs.

But considerations of weight and bulk put these ratios out of the question, and experience has shown that a ratio of about 1 : 3·25, regardless of boiler pressure, makes the best compromise. The extra quantity of coal to be carried is of little consequence where the voyage is only of three or four hours duration.

The ratios of cylinders for **auxiliary engines** are also determined more from considerations of weight and bulk than of economy, and without regard to boiler pressures,—with the result that they are rarely made so great as in the case of the propelling engines, even when the auxiliary engine must run hour for hour with the main engines; in practice this ratio is generally about 1 : 2·5 to 1 : 3 for cylinders side by side, and 1 : 3·5 to 1 : 4 for tandem engines.

Three-stage or **Triple engines,** and **Four-stage** or **Quadruple engines.**—In the case of main or propelling engines for merchant or mail steamers, space and weight are, within ordinary limits, of very little consequence, whilst coal consumption is of the utmost importance, and L.P. cylinders are therefore usually made large enough to expand the steam down to the full economical limit,—of say, 6 lbs. absolute.

The following Table shows the ratios of cylinders necessary to effect this, with a cut-off in the H.P. of ·6 of the stroke:—

Table XI.—Ratios of Cylinders.

Absolute pressure,	125	135	145	155	165	175	185	195
Ratio $\dfrac{\text{M.P.}}{\text{H.P.}}$	2·04	2·11	2·19	2·26	2·33	2·40	2·47	2·54
Ratio $\dfrac{\text{L.P.}}{\text{H.P.}}$	5	5·4	5·8	6·2	6·6	7·0	7·4	7·8
Working pressure, (above atmosphere)	110	120	130	140	150	160	170	180

The values given in the above table for the ratio $\dfrac{\text{L.P.}}{\text{H.P.}}$ are such that the nominal rate of expansion is, in each case, $\dfrac{\text{absolute pressure}}{15}$; and those for ratio $\dfrac{\text{M.P.}}{\text{L.P.}}$ are calculated by the formula,—

$$\text{M.P. area} = \frac{\text{L.P. area}}{1 \cdot 1 \times \sqrt{\text{Ratio of H.P. to L.P.}}}$$

When the working pressure exceeds 175 lbs., the steam should be expanded in four stages, or, in other words, the engine should be of the Quadruple type.

For Quadruple engines, working at 175 lbs. to 200 lbs. pressure, the relative areas of cylinders may be about 1 : 1·8 : 3·6 : 7·9 ;

$$\text{or} \quad \frac{\text{1st M.P.}}{\text{H.P.}} = 1 \cdot 8$$

$$\frac{\text{2nd M.P.}}{\text{1st M.P.}} = 2$$

$$\text{and} \quad \frac{\text{L.P.}}{\text{2nd M.P.}} = 2 \cdot 2$$

It is, of course, possible to arrange the cylinders of large engines in many different ways, and the designer must always keep in view equality of strains, and of ranges of temperature in the various cylinders.

FIG. 8. TRIPLE EXPANSION ENGINES. 37

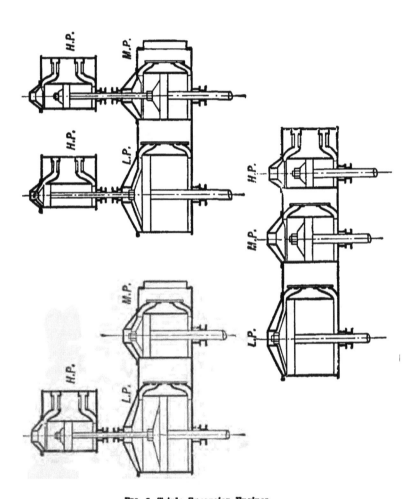

FIG. 3. Triple Expansion Engines.

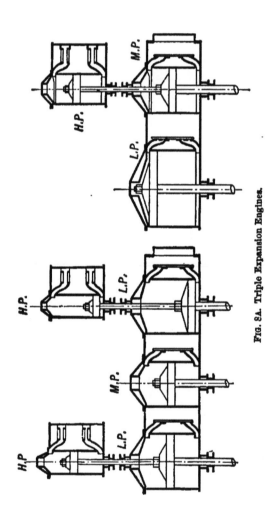

FIG. 8A. Triple Expansion Engines.

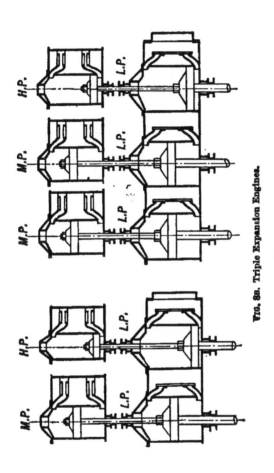

FIG. 8B. Triple Expansion Engines.

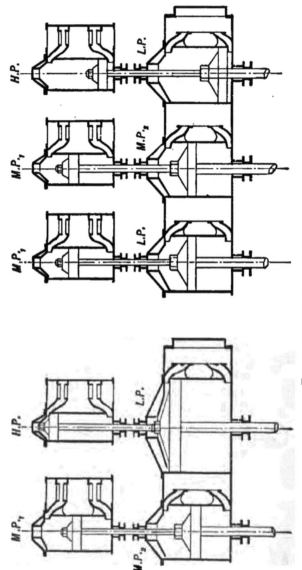

FIG. 8C. Quadruple Expansion Engines.

Figs. 8, 8A, 8B, and 8c show some of these arrangements.

It should be remarked, however, that, wherever possible, three cranks should be employed, and the tandem arrangement should only be resorted to in cases where there would otherwise be a low-pressure cylinder of very large size,—say over 100 inches diameter.

Naval engines.—In war-vessels economy is only a very secondary consideration, whilst space and weight are of the first importance, and the L.P. cylinders of triple engines are therefore rarely made more than 4·7 to 5 times the high-pressures.

The ratio of M.P. to H.P. is usually 2 to 2·25.

The boiler pressure now used is almost invariably 155 lbs.

Cases will, of course, often occur, in which the relative importance of economy, and of weight and bulk, will lie between the two extremes, of the cargo vessel on the one hand, and of the war-ship on the other, and the designer must meet them by adopting such intermediate ratios of cylinders as best suit the circumstances. In fast cargo steamers the ratio of H.P. to L.P. is commonly 1 to 5·5.

EQUIVALENT MEAN PRESSURES, ETC.

In estimating the mean pressure obtained in any multiple-stage engine, it is usual *to refer it to the L.P. cylinder, i.e.*, to calculate the pressure that would be required if the work of all three cylinders were to be done in the L.P. cylinder.

The value of this equivalent mean pressure, or Ep_m, for a triple engine is therefore,—

Rule 12.

$$Ep_m = \text{L.P. mean press.} + \frac{\text{M.P. mean press.}}{\text{Ratio of L.P. to M.P.}} + \frac{\text{H.P. mean press.}}{\text{Ratio of L.P. to H.P.}}$$

Records of the Ep_m obtained with each engine should be kept,—together with notes of the various conditions under which the trials were made,—such as pressure of steam at H.P. valve casing, degree of opening of throttle or regulator valve, amount gear is "linked-up," vacuum in condenser, &c.,—as they supply a ready means of checking calculations for sizes of cylinders, cuts-off, &c. of proposed engines of the same type, and of determining the exact values of the factors, approximate values of which are given in Table XIII., page 44.

Expansion of Steam: to determine Mean Pressures.

The mean pressure obtained, and consequently the work done, in any steam cylinder, or series of steam cylinders, depends mainly on the absolute pressure at which the steam is admitted, and the number of times it is expanded before release.

Let p_1 —absolute initial pressure.

p_m = ,, mean . ,,

r = number of expansions, or $\dfrac{\text{volume after expansion}}{\text{volume before expansion}}$.

$\dfrac{1}{r}$ = cut-off (reciprocal of r).

Then,—assuming that the steam is moderately moist, and that expansion takes place in accordance with Boyle and Mariotte's law, so that pressure varies inversely as volume (or pressure × volume = constant),—the expansion curve will be a hyperbola, and the p_m, or mean length of the ordinates to the curve will be given by the formula,—

Rule 13. $\qquad p_m = p_1 \dfrac{1 + \text{hyp. log } r}{r}$.

The following Table gives the values of $\dfrac{1 + \text{hyp. log } r}{r}$ for various numbers of expansions, and will be found useful in calculations relating to mean pressures, &c. :—

Table XII.—Expansion of Steam.

r	$\dfrac{1}{r}$	$\dfrac{1+\text{hyp.log }r}{r}$	$\dfrac{17-16r^{-\frac{1}{16}}}{r}$	r	$\dfrac{1}{r}$	$\dfrac{1+\text{hyp.log }r}{r}$	$\dfrac{17-16r^{-\frac{1}{16}}}{r}$
20	·050	·1998	·186	5·00	·200	·5218	·506
18	·055	·2161	...	4·44	·225	·5608	...
16	·062	·2358	...	4·00	·250	·5965	·582
15	·066	·2472	...	3·63	·275	·6308	...
14	·071	·2599	...	3·33	·300	·6615	·648
13·33	·075	·2690	·254	3·00	·333	·6993	...
13	·077	·2742	...	2·86	·350	·7171	·707
12	·083	·2904	...	2·66	·375	·7440	...
11	·091	·3089	...	2·50	·400	·7664	·756
10	·100	·3303	·314	2·22	·450	·8095	·800
9	·111	·3552	...	2·00	·500	·8465	·840
8	·125	·3849	·370	1·82	·550	·8786	·874
7	·143	·4210	...	1·66	·600	·9066	·900
6·66	·150	·4347	·417	1·60	·625	·9187	...
6·00	·166	·4653	...	1·54	·650	·9292	·926
5·71	·175	·4807	...	1·48	·675	·9405	...

Absolute initial pressure should be taken at 3 to 5 lbs. below the absolute boiler pressure, and back pressure at about 4 lbs. ; also, number of expansions is to be understood as the *nominal* number, or $\dfrac{\text{L.P. capacity}}{\text{H.P. capacity at cut-off}}$, without regard to clearances, or to the occurrence of release in final cylinder before end of stroke.

Then, to determine the p_m, when the boiler pressure is fixed, and the best number of expansions agreed upon,—multiply the p_1 by the quantity found in column 3 of the above Table opposite the proper number of expansions, and subtract back pressure from the product.

Column 4 gives the multipliers for *dry saturated* steam, calculated by Rankine's formula $\dfrac{p_m}{p_1} = \dfrac{17 - 16r^{-\frac{1}{16}}}{r}$ which is based on the assumption that $pv^{\frac{17}{16}} = \text{constant}$. It will be seen however, that, excepting at very high rates of expansion, there is no very great difference between column 3 and column 4.

As values of mean pressures, found by aid of above table, are determined without taking into consideration such disturbing elements as clearance, compression, radiation, friction of ports and passages, receiver "drop," initial condensation and subsequent re-evaporation, &c., it is necessary to allow for these in some way; but, the calculations for their values being laborious, and involving so many assumptions as to be of doubtful accuracy when finished, it is best to allow for them all at once by multiplying by a single factor, the value of which is derived from experience.

Then $p_m \times$ factor $= Ep_m$, or equivalent mean pressure,—(*q.v.* page 41).

The following Table gives the average value of this factor for the various types of engine, and under the various conditions named :—

Table XIII.—Ratios of Actual Mean Pressures to Theoretical.

Description of Engine.		Jacketed.	Unjacketed.
(1) Expansion taking place all in one cylinder; ports of average size; and ordinary slide valve driven by ordinary eccentric;—as in various auxiliary engines, . .		.	{ ·76 to ·80 MEAN ·78
(2) Expansion taking place all in one cylinder; ports of average size: and ordinary slide valves driven by ordinary eccentric gear;—as in low-pressure paddle engines, .		.	{ ·68 to ·75 MEAN ·71
(3) Expansion taking place in two cylinders with receiver between; ports of average size; ordinary slide valves (H.P. single-ported, and L.P. double-ported); and ordinary eccentric gear;—as in paddle engines, small screw engines, and some auxiliary engines, ...	Screw Paddle Auxly.	{ ·67 to ·73 MEAN ·70 . .	·58 to ·68 MEAN ·63 ·55 to ·65 MEAN ·60 ·45 to ·50 MEAN ·47
(4) Expansion taking place in two cylinders, placed in same line, and exhausting direct from one to the other; ports, slide valves, and gear as in (3);—as occasionally used, ...		.	{ ·71 to ·75 MEAN ·72
(5) Expansion taking place in three cylinders, placed side by side, with receivers between them; ports of average size; ordinary slide valves (say piston valve for H.P. and double-ported flat slides for M.P. and L.P.); and ordinary eccentric gear;—as used in most merchant steamers, . . .		{ ·64 to ·68 MEAN ·66	·60 to ·66 MEAN ·63
(6) Engines same as (5); but ratio of H.P. to L.P. not exceeding 1:5; and running at 90 to 140 revolutions;—as in ironclads and cruisers, ...		{ ·55 to ·65 MEAN ·60	.
(7) Engines same as (5); but ratio of H.P. to L.P. not exceeding 1:5; and running at 200 to 350 revolutions;—as in torpedo gunboats, &c., ...		.	{ ·60 to ·67 MEAN ·63

In the case of any multiple-stage engine, where the power is divided equally amongst the cylinders, whose number is N, and where the equivalent mean pressure is Ep_m, the mean pressures in the various cylinders are as follows :—

L. P. mean pressure $= \dfrac{Ep_m}{N}$.

M.P.   ,, $= \dfrac{Ep_m}{N} \times \dfrac{\text{L. P. area}}{\text{M.P. area}}$.

H.P.   ,, $= \dfrac{Ep_m}{N} \times \dfrac{\text{L. P. area}}{\text{H.P. area}}$.

PISTON SPEED.

Usually reckoned in feet per minnte.

Mean value $= 2 \times$ stroke in feet $\times$ revolutions per minute.

Experience with naval machinery has shown that the largest pistons (*i.e.*, up to 100 inches diameter) may be safely run at speeds up to 850 and 900 feet per minute.

With the lighter machinery of torpedo gunboats speeds of 1000 to 1100 feet per minute have given perfectly satisfactory results.

In the case of torpedo boats, the speeds usually employed are from 800 to 1200 feet per minute.

The standard practice in cargo steamers is very fairly represented by the expression $144 \sqrt[3]{S}$, where S is the stroke in inches. (North-East Coast rule—*sec* page 2.)

In large passenger and mail steamers the speed is now commonly 700 to 800 feet per minute.

In the case of paddle engines, and espécially of oscillating paddle engines, the available range of piston speeds is very closely limited by the other conditions of the case, and has not often in practice exceeded,—

For vertical oscillating engines,   .   .   450 feet per minute.
,,   inclined   ,,   .   .   520   ,,
,,   ,,   engines with connecting rods,   550 to 600 ,,

As the powers obtained from a given weight and bulk of engine vary almost directly as the piston speed, the efforts of engineers are constantly directed to obtain higher speeds, without undue increase of wear and tear, and there is therefore every probability of advance in this direction.

By increasing piston speed a perceptible reduction can also be made in the percentage of heat lost by radiation.

There are two ways of getting increased piston speed, viz.:—

(1) Increase length of stroke and let revolutions remain same ;
(2) Increase revolutions and let length of stroke remain same ;
but there is this important difference in the result obtained,—

In the first case, the cost, bulk, and weight of engines all increase directly as piston speed ; whilst in the second case, there is practically no increase under any of these heads ; and it is therefore in this second direction that the designer must go when limited as regards space and weight.

On the other hand, it must not be forgotten that, for a given cylinder capacity, the long stroke and small diameter cylinder has a distinct advantage over its rival, in its smaller piston area, and consequently reduced pressures and strains.

The most evident limits to increase of piston speed (and of revolutions) are,—the size of ports and consequent size and weight of valves, and the inertia of the reciprocating parts.

REVOLUTIONS.

In the case of paddle engines, the revolutions are, of course, strictly limited by the diameter of wheel and speed of vessel, and increased power must therefore be obtained either by increased stroke, pressure, or diameter of cylinders.

The following Table gives the maximum numbers of revolutions per minute usual in the classes of vessels named :—

Table XIV.—Revolutions.

Description or Class.	Revolutions.	Indicated Horse-power.
Battle Ships and First Class Cruisers, ...	90 to 110	Per set of engines. 10,000 to 4000
Second Class Cruisers, . .	120 to 140	Per set of engines. 4500 to 3000
Torpedo Gunboats, . . .	220 to 350	Per set of engines. 2500 to 1500
Torpedo Boats, . . .	350 to 600	2000 to 800
Large Mail Steamers, . .	65 to 75	16,000 to 8000
Fast Passenger Steamers, . .	90 to 140	3000 to 1500
Ordinary Cargo Steamers, .	65 to 100	2500 to 300
Paddle Steamers, . . .	25 to 55	...

For ordinary cargo steamers, the relation between length of stroke and revolutions given by the rule—piston speed = 144 $\sqrt[2]{S}$ (see page 2) shown in the following Table :—

elation between Stroke and Revolutions,
Cargo Steamers.

ions ute.	Stroke in inches.	Revolutions per minute.	Stroke in inches.	Revolutions per minute.
	33	84	48	65
	36	79	51	63
	39	75	54	61
	42	71	57	58
	45	68	60	56

STROKE OF PISTON.

The following are the lengths of stroke usual in the various classes of vessels named :—

Table XVI.—Stroke of Piston.

Battle Ships and First Class Cruisers,	45 to 51 inches.
Second Class Cruisers,	36 to 42 ,,
Torpedo Gunboats,	18 to 21 ,,
Torpedo Boats,	9 to 18 ,,
Large Mail Steamers,	60 to 78 ,,
Fast Passenger Steamers,	30 to 48 ,,
Ordinary Merchant Steamers,	18 to 60 ,,
Paddle Steamers,	30 to 102 ,,

When engines of the overhead beam type are used in paddle vessels the stroke is sometimes as much as 10 feet and even 12 feet.

In ordinary cargo steamers the length of stroke is usually about 66 × diameter of L.P. cylinder. (*See* Table I., page 4.)

TO CALCULATE DIAMETER OF CYLINDER FOR A GIVEN POWER.

The following rules apply to L.P. cylinders only,—the sizes of M.P. and H.P. cylinders being supposed to be fixed from those of the L.P. by the rules given in section on "Ratio of cylinders," pages 35–41.

For rapid calculations, use the formula for Estimated Horse Power given on page 1, and here repeated ; the result will usually be within 5 per cent.

Rule 14.
$$E.H.P. = \frac{D^2 \times \sqrt{p} \times R \times S}{8500}$$

where D = diameter of L.P. cylinder in inches.
p = absolute boiler pressure.
R = revolutions per minute.
S = stroke in feet.

If the size of cylinder is required for a given N.H.P., on the basis of a rule of the form,—

$$N.H.P. = \frac{d_1^2 + d_2^2 + d_3^2}{F}$$

where d_1, d_2, d_3, are respectively diameters of H.P., M.P., and L.P. cylinders, and F is some factor representing number of circular inches of piston per N.H.P.—put r for ratio $\dfrac{H.P.\ area}{L.P.\ area}$, and R for ratio $\dfrac{M.P.\ area}{L.P.\ area}$; then,—

Rule 15.   Diameter of L.P. $= \sqrt{\dfrac{N.H.P. \times F}{r + R + 1}}$

If the rule takes the form,—

$$N.H.P. = \frac{d_1^2 + d_2^2 + d_3^2}{F_1} \times \sqrt[3]{S}$$

where S is the stroke in inches; then,—

Rule 16.   Diameter of L.P. $= \sqrt{\dfrac{N.H.P. \times F_1}{(r + R + 1)\sqrt[3]{S}}}$.

If the size of cylinder for a given **Indicated Horse-power** is required, let S stand for piston speed in feet per minute, and Ep_m for the equivalent mean pressure in pounds per square inch (calculated by rules given on pages 41–45); then,—

Rule 17.   Area of L.P. cylinder $= \dfrac{I.H.P. \times 33,000}{Ep_m \times S}$;

or, if piston speed is required,—

Rule 18.   **Piston speed** $= \dfrac{I.H.P. \times 33,000}{L.P.\ area \times Ep}$;

or, again, if equivalent mean pressure is required,—

Rule 19. $Ep = \dfrac{I.H.P. \times 33,000}{L.P.\ area \times S}$;

and lastly, if I.H.P. i required,—

Rule 20.   I.H.P. $= \dfrac{L.P.\ area \times S \times Ep_m}{33,000}$.

CYLINDER PORTS, PIPES, AND PASSAGES.

In fixing the sizes of cylinder ports, &c., it is necessary—as in so many other cases—to discover the best compromise. By increasing the port areas, for instance, "wire-drawing" is diminished, a freer exhaust is obtained, and the resulting indicator diagram is fuller; but, at the same time, the size and weight of cylinder, of slide-valve, and of valve-gear, are increased, and the loss may easily balance the gain.

The following figures give such speeds of steam as are usual in triple engines of the best class; but it must be understood that in dealing with very high piston speeds (say over 900 feet per minute), it is not always either possible or advisable to give such large areas:—

Table XVII.—Speeds of Steam.

Main steam pipe, 8000 ft. per minute; or, say 8100 ft., then—

$$\text{Diameter} = \frac{\text{Dia. of H.P. cyl.}}{90} \times \sqrt{\text{Mean piston speed.}}$$

Mean of maximum valve openings,
- H.P. — 7,500 ft. per min.
- M.P. — 9,000 ,,
- L.P. —12,000 ,,

Ports (during exhaust),
- H.P. — 5,800 ,,
- M.P. — 7,200 ,,
- L.P. — 8,600 ,,

Nearly equivalent to 40, 50, and 60 c. ft. of cylinder per sq. inch of port per minute.

Exhaust pipe or passage from one cylinder to next or to condenser,
- H.P. — 4,500 ,,
- M.P. — 5,500 ,,
- L.P. — 7,000 ,,

Ports (during exhaust) in light, high-speed engines,
- H.P. — 5,800 ,,
- M.P. — 8,600 ,,
- L.P. —11,500 ,,

Nearly equivalent to 40, 60, and 80 c. ft. of cylinder per sq. inch of port per minute.

For **Two-stage** or **Compound** engines use the above figures, only omitting those referring to M.P. cylinders.

The following Table is based on the above figures, and gives the proportions at a glance,—where A is the area of the cylinder :—

4

Table XVIII.—Divisors for Ports, &c.

Piston speed, feet per minute.	H.P.A. ÷ Area of main steam.	H.P.A. ÷ Mean area of H.P. max. openings.	M.P.A. ÷ Mean area of M.P. max. openings.	L.P.A. ÷ Mean area of L.P. max. openings.	H.P.A. ÷ Area of H.P. ports.	M.P.A. ÷ Area of M.P. ports.	L.P.A. ÷ Area of L.P. ports.	H.P.A. ÷ Area of pipe from H.P. to M.P.	M.P.A. ÷ Area of pipe from M.P. to L.P.	L.P.A. ÷ Area of main eduction.
200	40·0	37·5	45·0	60·0	29·0	36·0	43·0	22·5	27·5	35·0
250	32·0	30·0	36·0	48·0	23·2	28·8	34·4	18·0	22·0	28·0
300	26·6	25·0	30·0	40·0	19·3	24·0	28·6	15·0	18·3	23·3
350	22·8	21·4	25·7	34·3	16·6	20·6	24·6	12·8	15·7	20·0
400	20·0	18·7	22·5	30·0	14·5	18·0	21·5	11·2	13·7	17·5
450	17·7	16·6	20·0	26·6	12·9	16·0	19·1	10·0	12·2	15·5
500	16·0	15·0	18·0	24·0	11·6	14·4	17·2	9·0	11·0	14·0
550	14·5	13·6	16·3	21·8	10·5	13·1	15·6	8·2	10·0	12·7
600	13·3	12·5	15·0	20·0	9·6	12·0	14·3	7·5	9·1	11·6
650	12·3	11·5	13·9	18·4	8·9	11·1	13·2	6·9	8·4	10·7
700	11·4	10·7	12·8	17·1	8·3	10·3	12·3	6·4	7·8	10·0
750	10·6	10·0	12·0	16·0	7·7	9·6	11·5	6·0	7·3	9·3
800	10·0	9·4	11·2	15·0	7·2	9·0	10·7	5·6	6·9	8·7
850	9·4	8·8	10·6	14·1	6·8	8·5	10·1	5·3	6·5	8·2
900	8·9	8·3	10·0	13·3	6·4	8·0	9·5	5·0	6·1	7·8
950	8·4	7·9	9·5	12·6	6·1	7·6	9·0	4·7	5·8	7·4
1000	8·0	7·5	9·0	12·0	5·8	7·2	8·6	4·5	5·5	7·0

To determine the area of any port, pipe, passage, &c., for any given speed of piston,—Divide the area of the cylinder by the number found in the proper column opposite the given piston speed, and quotient will be area required.

GENERAL REMARKS ON STRENGTHS, &c.

The dimensions and proportions given for the various details of engines, in this and succeeding sections, are in all cases such as experience has shown to give satisfactory results, both as regards stiffness and strength, and also as regards durability. Where weight is not a consideration of importance, the various parts may be made heavier, or larger, but, with good material this is unnecessary ; on the other hand, they may be, and are sometimes, made smaller or lighter,— as in some Naval vessels for example,—but in such cases, the increased risk of breakdown, and the increased wear and tear when working at full power, are understood, and accepted as the price that must be paid for reduction in weight.

CYLINDER BARRELS.

Cylinder barrels should be made of good sound cast-iron, at least twice melted, and of the thicknesses given by the following rules :—

Rule 21. $T = \dfrac{D}{6000}(p + 50) + \cdot 2$; for cylinders fitted with liners.

Rule 22. $T = \dfrac{D}{6000}(p + 50) + \cdot 4$; for cylinders without liners.

Where T is thickness of barrel, and D diameter of cylinder, both in inches ; and p is maximum pressure in cylinder,—the values of which are assumed to be as follows :—

For H.P. cylinders of Triple and Quadruple

		engines,	$p =$ boiler pressure.
H.P.	,,	Compound	,,	$p =$ boiler pressure + 20.
M.P.	,,	Triple	,,	$p = \cdot 6 \times$ boiler pressure.
M.P.₁	,,	Quadruple| ,,	$p = \cdot 7 \times$,,
M.P.₂	,,	,,	,,	$p = \cdot 45 \times$,,
L.P.	,,	Compound	,,	$p = \cdot 5 \times$ (boiler pressure + 20.)
L.P.	,,	Triple	,,	$p = \cdot 37 \times$ boiler pressure.
L.P.	,,	Quadruple| ,,	$p = \cdot 3 \times$,,

It is also assumed, as previously stated, that compound engines work at pressures between 70 lbs. and 120 lbs.; Triple engines between 120 lbs. and 175 lbs.; and Quadruple engines between 175 lbs. and 200 lbs.

For working pressure of 160 lbs. the formulæ may be reduced to the forms given at the heads of the columns in the following Table, which is calculated in accordance with them, for 160 lbs. :—

Table XIX.—Cylinder Barrels of Triple Engines (160 lbs.).

Dia. of Cylinder in ins.	Thickness in inches, WITH LINERS						Thick. in inches, WITHOUT LINERS					
	H.P. $\frac{D}{23.5}+.2$		M.P. $\frac{D}{41}+.2$		L.P. $\frac{D}{55}+.2$		H.P $\frac{D}{23.5}+.4$		M.P. $\frac{D}{41}+.4$		L.P. $\frac{D}{55}+.4$	
10							·75	3/4			·58	5/8
15							·93	15/16			·67	11/16
20	·90	15/16					1·10	1 1/8			·76	13/16
25	1·07	1 1/16	·81	13/16			1·27	1 1/4	1·01	1	·85	7/8
30	1·25	1 1/4	·93	15/16			1·45	1 7/16	1·13	1 1/8	·94	1
35	1·42	1 7/16	1·05	1 1/16	·83	13/16	1·62	1 5/8	1·25	1 1/4	1·03	1 1/16
40	1·60	1 5/8	1·17	1 5/16	·93	15/16	1·80	1 13/16	1·37	1 3/8	1·13	1 1/8
45	1·77	1 3/4	1·29	1 5/16	1·02	1 1/16	1·97	2	1·49	1 1/2	1·22	1 1/4
50	1·95	1 15/16	1·42	1 7/16	1·11	1 1/8	2·15	2 1/8	1·62	1 5/8	1·31	1 5/16
55	2·12	2 1/8	1·54	1 9/16	1·20	1 1/4	2·32	2 5/16	1·74	1 3/4	1·40	1 7/16
60	2·30	2 5/16	1·66	1 11/16	1·29	1 5/16	2·50	2 1/2	1·86	1 7/8	1·49	1 1/2
65			1·78	1 13/16	1·38	1 3/8			1·98	2	1·58	1 9/16
70			1·90	1 15/16	1·47	1 1/2			2·10	2 1/8	1·67	1 11/16
75			2·03	2 1/16	1·56	1 9/16			2·23	2 1/4	1·76	1 3/4
·80			2·15	2 3/16	1·65	1 11/16			2·35	2 3/8	1·85	1 7/8
85					1·74	1 3/4					1·94	1 15/16
90					1·83	1 7/8					2·03	2 1/16
95					1·93	1 15/16					2·13	2 1/8
100					2·02	2 1/16					2·22	2 1/4
105					2·11	2 1/8					2·31	2 5/16
110					2·20	2 1/4					2·40	2 7/16

The thicknesses given by the rule for the cylinders of Compound engines working at 100 lbs. pressure are exhibited in Table XX.

Table XX.—Cylinder Barrels of Compound Engines (100 lbs.).

Dia. of Cylinder in ins.	Thickness in inches, WITH LINERS.				Thick. in inches, WITHOUT LINERS.			
	H.P. $\frac{D}{35\cdot3}+\cdot2$		L.P. $\frac{D}{55}+\cdot2$		H.P. $\frac{D}{35\cdot3}+\cdot4$		L.P. $\frac{D}{55}+\cdot4$	
10					·68	11/16	·58	5/8
15					·82	13/16	·67	11/16
20	·76	3/4			·96	1	·76	13/16
25	·91	15/16			1·11	1 1/8	·85	7/8
30	1·05	1 1/16			1·25	1 1/4	·94	1
35	1·19	1 3/16	·83	13/16	1·39	1 3/8	1·03	1 1/16
40	1·33	1 5/16	·93	15/16	1·53	1 1/2	1·13	1 1/8
45	1·47	1 1/2	1·02	1 1/16	1·67	1 11/16	1·22	1 1/4
50	1·61	1 5/8	1·11	1 1/8	1·81	1 13/16	1·31	1 5/16
55	1·75	1 3/4	1·20	1 1/4	1·95	1 15/16	1·40	1 7/16
60	1·90	1 7/8	1·29	1 5/16	2·10	2 1/8	1·49	1 1/2
65	2·04	2 1/16	1·38	1 3/8	2·24	2 1/4	1·58	1 9/16
70			1·47	1 1/2			1·67	1 11/16
75			1·56	1 9/16			1·76	1 3/4
80			1·65	1 11/16			1·85	1 7/8
85			1·74	1 3/4			1·94	1 15/16
90			1·83	1 7/8			2·03	2 1/16
95			1·93	1 15/16			2·13	2 1/8
100			2·02	2 1/16			2·22	2 1/4
105			2·11	2 1/8			2·31	2 5/16
110			2·20	2 1/4			2·40	2 7/16

For the very light machinery of torpedo gun-boats, &c., the thicknesses of cylinders for working pressure of 175 lbs. may be as shown in following Table :—

Table XXI.—Cylinder Barrels of Torpedo Gunboats.

Diameter of cylinder in inches.	H.P. Thickness in inches $\frac{D}{45}+\cdot35$.		M.P. Thickness in inches $\frac{D}{52}+\cdot3$.		L.P. Thickness in inches. $\frac{D}{60}+\cdot25$.	
10	·57	9/16	...	...		
15	·68	11/16	·58	9/16		
20	·79	13/16	·68	11/16	·58	9/16
25	·90	15/16	·78	13/16	·67	11/16
30	1·01	1	·88	7/8	·75	3/4
35	1·12	1 1/8	·97	1	·83	13/16
40			1·07	1 1/16	·91	7/8
45			1·16	1 3/16	1·00	1
50					1·08	1 1/16
55					1·16	1 3/16

The mixture of iron for cylinders of this type must, of course, be of a very special character, and should contain a good proportion of best cold-blast iron.

For thicknesses of cylinder barrels for oscillating engines see page 68.

The barrels of all cylinders are improved by the addition of external stiffening ribs or rings; these may have a thickness of 1·5 × thickness of barrel, and may stand ·75 × thickness of barrel above the surface, whilst they may be pitched about 12 × thickness of barrel apart.

CYLINDER LINERS.

Cast-iron cylinder liners should be of thickness given by the rule,—

Rule 23.
$$T = \cdot 8 \left\{ \frac{D}{6000}(p + 50) \right\} + \cdot 35$$

where T is thickness, and D diameter, both in inches, and p has the same values as given in section on cylinder barrels, above.

The following Table is calculated by means of the above formula, for liners of triple engines working at 160 lbs. pressure :—

Table XXII.—Cast-iron Cylinder Liners (160 lbs.).

Diameter in inches.	Thickness in inches. H.P.		M.P.		L.P.	
10	...	...	...	...	...	...
15	...	...	...	...	...	...
20	0·91	15/16	...	...	...	...
25	1·05	1 1/16	0·83	13/16	...	...
30	1·19	1 3/16	0·93	15/16	...	...
35	1·33	1 5/16	1·03	1	0·85	7/8
40	1·47	1 1/2	1·13	1 1/8	0·92	15/16
45	1·61	1 5/8	1·22	1 1/4	1·00	1
50	1·75	1 3/4	1·32	1 5/16	1·08	1 1/16
55	1·89	1 7/8	1·42	1 7/16	1·15	1 1/8
60	2·03	2	1·52	1 1/2	1·22	1 1/4
65	...	...	1·62	1 5/8	1·20	1 3/16
70	...	...	1·71	1 11/16	1·36	1 3/8
75	...	...	1·81	1 13/16	1·44	1 7/16
80	...	...	1·91	1 15/16	1·51	1 1/2
85	...	...	...	...	1·59	1 9/16
90	...	...	...	...	1·66	1 5/8
95	...	...	...	...	1·73	1 3/4
100	...	...	...	...	1·80	1 13/16
105	...	...	...	...	1·88	1 7/8
110	...	...	...	...	1·95	1 15/16

When of forged steel, cylinder liners should be of thickness given by

Rule 24.
$$T = \cdot 65 \left\{ \frac{D}{6000}(p + 50) \right\} + \cdot 3$$

where all symbols have same meanings as above.

Table XXIII. is calculated by means of this formula, for liners of triple engines working at 160 lbs. pressure.

Table XXIII.—Forged steel Cylinder Liners (160 lbs.).

Diam. of cylinder in inches.	Thickness in inches.					
	H.P.		M.P.		L.P.	
20	·75	3/4				
25	·87	7/8				
30	·98	1	·77	13/16		
35	1·10	1 1/8	·85	7/8		
40	1·21	1 1/4	·93	15/16		
45	1·33	1 5/16	1·01	1	·83	7/8
50	1·44	1 7/16	1·09	1 1/8	·89	15/16
55	1·55	1 9/16	1·17	1 3/16	·95	1
60	1·66	1 11/16	1·25	1 1/4	1·01	1 1/16
65			1·33	1 5/16	1·07	1 1/16
70			1·41	1 7/16	1·12	1 1/8
75			1·49	1 1/2	1·18	1 3/16
80			1·57	1 9/16	1·24	1 1/4
85					1·30	1 5/16
90					1·36	1 3/8
95					1·42	1 7/16
100					1·48	1 1/2
105					1·54	1 9/16
110					1·60	1 5/8

For liners of horizontal cylinders the above thicknesses may be increased by 1/16-inch.

For steel liners the piston packing rings should be of hard bronze, as cast-iron will not work satisfactorily with forged steel.

CYLINDER ENDS AND COVERS.

In most engines the cylinder end has, in addition to supporting the uniformly distributed load due to the steam pressure, to take the more or less locally applied pull of the frames or columns, and distribute it to the barrel; and the various ribs or webs must therefore be carefully arranged to effect this.

When the L.P. diameter is over 70 inches, it is desirable to make the end double, and this must of course be done for any size when steam jackets are required.

The double bottom adds considerably to the weight of a cylinder, as the inner metal must still be made strong enough to stand all local shock and strain; and it increases the risk of unsoundness in the casting.

In Naval work the double bottom is frequently dispensed with, on account of its weight; but owing to the use of the conical or "dished" steel piston, the cylinder end can also be "dished," and is thus given such additional structural strength that L.P. cylinders are commonly made with single ends up to 90 inches diameter.

Table XXIV. gives the thicknesses of single and double cylinder ends for triple engines working at 160 lbs. pressure, the following conditions being assumed : —

(a) The over all depth or thickness of a double end is not less than five times that of the metal given in the table.

(b) In a single end, the total depth of the central ring (forming the hole for boring-bar) is not less than 5½ times the thickness of end given in the table.

(c) The number, strengths, and positions of webs, &c., are as customary in "average" practice for two, three, or four columns or frame attachments.

As much depends on (c) the Table and rules from which it is calculated must be used with judgment, and checked by the rule for flat surfaces given on page 59.

The rules from which Table XXIV. is calculated are, —

Rule 25. Thickness of single cylinder end = ·85 × f.

Rule 26. Thickness of double cylinder end = ·7 × f.

where f is thickness of barrel (columns 8, 10, and 12, Table XIX.) + ·25 in.

Table XXIV.—Cylinder Ends.

Diam. of cylin. in inches	H.P.—Thick. in inches.				M.P.—Thick. in inches.				L.P.—Thick. in inches.			
	Single end.		Double end.		Single end.		Double end.		Single end.		Double end.	
10	·85	7/8	·70		·76	3/4	·63		·70	3/4	·58	
15	1·00	1	·83		·87	7/8	·71		·78	13/16	·64	
20	1·15	1 3/16	·94		·97	1	·80		·86	7/8	·71	
25	1·29	1 5/16	1·06		1·07	1 1/16	·88		·94	15/16	·77	
30	1·44	1 7/16	1·19	1 3/16	1·17	1 3/16	·96		1·01	1	·83	
35	1·59	1 9/16	1·31	1 5/16	1·27	1 1/4	1·05		1·09	1 1/8	·89	
40	1·74	1 3/4	1·43	1 7/16	1·38	1 3/8	1·13	1 1/8	1·17	1 3/16	·96	
45	1·89	1 7/8	1·55	1 9/16	1·48	1 1/2	1·22	1 1/4	1·25	1 1/4	1·03	
50	2·04	2 1/16	1·67	1 11/16	1·59	1 5/8	1·31	1 5/16	1·32	1 5/16	1·09	
55	2·19	2 3/16	1·80	1 13/16	1·69	1 11/16	1·40	1 7/16	1·40	1 3/8	1·15	
60	2·34	2 5/16	1·92	1 15/16	1·79	1 13/16	1·48	1 1/2	1·48	1 1/2	1·22	
65					1·89	1 7/8	1·54	1 9/16	1·56	1 9/16	1·28	1 5/16
70					2·00	2	1·60	1 5/8	1·63	1 5/8	1·35	1 3/8
75					2·11	2 1/8	1·66	1 11/16	1·71	1 3/4	1·41	1 7/16
80					2·21	2 3/16	1·82	1 13/16	1·78	1 13/16	1·47	1 1/2
85									1·86	1 7/8	1·53	1 9/16
90									1·94	1 15/16	1·60	1 5/8
95									2·02	2	1·66	1 11/16
100									2·10	2 1/8	1·73	1 3/4
105									2·18	2 3/16	1·79	1 13/16
110									2·25	2 1/4	1·85	1 7/8

There is something to be said in favour of the practice of making the inner metal of double bottoms and covers a little heavier than the outer in large cylinders, and if it is desired to do this, the inner metal may be made $\frac{1}{16}$-inch thicker than shown in table, and the outer $\frac{1}{16}$-inch lighter than table.

Single cylinder covers should not be used for larger sizes than about 12 inches H.P., 24 inches M.P., and 40 inches L.P.; and in vertical engines it is better to make all but the very smallest of the double type. Single covers may be much strengthened by adding small T or bulb heads to the ribs. Single covers should be of thickness given by,—

Rule 27.   **Thickness of single cylinder cover = ·77 × f ;**

and for double covers,—

Rule 28.   **Thickness of double cylinder cover = ·65 × f ;**

similar general conditions being assumed, as in the case of cylinder ends.

A sufficiently close approximation to these thicknesses, for triple engines working at 160 lbs. pressure, may be obtained by multiplying the thicknesses given in Table XXIV. for cylinder ends by ·9.

In large cylinder covers and bottoms, or ends, care should be taken that there is no lack of strength in the central ring which surrounds the stuffing-box or manhole in the one case, and the boring hole in the other; it should have the same thickness as the metal of the cover or end, and good radii should be used in the corners formed by its junction with the metals of the bottom or cover, and of the various webs.

Very light cylinders, of the type referred to in Table XXI., are better made with loose covers for both ends, the thicknesses of which should be as follows :—

Rule 29.   **Thickness of end covers = ·5 × f_1 ;**

where f_1 is the thickness of cylinder barrel (from Table XXI.) + ·25 inch.

They should be made of the same special iron as the barrels, and should each have from 6 to 12 deep stiffening ribs, the mean thickness of which (they should have $\frac{1}{16}$-inch taper for moulding) may be $\frac{1}{16}$-inch less than that of the cover.

CYLINDER VALVE BOXES, PORTS, ETC.

The casings, ports, passages, &c., for piston slide-valves should be made of thicknesses given by,—

Rule 30.   **Thickness of piston valve casing = ·8 × f ;**

and when flat slide-valves are used the thicknesses of casings may be,—

Rule 31.   **Thickness of valve casing $= \cdot 7 \times f$;**

f having, in both cases, the same value as above.

It will be noted that the latter value ($\cdot 7 \times f$) is the same as for metal of double ends, the thicknesses of which (for triple engines working at 160 lbs. pressure) are given in Table XXIV.

It must be remembered that in the case of valve casings, as in that of cylinder ends, a very great deal depends on the number, size, and arrangement of stiffening webs, and the following rule should therefore be used in conjunction with the Table :—

Rule for flat surfaces.—All flat surfaces should be stiffened by webs or stays of some form, whose distance apart should not exceed,—

Rule 32.   **Pitch in inches $= \sqrt{\dfrac{t^2 \times C}{p}}$;**

where t is the thickness in 16ths of an inch, p, the pressure in pounds per square inch on the surface, and C equals 50 for cast iron, and 120 for cast steel.

These webs should be of the same thickness as the flat surface, and their depth at least 2·5 times the thickness.

When flat slide-valves are used the thickness of the metal which separates the steam ports from the steam chest or receiver, and from the exhaust port, should be in accordance with the following rule,—it being understood that the ports of large cylinders are so divided by webs that no section or compartment exceeds 20 inches in width :—

Rule 33.   **Thickness of port metal $= \cdot 65 \times f$,**

where f has the same value as before.

These thicknesses (for triple engines working at 160 lbs. pressure) may be obtained with sufficient accuracy by multiplying the figures found in columns 8 and 12 of Table XXIV. by ·9.

The slide-valve faces of cylinders, when a false face is fitted, should be made of thickness given by the rule,—

Rule 34.   **Thickness of cylinder valve-face (when a false face is fitted) $= \cdot 8 \times f$,**

and when no false face is used, the rule becomes,—

Rule 35.   **Thickness of cylinder valve-face (when no false face is used) $= \cdot 9 \times f$.**

False faces of hard cast-iron may have a thickness,—

Rule 36.   **Thickness of cast-iron false face $= \cdot 75 \times f$.**

CAST-STEEL CYLINDER COVERS.

For a flat plate secured round the edge after the manner of a cylinder cover, but yet not quite encastré, the formula may be written,—

$$t = \sqrt{\frac{3r^2 p}{4f}} \qquad . \qquad . \qquad . \qquad (a)$$

where t is the thickness in inches; r the radius of the cylinder,—also in inches; p the maximum pressure on cover in lbs. per square inch above atmosphere; and f the greatest permissible stress per square inch on the material. When a suitable value of f has been inserted, and a small constant quantity (necessary for practical reasons) added, this becomes,—

Rule 37. $t = \sqrt{\dfrac{3r^2 p}{160,000}} + \cdot 25,$

which gives the thickness (very nearly) of steel cylinder covers of the single type, ribbed as directed below, and coned to suit steel pistons, made in accordance with formulæ and Table given on pages 76–79.

For steel covers of triple engines, working at 160 lbs. pressure, this formula (after reduction and further slight adjustment of constant) takes the forms given at heads of columns in following Table :—

Table XXV.—Cast-Steel Cylinder Covers.

Dia. of cylinder in inches.	H.P. Thick. in inches $\dfrac{D}{36} + \cdot 24.$		No. of ribs $\dfrac{D+20}{6}$	M.P. Thick. in inches $\dfrac{D}{56} + \cdot 34.$		No. of ribs $\dfrac{D+20}{7\cdot5}$	L.P. Thick. in inches $\dfrac{D}{100} + \cdot 42.$		No. of ribs $\dfrac{D+20}{9}$
10	·52	$\frac{9}{16}$	5	·52	$\frac{9}{16}$	4	·52	$\frac{9}{16}$	4
20	·79	$1\frac{5}{16}$	7	·70	$\frac{3}{4}$	6	·62	$\frac{5}{8}$	5
30	1·07	$1\frac{1}{16}$	9	·87	$\frac{7}{8}$	7	·72	$\frac{3}{4}$	6
40	1·35	$1\frac{3}{8}$	10	1·05	$1\frac{1}{16}$	8	·82	$\frac{7}{8}$	7
50	1·63	$1\frac{5}{8}$	12	1·23	$1\frac{1}{4}$	10	·92	$\frac{15}{16}$	8
60	1·91	$1\frac{15}{16}$	14	1·41	$1\frac{7}{16}$	11	1·02	$1\frac{1}{16}$	9
70	...	...	...	1·59	$1\frac{5}{8}$	12	1·12	$1\frac{1}{8}$	10
80	...	...	...	1·77	$1\frac{13}{16}$	14	1·22	$1\frac{1}{4}$	12
90	...	...	...	...	...	...	1·32	$1\frac{3}{8}$	13
100	...	...	...	...	...	...	1·42	$1\frac{7}{16}$	14
110	...	...	...	...	...	...	1·52	$1\frac{9}{16}$	15

Where the size of the cover permits it, and there is no central stuffing-box, a very strong form may be obtained by fitting a central manhole (say 15 inches diameter), and surrounding it by a good stout ring of metal to form an abutment for the radial ribs.

The thickness of the ribs may be $\cdot 9 \times$ thickness of cover.

The steeper cone of the H.P. cover undoubtedly gives it considerable additional strength, but its greater liability to shock and strain from the presence of water may be considered as almost balancing the account.

CAST-IRON VALVE BOX COVERS.

Where these are circular, as in the case of covers for piston-valve boxes, a modification of the formula (*a*) given above may be again employed ; the nature of the material renders the addition of a larger constant quantity necessary, and provision must be made for stiffness in L.P. covers by adding a constant to the pressure also. The formula may then be written,—

Rule 38. $\qquad t = \sqrt{\dfrac{3 \times r^3 \times (p + 15)}{4 \times 45{,}000}} + \cdot 5.$

This will be found to give practically the same results as Rule 28 and Table XXIV., and may be used in their place for cylinder covers, if preferred.

These covers should be stiffened by radial ribs which may have a thickness of $\cdot 9 \times$ thickness of cover, and a depth at centre of $3\cdot5 \times$ thickness of cover,—the outer edge of rib being a curve of the parabolic type when the cover is flat.

It will often happen that a cover can be made more or less curved, or dished one way or the other, and this should always be done where possible,—as the whole cover may then be made lighter.

The number of ribs may be the same as for cast-iron pistons (see Table XXXVI., pages 82 and 83).

Where weigh-shaft brackets, or slide-rod guides, are attached to these covers, they must, of course, be made proportionately stronger.

For rectangular doors consisting of a single flat plate of metal of thickness (*t*), stiffened by ribs, the greatest section of which is not less than $(3 \times t) \times (\cdot 9t)$, and which are spaced in accordance with Rule 32 (page 59), the formula is,—

Rule 39. $\qquad t = \sqrt{\dfrac{K \times b^3 \times (p + 15)}{c}} + \cdot 5 \qquad . \qquad . \qquad (b)$

where $t =$ thickness in inches.

$\qquad b =$ breadth within flanges of valve-boxes, in inches.

$\qquad K = \dfrac{\text{Length}^4}{\text{Length}^4 + \text{breadth}^4}$

$\qquad p =$ max. pressure in lbs. per sq. inch to which door is subject.

$\qquad c = 70{,}000$ for single doors.

$\qquad c = 450{,}000$ for double doors (see next paragraph).

When the door is rectangular and of the double type—consisting of two equal thicknesses of metal tied together and locally stiffened by ribs, in conformity with Rule 32—the overall thickness of the door at centre should be not less than $\dfrac{\text{Shortest span}}{5}$, and the outer surface may be curved in both directions, like a "hog-back" girder. When these conditions are complied with, the formula (b), given above, will give the thickness of either the outer or inner metal if the value $c = 450,000$ be employed.

Large doors of this type, when fitted to naval engines, are commonly made in cast-steel—to save weight. When this material is used the overall thickness of door at centre may be $\dfrac{\text{Shortest span}}{6\cdot5}$, and the following variant of the above formula (b) may be used to determine thickness of metal :—

Rule 40. $t = \sqrt{\dfrac{K \times b^3 \times (p + 15)}{250,000}} + \cdot25.$

A convenient method of saving some space and weight, in connection with doors of this type, is to slightly increase the thickness of the inner metal, and then make sufficient openings in it, towards the ends, to allow of the door itself being utilized as a passage for the steam from one end of the valve-box to the other.

When L.P. valve-boxes and doors, &c., are to be subjected to high test pressures—such as 90 or 100 lbs. per sq. inch—they should be designed as for about half that pressure—45 to 50 lbs. instead of for the usual 25 or 30 lbs. to which the receiver safety-valve is loaded.

JOINTS OF CYLINDER COVERS, ETC.

As the widths and thicknesses of flanges should vary directly as the diameter of the stud employed, it is best when designing a joint to fix this diameter first.

In fixing the diameter of stud the three following conditions must be satisfied :—

(1) $\dfrac{\text{Total load on cover}}{\text{Effective area of 1 stud} \times \text{No. of studs}}$ must not exceed value given in Table XXXI., page 73.

Rule 41.—(2) The diameter of stud should not exceed $\cdot8 \times$ thickness of flange ; and the thickness of flange should not exceed $1\cdot5 \times$ thickness of metal to which it is attached.

Rule 42.—(3) The pitch of the studs should not be less than diameter $\times 3\cdot5$—(to give room for spanner)—nor more than diameter $\times 4\cdot75$.

When no water test is required a pitch of 6 diameters is often ·d for L.P. doors, &c.

The diameter is usually settled in the first instance by a rough process of trial and error.

Rule 43.—The width of joint may be from diameter × 2·3 to diameter × 3, but—unless weight is of the greatest importance—it should not be less than diameter × 2·8.

The thickness of cover flange may be from diameter × 1 to diameter × 1·25.

The limit which the pitch should in no case exceed is given by the rule,—

Rule 44. Maximum pitch in inches $= \sqrt{\dfrac{t \times 100}{p}}$

where t is the thickness of cover or door flange in sixteenths of an inch, and p the pressure in pounds per square inch, on the cover.

For covers of cast-steel the co-efficient in the above rule may be 150 in place of 100.

It is, of course, not at all necessary that studs should always carry the full load allowed by Table XXXI., and it is sometimes convenient to load them much more lightly—as, for instance, when it is desired to produce a more uniform effect by using the same size of stud for all the covers in connection with a set of cylinders.

Where bolts are used, as in the case of a steam joint between two cylinders, the thickness of the flanges should be about diameter of bolt × 1·25, but should not exceed 1·5 × (thickness of metal to which flange is attached) ; the width of the joint is, of course, also increased by an amount equal to about ·6 of the thickness of that metal.

CYLINDER RELIEF OR ESCAPE VALVES.

Rule 45.—For the L.P. cylinder these should have a diameter of about one-fifteenth the diameter of the cylinder ; for the H.P. cylinder, where priming water may have to be dealt with, the proportion may be one-eleventh of the cylinder diameter ; and for the M.P. cylinder, an intermediate proportion of say one-thirteenth.

In horizontal engines, there should be at least one valve at each end of each cylinder—(the Admiralty requirement is two valves at each end of each cylinder)—and in vertical engines there should be one at each end of the H.P. cylinders, and one at the bottom of each of the others.

The following Table of sizes of Relief-valves is calculated to give about the above named proportions, but the nearest half inch is given in most cases—as it is neither usual nor necessary to make patterns for every quarter of an inch.

Table XXVI.—Cylinder Relief Valves.

Dia. of cylinder.	Dia. of Relief Valve in Ins.			Dia. of cylinder.	Dia. of Relief Valve in Ins.		
	H.P.	M.P.	L.P.		H.P.	M.P.	L.P.
15	1½	...	...	65	6	5	4½
20	2	1¾	...	70	...	5½	4½
25	2¼	2	1¾	75	...	6	5
30	2½	2¼	2	80	...	6	5½
35	3	2½	2¼	85	...	...	5½
40	3½	3	2½	90	...	...	6
45	4	3½	3	95	...	...	6½
50	4½	4	3½	100	...	...	6½
55	5	4½	3½	105	...	...	7
60	5½	4½	4	110	...	...	7½

The H.P. and L.P. columns are equally applicable to the cylinders of Compound, Triple, and Quadruple engines.

It is very desirable that all relief-valves should be fitted with guard rings, or other suitable appliances, to prevent thoughtless screwing up of the springs whenever a slight leakage occurs.

In confined engine-rooms it is also desirable to have light pipes fitted to convey any water that may escape down to the crank-pits or bilge.

For methods of calculating strengths and proportions of springs see page 256.

RECEIVER SAFETY VALVES.

For small and medium sized cylinders (up to 60 inches diameter say), these may be of the same size and pattern as the cylinder relief-valves,—*i.e.*, L.P. receiver valve same as L.P. cylinder valve, and so on; but this rule makes the valves unnecessarily large for larger cylinders, and more appropriate sizes are given by the rule,—

Rule 46.

$$\textbf{Diameter of Receiver Safety Valve} = \frac{\sqrt{\text{Diameter of cylinder.}}}{2}$$

DRAIN VALVES, OR COCKS, FOR CYLINDERS, RECEIVERS, ETC.

These should be of the proportions given by the following Table :—

TABLE XXVII.—DRAIN VALVES.   65

Table XXVII.—Drain Valves.

Diameter of Cylinder or L.P. Cylinder, if Compound, Triple, &c.	Diameter of valve or cock.	Diameter of Cylinder or L.P. Cylinder, if Compound, Triple, &c.	Diameter of valve or cock.
Up to 10″	½″	46 to 60″	1¼″
11 to 15″	⅝″	61 to 75″	1½″
16 to 30″	¾″	76 to 90″	1¾″
31 to 45″	1″	Above 90″	2″

In Compound, Triple, and Quadruple engines, the valves or cocks in connection with the H.P. and M.P. cylinders, should be of the same size as those for L.P. cylinder.

All pipes from H.P. and M.P. drain valves should be led to hot-well or feed tank, and should be fitted with self-acting non-return valves to prevent water getting back into cylinders. The pipes from L.P. drain valves should be led to the condenser, and also be fitted with non-return valves.

The drainage from the steam jackets should be collected in suitable vessels (each fitted with a gauge glass), and then led away to the hot-well or feed tank. The pipes should be fitted with adjustable screw-down valves, placed in sight of the gauge glasses, and within easy reach of the attendants.

Reversing engine drains should be led to the condenser, and fitted with non-return valves ; but the drainage from other auxiliary engines is better led into the bilges, as it always contains a large percentage of oil.

Each steam jacket must have its own separate adjustable drain-valve, but the cylinder and receiver drains may be, to a certain extent, grouped, or led into common pipes so as to reduce the number of valves on hot-well or feed tank, and condenser.

Drain cocks of suitable sizes should be fitted to every pipe, passage, or place where water can lodge, so that, when engines are cold, they may be quite free from water.

STARTING, PASS, OR AUXILIARY VALVES.

In small Paddle engines it is usual to arrange the slide-valves to be worked by hand, and starting valves are therefore unnecessary.

In larger low-pressure Paddle engines, where the valves are driven by double eccentrics and links, the cuts-off are so late that starting-valves are not required.

Compound engines, with cuts-off at about six-tenths of the stroke, would occasionally be unhandy without some means of admitting steam to one of the cylinders later than the main slide allows, and starting valves should therefore be fitted to the L.P. cylinders so that steam may be admitted into either top or bottom port.

Triple engines, with three cranks, require only a small valve to admit steam to each receiver; the valves should be raised from their seats by means of levers acting on the spindles, and should be held shut by small spiral springs in addition to the pressure of the steam. The steam supply should be taken from the boiler side of the regulator valve, and may be about one-fifth the diameter of the main steam-pipe.

JACKET PIPES AND VALVES.

If the cylinder barrels are to be jacketed, it is very desirable to use double covers and bottoms, or ends, and admit steam to these as well. Great care should be taken to place the drains at the lowest points, and to see that water can get freely to them from all parts. When possible, the internal ribs should be arranged so as to prevent the steam rushing straight through from the supply to the drain; it should be compelled to circulate as much as possible.

H.P. cylinders should be jacketed with steam of boiler pressure; M.P. and L.P. cylinders with steam of not less than 15 lbs. above the highest pressure usually found in their respective receivers.

As it is necessary to fit reducing valves to secure these pressures in the jackets, and as reducing valves are notoriously uncertain in their action, great care should be taken,—by fitting thoroughly efficient safety-valves,—to guard against the presence of undue pressures in the jackets.

Supply and drain pipes need not exceed 1¼ inches in diameter, even in the largest engines.

It is customary to connect the cover, barrel, and end jackets, either by means of small external copper pipes, or by passages in the castings, and so have only one supply and one drain for each cylinder. For drainage of jackets see also page 65.

CAPACITY OF RECEIVERS.

For Compound engines with cranks set at from 90° to 105°, the receiver capacity should be 1 to 1·5 times capacity of H.P. cylinder; when cranks are opposite one another this capacity may be very much reduced.

Triple engines, with cranks at 120°, require only very small receivers; the exhausts from one cylinder to the next may be as short and direct as possible, and only large enough to keep down the mean speed of steam to the limit given in Table XVII., page 49; whilst the spaces around the valves need not exceed those actually required for the passage of the steam to the cylinders. When the M.P. crank leads the H.P., an even smaller receiver will suffice.

COLUMN FEET AND BOLTS.

Great care should be exercised in designing these feet, for through them the whole force of the steam on the cover is transmitted ; and, as the strain is always applied suddenly, very ample section of metal should be provided to sustain it.

Rule 47.—The area of section through these feet should be such that the strain does not exceed 600 lbs. per square inch.

The webs from the flanges of the feet should be well spread over the cylinder bottom and towards the sides, so as to distribute the strain.

Rule 48.—The bolts connecting the cylinder to the columns or frames should be such that the strain on them does not exceed 4000 lbs. per square inch of area at the bottom of the thread, and when there are a large number, of comparatively small size, it should not exceed 3000 lbs. per square inch.

See also Table XXXI., page 73, for loads that bolts may carry. The maximum or initial pressure should be used in calculating these stresses.

The feet should always be formed so as to permit of bolts being used (not studs), and two at least, in each foot, should be a driving it.

GENERAL REMARKS ON CYLINDERS.

Where weight is of great importance, as in Naval machinery, it is a common practice to shorten the cylinder, and form the port in the cover, as shown in Fig. 9,—the more ordinary plan being shown by Fig. 9A.

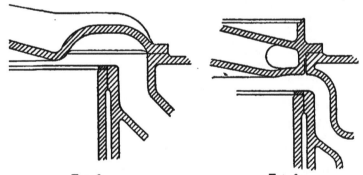

FIG. 9.   FIG. 9A.

Another method of saving weight, sometimes practised in connection with the very light machinery of torpedo gun-boats, is to lead the exhaust steam through the middle of the relief-rings on the backs of the flat slides, thus doing away with the exhaust port in the cylinder face, and shortening the slide-valves and casings to that extent.

The corners of ports, both in false faces and in cylinder faces, should be well rounded, as the castings are very apt to crack if they are made quite square.

Steel screw-stays may be used with advantage to strengthen the various ports and passages in cylinder castings.

Where a false face is secured by the usual "cheese-headed" screws, a certain measure of relief may be given to the slide-valve, and the lubrication may also be improved, by connecting the recesses in which the screw heads are sunk, by small grooves cut in the face.

Horizontal Cylinders.

Horizontal cylinder barrels should be stiffened by rings, as recommended on page 54 ; but they require something more than this, and should have—especially when the valve-box is on top—two strong ribs running up from each foot to the flange of the valve-box door, to prevent deformation of those portions of the barrel lying between the feet brackets and the valve-box ; these ribs may have a thickness of ·8 × thickness of barrel, and may stand out about 2 × thickness of barrel. Similar ribs should also be carried round under the bottom of the cylinder from foot to foot.

In bolting horizontal cylinders down to the seatings, fitted bolts should be used at the front ends only—sufficient clearance being allowed in the other bolt holes to permit the expansion of the cylinder.

Oscillating Cylinders.

In cases where oscillating engines are employed, weight is generally of great importance, and the cylinder barrels are commonly made rather lighter than would be the case if the thickness were determined by the rules previously given (page 51). This reduction is rendered possible, partly by the absence of any strains communicated from the framing to the cylinder, and partly by the great stiffness imparted to the barrel by the steam and exhaust belts, valve-boxes, and ports, and their accompanying ribs. The H.P. cylinders of compound oscillating engines should be of thickness given by the following formula :—

Rule 49. Thickness of H.P. cylinder barrel $= \dfrac{D}{6200}\,(p+50)+ \cdot45$

inch, where D is diameter of cylinder in inches, and p boiler pressure.

The formula also gives the thickness of barrel for the L.P. cylinders of compound engines, and for the cylinders of simple engines working at 30 to 35 lbs. per square inch, if p be taken as 25. The above thicknesses are for cylinders without liners.

The trunnions should be all of the same diameter in the same pair of engines, whether engines are simple or compound, and must, of course, be large enough to allow proper area for exhaust. This plan allows of an annular air space between the steam pipe and the outer, or working surface of the trunnion, and so gives a comparatively cool bearing.

The area of section where the trunnion joins the cylinder barrel should be such that the value of the expression,—

Rule 50.

$$\frac{\text{Max. effective pressure} \times \text{area of piston} + \text{weight of cylinder in lbs.}}{2 \times \text{area of section}}$$

does not exceed 500 ; *i.e.*, the shearing stress should not exceed 500 lbs. per square inch.

A length of trunnion of about ·4 of the diameter will usually give a satisfactory amount of surface, but the area should be such that the pressure per square inch (excluding pressure due to weight of cylinder) does not exceed 350 lbs. A safe proportion, which should always be obtained where possible, is given by,—

Rule 51. $\dfrac{\text{Max. effective pressure} \times \text{area of piston}}{2 \times \text{diameter of trunnion} \times \text{length of same}} = 300.$

A very strong attachment of trunnion is obtained by making the outer cylinder of barrel form, and arranging the steam and exhaust belts between it and the inner cylinder or liner.

The valve faces should not be at right angles to the line joining valve-spindle and piston-rod centres, but angled so that the side next the steam entrance is nearer to the cylinder barrel than the opposite or exhaust side ; this will bring the centre of valve-spindle as close in to the cylinder as possible, whilst allowing free way for the exhaust.

The receivers of compound oscillating engines are commonly formed around the H.P. cylinder, between the inner cylinder or liner, and the outer shell to which the trunnions are attached. This method has the accompanying advantage of making the outer cylinders more nearly of a size.

CLEARANCE OF PISTON.

The following Table shows the clearances required in cylinders of the various diameters mentioned :—

Table XXVIII.—Piston Clearances.

Diameter of Cylinder.	Clearances.		Diameter of Cylinder.	Clearances.	
	Crank end.	Cover end.		Crank end.	Cover end.
Up to 14″	3/8″	1/4″	61 to 80″	7/8″	5/8″
15 to 20″	1/2″	3/8″	81 to 100″	1 5/16″	1 1/16″
21 to 40″	5/8″	1/2″	Above 100″	1″	3/4″
41 to 60″	3/4″	9/16″	...	...	...

STUFFING BOXES, &c.

Table XXIX.—Stuffing-boxes for elastic packing. (Fig. 10.)

A Diameter of Rod.	B Width of packing space.	C Diameter of Box. 1·25×A+·55	D_1 50 lbs. Depth of packing. ·9×A+·75	D_2 100 lbs. Depth of packing. 1·15×A+·75	D_3 150 lbs. Depth of packing. 1·4×A+·75	E Overall depth of neck-ring. ·45A+·35	F Thickn's of bushes, gland, and neck. ·045×A+·18	Number and diameter of studs.
3/4"	3/8"	1 1/2"	1 3/8"	1 5/8"	1 3/4"	3/4"	3/16"	2- 3/8"
7/8	3/8	1 5/8	1 1/2	1 3/4	2	3/4	3/16	2- 3/8
1	7/16	1 7/8	1 5/8	2	2 1/4	7/8	1/4	2- 1/2
1 1/8	7/16	2	1 3/4	2 1/8	2 3/8	7/8	1/4	2- 1/2
1 1/4	7/16	2 1/8	1 7/8	2 1/4	2 1/2	7/8	1/4	2- 1/2
1 3/8	7/16	2 1/4	2	2 3/8	2 5/8	1	1/4	2- 5/8
1 1/2	1/2	2 1/2	2 1/8	2 1/2	2 7/8	1	1/4	2- 5/8
1 3/4	1/2	2 3/4	2 1/4	2 3/4	3 1/4	1 1/8	1/4	2- 3/4
2	9/16	3 1/8	2 1/2	3 1/8	3 1/2	1 1/4	1/4	2- 3/4
2 1/4	9/16	3 3/8	2 3/4	3 1/2	3 7/8	1 3/8	5/16	2- 7/8
2 1/2	5/8	3 3/4	3	3 3/4	4 1/4	1 1/2	5/16	2- 7/8 or 3- 3/4
2 3/4	5/8	4	3 1/4	4	4 5/8	1 5/8	5/16	2-1 or 3- 3/4
3	11/16	4 3/8	3 1/2	4 1/4	5	1 3/4	5/16	2-1 or 3- 7/8
3 1/4	11/16	4 5/8	3 3/4	4 3/4	5 3/8	1 7/8	3/8	2-1 1/8 or 3- 7/8
3 1/2	3/4	5	4	5	5 5/8	2	3/8	2-1 1/8 or 3-1
3 3/4	3/4	5 1/4	4 1/4	5 1/4	6	2 1/8	3/8	2-1 1/4 or 3-1
4	13/16	5 5/8	4 1/2	5 1/2	6 3/8	2 1/8	3/8	2-1 1/4 or 3-1
4 1/2	7/8	6 1/4	4 3/4	6	7	2 3/8	7/16	2-1 3/8 or 3-1 1/8
5	15/16	6 7/8	5	6 1/2	7 3/4	2 5/8	7/16	2-1 3/8 or 3-1 1/8
5 1/2	1	7 1/2	5 5/8	7	8 1/2	2 7/8	7/16	4-1 1/8
6	1 1/8	8 1/4	6 1/4	7 5/8	9 1/4	3	7/16	4-1 1/8
6 1/2	1 1/8	8 3/4	6 5/8	8 1/8	9 7/8	3 1/4	1/2	4-1 1/8
7	1 1/4	9 1/2	7	8 3/4	10 1/2	3 1/2	1/2	4-1 1/4
7 1/2	1 1/4	10	7 1/2	9 3/8	11 1/4	3 3/4	1/2	4-1 1/4
8	1 3/8	10 3/4	8	10	12	4	9/16	4-1 1/4
8 1/2	1 3/8	11 1/4	8 1/2	10 1/2	12 3/4	4 1/4	9/16	4-1 3/8
9	1 1/2	12	8 7/8	11	13 3/8	4 3/8	9/16	4-1 3/8
9 1/2	1 1/2	12 1/2	9 3/8	11 5/8	14 1/8	4 5/8	5/8	4-1 3/8
10	1 1/2	13	9 3/4	12 1/4	14 3/4	4 7/8	5/8	4-1 3/8

In all cases, the amount that the gland can enter (G) should be one-half the depth of the stuffing box (D_1, D_2, or D_3).

The width of packing space (B) given in above Table assumes that all glands are of gun-metal, as is usual in naval practice; if it is desired to use cast-iron glands bushed with gun-metal the width of space must be increased ⅛ inch in each case.

A ³⁄₁₆″ bush for a cast-iron gland may be used up to 1¼″ dia. of rod.

¼ ″	,,	,,	,,	2¾″	,,
⁵⁄₁₆″	,,	,,	,,	3¾″	,,
⅜ ″	,,	,,	,,	6 ″	,,
⁷⁄₁₆″	,,	,,	,,	7½″	,,
½ ″	,,	,,	,,	9 ″	,,
⁹⁄₁₆″	,,	,,	,,	10½″	,,

Where two numbers and sizes of studs are given, the larger studs are intended for use with pinion nuts and toothed rings.

The sizes of studs for rods of 5½ inches diameter and upwards are fixed on the assumption that pinion nuts and toothed rings will always be used.

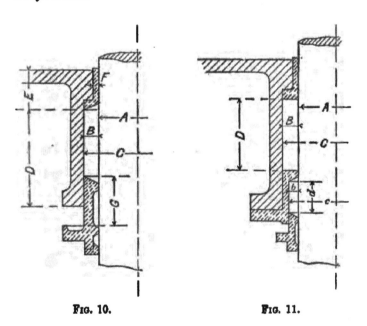

FIG. 10.   FIG. 11.

Table XXX.—Stuffing-boxes for Metallic Packing. (Fig. 11.)

A. Diameter of rod.	B. Width of packing space.	C. Diameter of box. $1\cdot2A+1\cdot25$	D. Depth of metallic packing. $2\sqrt{A}+\cdot5$.	b. Width of packing space.	c. Diameter of box.	d. Depth of elastic packing. $\frac{D}{3}+1\cdot15$.
3	1	5	4	¾	4½	2⅛
3¼	1	5¼	4⅛	¾	4¾	2⅛
3½	1	5½	4¼	¾	5	2¼
3¾	1	5¾	4⅜	¾	5¼	2¼
4	1	6	4½	¾	5½	2⅜
4¼	1	6¼	4⅝	¾	5¾	2½
4½	1⅛	6¾	4¾	¾	6	2⅝
5	1⅛	7¼	5	¾	6½	2¾
5½	1⅛	7¾	5¼	¾	7	2⅞
6	1¼	8½	5½	¾	7½	3
6½	1¼	9	5⅝	¾	8	3⅛
7	1¼	9½	5¾	¾	8½	3¼
7½	1⅜	10¼	6	⅞	9¼	3½
8	1⅜	10¾	6¼	⅞	9¾	3⅝
8½	1⅜	11¼	6⅜	⅞	10¼	3¾
9	1½	12	6½	1	11	3⅞
9½	1½	12½	6¾	1	11½	4
10	1½	13	7	1	12	4⅛

For the sake of uniformity the H.P., M.P., and L.P. stuffing boxes of triple engines are commonly made all of the same depth.

STRENGTH, ETC., OF STUDS AND BOLTS.

When a nut is screwed up tightly, as in making a joint, the stud or bolt is subjected to two distinct stresses, viz.:—tension, and torsion; and the tensile or axial stress cannot be set up, without at the same time producing the twisting stress. The twisting moment,—since it is due to the frictional resistances acting at the surfaces of the threads,—is probably very nearly proportional to the square of the diameter, whilst the moment of resistance of the bolt varies as the cube of the diameter. If, therefore, two bolts be screwed up until the *axial* stress per square inch of section is the same in both, the *total* stress will always be proportionally greater in the smaller of the two; and every day experience shows that the tightening up of small studs (below ⅝ inch),—especially if they are of such easily sheared material as muntz-metal,—must only be entrusted to skilled hands.

The readiest way of allowing for this twisting stress is to assume progressively lower values for the *working* stress as the bolts diminish in size, and the Table below gives values which should not be exceeded.

The Table is based on the relation,—

Working stress per sq. in. = (Area at bottom of thread)$^{\frac{5}{12}}$ × C;

where **C** = 5000 for iron or mild steel, and 1000 for muntz or gun-metal. For iron or steel bolts above 2 inches in diameter, and gun-metal or bronze ones above 3½ inches diameter, the moment of the twisting stress is so small, proportionately, that it may be neglected.

Table XXXI.—Strengths of Studs and Bolts.

Dia. of stud or bolt.	Iron or Mild steel.		Muntz or Gun-metal.	
	Working stress in lbs. per square inch.	Effective strength of 1 stud or bolt in lbs.	Working stress in lbs. per square inch.	Effective strength of 1 stud or bolt in lbs.
½	2000	250	400	50
⅝	2500	500	500	100
¾	3000	900	600	180
⅞	3400	1,450	680	290
1	3900	2,150	780	430
1⅛	4300	3,000	860	600
1¼	4700	4,200	940	840
1⅜	5100	5,400	1020	1,080
1½	5500	7,100	1100	1,420
1⅝	5800	8,500	1160	1,700
1¾	6300	11,000	1260	2,200
1⅞	6600	13,100	1320	2,620
2	7000	16,100	1400	3,220
2¼	7000	20,400	1560	4,560
2½	7000	26,100	1730	6,450
2¾	7000	31,200	1860	8,300
3	7000	38,100	2030	11,000
3¼	7000	44,800	2170	13,900
3½	7000	53,000	2350	17,800
3¾	7000	60,500	2500	21,600
4	7000	70,100	2500	25,000
4¼	7000	79,500	2500	28,400
4½	7000	90,300	2500	32,200
4¾	7000	100,800	2500	36,000
5	7000	113,000	2500	40,300
5¼	7000	124,600	2500	44,500
5½	7000	138,000	2500	49,200

Studs and bolts may be loaded to the figures given in the above table whether the load is "dead" (as in the case of a joint), or "live" (as in the case of a connecting-rod bolt), as in the latter case mild steel will always be used, and the shearing stress due to tightening up is practically absent.

Mild steel studs and bolts should always be fitted with iron nuts, as steel ones have a much greater tendency to "seize," and so greatly increase the twisting stress; for the same reason muntz-metal or naval brass studs, &c. should always have iron nuts if possible.

Gun-metal, and the various bronzes are unsatisfactory materials for small studs and bolts, not because of any lack of tensile strength,—which is often high,—but because of their very low elastic limit under a shearing stress.

Gun-metal, muntz-metal, &c., studs and bolts are not suitable for joints of covers, &c., on which the pressure exceeds 50 lbs. per square inch.

When iron or steel studs are used in connection with gun-metal steam or water valves, &c., they must not be allowed to penetrate into the steam or water space, or they will rapidly corrode and come loose.

The part of a stud that is screwed into the work should be :—

Rule 52. Not less than 1¼ diameters long when screwed into cast-iron, and 1½ diameters when not inconvenient.

Not less than 1 diameter long when screwed into gun-metal, wrought-iron, or cast-steel.

The general dimensions, numbers of threads, &c., for bolts and nuts are given in Table XXXII.; the numbers of threads, and dimensions depending thereon are in accordance with the Whitworth Standard, and the sizes over flats and angles are the nearest sixteenth to the same.

The proportions of the Whitworth Standard thread are shown by Fig. 12.

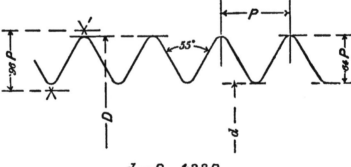

$$d - D - 1{\cdot}28P$$

FIG. 12.

TABLE XXXII.—DIMENSIONS OF NUTS, BOLTS, ETC.   75

Table XXXII.—Dimensions of Nuts, Bolts, etc.

Dia. of Bolt.	Head and nut over flats.	Head and nut over angles.	Height of nut.	Height of head for screws and bolts.	No. of threads per inch.	Area at bottom of thread in sq. inches.	Thick. of check-nut.	Size of split-pin L. S. G.
1/4	1/2	5/8	1/4	3/16	20	·027	3/16	No. 14
3/8	1 1/16	1 3/16	3/8	5/16	16	·068	1/4	,, 13
1/2	1 3/16	1 1/16	1/2	7/16	12	·121	3/8	,, 12
5/8	1 1/8	1 1/4	5/8	9/16	11	·203	7/16	,, 11
3/4	1 5/16	1 1/2	3/4	11/16	10	·303	9/16	,, 10
7/8	1 1/2	1 11/16	7/8	3/4	9	·421	5/8	,, 9
1	1 11/16	1 15/16	1	7/8	8	·554	3/4	,, 8
1 1/8	1 7/8	2 1/8	1 1/8	1	7	·697	1 3/16	,, 7
1 1/4	2 1/16	2 3/8	1 1/4	1 1/16	7	·894	1 5/16	,, 6
1 3/8	2 3/16	2 9/16	1 3/8	1 3/16	6	1·059	1 1/16	,, 5
1 1/2	2 7/16	2 13/16	1 1/2	1 5/16	6	1·300	1 1/8	,, 4
1 5/8	2 9/16	3	1 5/8	1 7/16	5	1·471	1 1/4	,, 3
1 3/4	2 3/4	3 3/16	1 3/4	1 9/16	5	1·752	1 5/16	,, 2
1 7/8	3	3 1/2	1 7/8	1 5/8	4 1/2	1·986	1 3/8	,, 1
2	3 1/4	3 5/8	2	1 3/4	4 1/2	2·311	1 1/2	,, 1
2 1/4	3 9/16	4 1/16	2 1/4	2	4	2·925	...	5/16
2 1/2	3 7/8	4 1/2	2 1/2	2 3/16	4	3·732	...	5/16
2 3/4	4 3/16	4 15/16	2 3/4	2 7/16	3 1/2	4·463	...	3/8
3	4 1/2	5 1/4	3	2 5/8	3 1/2	5·449	...	3/8
3 1/4	4 7/8	5 5/8	3 1/4	2 13/16	3 1/4	6·406	...	3/8
3 1/2	5 3/16	6	3 1/2	3 1/16	3 1/4	7·572	...	7/16
3 3/4	5 9/16	6 3/8	3 3/4	3 1/4	3	8·656	...	7/16
4	5 15/16	6 7/8	4	3 1/2	3	10·026	...	1/2
4 1/4	6 3/8	7 3/8	4 1/4	3 3/4	2 7/8	11·480	...	1/2
4 1/2	6 15/16	7 7/8	4 1/2	3 15/16	2 7/8	12·784	...	9/16
4 3/4	7 1/4	8 7/16	4 3/4	4 3/8	2 3/4	14·413	...	9/16
5	7 15/16	9	5	4 3/8	2 3/4	16·145	...	5/8

The heads of all ordinary screws and bolts should be made hexagonal, and of the same size over flats as the corresponding nuts.

All check-nuts should be chamfered on both sides.

Set-screws that are frequently handled, such as those in the lock-rings on the various large nuts, should have *square* heads, and should either be of hard steel or should have their heads properly case-hardened.

Wherever a square is used about the engines in place of a hexagon care should be taken that it is made of the same size as some nut over flats, as otherwise much trouble and annoyance are often caused.

Spanners or Screw Keys.

Excellent proportions for these are given by the diagram below and Table on next page. They should be of forged steel, and the jaws should be case-hardened.

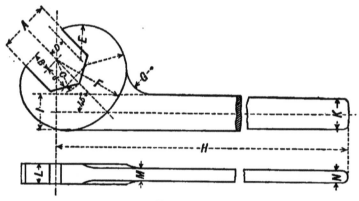

Fig. 12a.

For nuts above 2½ inches it is better to have a short-handled heavy or "striking" spanner, and a very light, or "handling" one.

Where "check" nuts are used, care should be taken that suitable thin spanners are provided for handling them.

For very heavy nuts, such as those on the piston-rods, and on the end of the "tail" shaft, the spanners should be of the closed or "ring" type.

PISTONS.

For the case of a circular flat plate, supported at centre, and uniformly loaded, the formula may be written,—

$$t = C \times D \sqrt{p} \, ;$$

where $t =$ thickness, and D diameter,—both in inches; $p =$ effective absolute pressure (or greatest difference of absolute pressures on the two sides); and C = co-efficient.

Cast steel pistons. —These are made of a single thickness of metal, and should be coned, or "dished," to get the necessary rigidity. The three pistons of an ordinary triple engine should all be made of the same total depth, which should be such that the slope of the face of P. piston next crankshaft is about 1 in 3 ; this will give a perfectly

TABLE XXXIII.—PROPORTIONS OF SPANNERS. 77

Table XXXIII.—Proportions of Spanners.

Diam.	A	B	C	D	E	F	G	H	I	J	K	L	M	N
"	"	"	"	"	"	"	"	"	"	"	"	"	"	"
3/8	1 1/16	1/4	1/2	1/4	9/16	1 1/16	5/8	7 3/8	3/4	3/8	3/4	7/16	3/4	1/4
1/2	1 5/16	3/8	11/16	7/16	3/4	1 1/8	5/8	9 3/8	15/16	9/16	13/16	1/2	3/8	1/2
5/8	1 3/8	7/16	3/4	9/16	13/16	1 3/8	3/4	10 1/4	1 1/16	9/16	7/8	5/8	3/8	5/16
3/4	1 9/16	1/2	7/8	5/8	7/8	1 15/16	3/4	11 1/4	1 3/16	5/8	1	5/8	3/8	5/16
7/8	1 11/16	9/16	1	7/8	1	2 3/16	7/8	1–1"	1 5/16	11/16	1 1/8	11/16	7/16	3/8
1	1 7/8	5/8	1 1/4	1	1 1/8	2 3/8	1	1–1 3/4	1 3/8	3/4	1 1/8	13/16	1/2	7/16
1 1/8	2 1/8	11/16	1 3/8	1 1/8	1 3/8	2 5/8	1 1/8	1–3 7/8	1 1/2	13/16	1 1/4	3/4	1/2	7/16
1 1/4	2 1/4	3/4	1 5/8	1 3/8	1 5/8	3 1/4	1 1/4	1–5 5/8	1 5/8	7/8	1 3/8	7/8	9/16	1/2
1 3/8	2 5/8	3/4	1 3/4	3/4	1 5/8	3 1/2	1 3/8	1–6 3/4	1 11/16	13/16	1 3/8	15/16	9/16	7/8
1 1/2	2 13/16	13/16	1 7/8	7/8	1 3/4	3 3/4	1 1/2	1–7 3/4	1 3/4	7/8	1 7/16	1 5/16	9/16	1/2
1 5/8	3 1/16	7/8	2	1 1/8	1 13/16	4	1 5/8	1–9 3/4	1 13/16	1	1 1/2	1	9/16	1/2
1 3/4	3 3/16	15/16	2 1/16	1 1/8	1 15/16	4 1/4	1 1/2	1–10 1/8	1 7/8	1 1/8	1 5/8	1 1/8	5/8	1/2
1 7/8	3 3/8	7/8	2 3/16	1 5/16	2 1/16	4 1/2	1 5/8	1–10 3/4	2 1/16	1 1/8	1 3/4	1 1/8	5/8	9/16
2	3 5/8	15/16	2 3/8	1 1/4	2 7/16	4 3/4	1 3/4	2–1 3/4	2 1/4	1 1/4	1 3/4	1 3/16	5/8	9/16
2 1/8	3 3/4	1	2 3/8	1 1/2	2 9/16	5 1/8	1 5/8	2–2 3/8	2 9/16	1 5/16	1 7/8	1 1/4	13/16	5/8
2 1/4	3 15/16	1 1/16	2 7/8	1 3/4	2 3/8	5 5/8	1 3/4	2–4 1/2	2 9/16	1 3/8	1 15/16	1 5/16	1 1/16	5/8
2 3/8		1 1/8	2 5/8	1 15/16	2 5/8	5 15/16	1 7/8	2–7 1/4	2 3/4	1 7/8	2	1 1/4	1 1/16	5/8
2 1/2		1 1/4	2 5/8	1 15/16		5 1/2	2	2–9 1/4	2 15/16	1 11/16		1 1/2	1 11/16	5/8

rigid piston, and will also give room for a properly proportioned piston-rod stuffing-box.

The thickness near boss, for a piston of this type, is given very nearly by the above formula, when $C = \cdot 0046$; that is,—

Rule 53. $\qquad t = \cdot 0046\ D\sqrt{p}$ . . . . (a).

For practical reasons, however, the addition of a small constant quantity, and other slight modifications are necessary, and the formulæ for the pistons of triple engines working at about 160 lbs. pressure, and of compound engines working at about 100 lbs. pressure (since p will have practically the same values in both cases) will be,—

For H.P. pistons, $t = \dfrac{D}{23} + \cdot 24$ (value of p about 90).

For M.P. pistons, $t = \dfrac{D}{34} + \cdot 40$ (value of p about 60).

L.P. do., $\quad t = \dfrac{D}{50} + \cdot 48$ (value of p about 30).

The thicknesses near the edge, or rim, should be,—

Rule 54. For H.P. pistons, $\cdot 65 \times$ thickness near boss ;
M.P. do., $\quad \cdot 60 \times \quad$ do. do.
L.P. do., $\quad \cdot 55 \times \quad$ do. do.

For the pistons of quadruple, or other engines, similar modifications of the general formula (a) above may be employed.

Table **XXXIV.** is calculated by means of the above formulæ, and may be used for compound engines by simply omitting the M.P. columns.

Cast-steel pistons for the lighter machinery of torpedo gun-boats give perfectly satisfactory results, even up to 300 revolutions per minute, when made of thickness, near boss, given by,—

Rule 55. $\qquad t = \cdot 0042\ D\sqrt{p} + \cdot 125$

where the symbols have the same values as above.

The slope of the underside of L.P. piston should not be less than 1 in 6·5. The thickness near edge should be half that near boss ; and the depth of rim may be $\dfrac{\text{L. P. Diameter}}{12}$. They are sometimes fitted with Ramsbottom rings of hard gun-metal, and sometimes with ordinary cast-iron packing rings.

They are, as a matter of course, turned all over on both sides.

Table **XXXV.** is calculated by means of the above formula for boiler pressure of about 175 lbs.

TABLE XXXIV.—CAST STEEL PISTONS.   79

Table XXXIV.—Cast Steel Pistons.

Diam. of piston in inches.	H.P. Thick. in inches.				M.P. Thick. in inches.				L.P. Thick. in inches.			
	Near boss.		Near edge.		Near boss.		Near edge.		Near boss.		Near edge.	
15	·89	1 5/16	·58	9/16	·84	7/8	·50	1/2	·78	1 3/16	·43	7/16
20	1·11	1 1/8	·72	3/4	·99	1	·59	5/8	·88	1 5/16	·48	1/2
25	1·33	1 3/8	·86	7/8	1·14	1 3/16	·68	11/16	·98	1	·54	9/16
30	1·54	1 9/16	1·00	1	1·28	1 5/16	·77	13/16	1·08	1 1/16	·59	5/8
35	1·76	1 3/4	1·14	1 3/16	1·43	1 7/16	·86	7/8	1·18	1 3/16	·65	11/16
40	1·98	2	1·29	1 5/16	1·57	1 9/16	·94	15/16	1·28	1 5/16	·70	3/4
45	2·20	2 1/4	1·43	1 9/16	1·72	1 3/4	1·03	1 1/16	1·38	1 3/8	·76	13/16
50	2·41	2 7/16	1·57	1 9/16	1·87	1 7/8	1·12	1 1/8	1·48	1 1/2	·81	15/16
55	2·63	2 5/8	1·71	1 3/4	2·02	2 1/16	1·21	1 1/4	1·58	1 5/8	·87	7/8
60	2·85	2 7/8	1·85	1 7/8	2·16	2 3/16	1·29	1 5/16	1·68	1 11/16	·92	15/16
65	...	...	...	...	2·31	2 5/16	1·38	1 3/8	1·78	1 13/16	·98	1
70	...	...	...	...	2·46	2 1/2	1·47	1 1/2	1·88	1 7/8	1·03	1 1/16
75	...	...	...	...	2·61	2 5/8	1·56	1 9/16	1·98	2	1·09	1 1/8
80	...	...	...	...	2·75	2 3/4	1·65	1 11/16	2·08	2 1/16	1·14	1 3/16
85	...	...	...	...	...	...	...	...	2·18	2 3/16	1·20	1 1/4
90	...	...	...	...	...	...	...	...	2·28	2 5/16	1·25	1 1/4
95	...	...	...	...	...	...	...	...	2·38	2 3/8	1·31	1 5/16
100	...	...	...	...	...	...	...	...	2·48	2 1/2	1·36	1 3/8
105	...	...	...	...	...	...	...	...	2·58	2 9/16	1·42	1 7/16
110	...	...	...	...	...	...	...	...	2·68	2 11/16	1·47	1 1/2

Table XXXV.—Pistons of Torpedo Gun-Boats.

Diameter of piston in inches.	H.P. Thick. in inches.		M.P. Thick. in inches.		L.P. Thick. in inches.	
	Near boss.	Near edge.	Near boss.	Near edge.	Near boss.	Near edge.
10	·52 9/16	·26 5/16	...	...	...	...
15	·72 3/4	·36 3/8	·61 5/8	·31 5/16	...	...
20	·92 15/16	·46 7/16	·77 3/4	·39 7/16	·58 9/16	·29 5/16
25	1·12 1 1/8	·56 9/16	·93 15/16	·47 1/2	·69 11/16	·35 3/8
30	1·32 1 5/16	·66 11/16	1·09 1 1/16	·55 9/16	·80 13/16	·40 7/16
35	...	...	1·26 1 1/4	·63 5/8	·92 15/16	·46 7/16
40	...	...	1·42 1 7/16	·71 3/4	1·03 1 1/16	·52 1/2
45	...	...	...	...	1·15 1 1/8	·58 9/16
50	...	...	...	...	1·26 1 1/4	·63 5/8
55	...	...	...	...	1·38 1 3/8	·69 11/16

Cast-iron Pistons.—These must be made of thoroughly sound and good material,—at least twice melted.

A convenient unit on which to base the proportions of cast-iron box (or hollow) pistons is the thickness requisite, in a circular plate of cast-iron, to carry safely a given pressure per square inch when supported at centre and loaded uniformly.

Using again the general formula given above for such a case, viz.:—

$$t = C \times D\sqrt{p}$$

and inserting a value of c that will limit the strain on the material to 3000 lbs. per square inch it becomes,—

$$t = ·008 \, D\sqrt{p}$$

If the necessary constant quantity be then added, the units (x) for pistons of triple engines working at about 160 lbs., and of compound engines working at about 100 lbs., will be given by,—

Rule 56. For H.P. pistons, $x = ·008 \, D \sqrt{p} + ·3$

M.P.   ,, $x = ·008 \, D \sqrt{p} + ·45$

L.P.   ,, $x = ·008 \, D \sqrt{p} + ·6$

Other proportions are then as follows :—

Rule 57. Depths at centre, ... $3·2 \times x$.

,, **57a.** Thickness of metal of faces $\begin{cases} \text{near boss } ·33 \times (x + ·5). \\ \text{near edge } ·3 \times (x + ·5). \end{cases}$

,, **57b.** Thickness of webs, ... $·32 \times (x + ·5)$.

,, **57c.** Thickness of boss round rod, ... $·7 \times x$.

,, **57d.** Depth of packing-ring, ... $·75 \times (\text{L.P.} x + 1·5)$.

,, **57e.** Thickness   do.   ...   ... $·22 \times (\text{L.P.} x + 1·5)$.

,, **57f.** Diameter of junk-ring bolts, ... $(·2 \times \text{L.P.} x) + ·5$.

Rule 58. Pitch of junk-ring bolts, $\begin{cases} \text{H.P.—7} \frac{1}{2} \text{ diameters.} \\ \text{M.P.—8} \frac{3}{4} \quad \text{do.} \\ \text{L.P.—10} \quad \text{do.} \end{cases}$

Rule 59. Spring pressures per square inch of packing-ring in contact with cylinder,... ... ... $\begin{cases} \text{H.P.—3} \frac{1}{2} \text{ lbs.} \\ \text{M.P.—3 lbs.} \\ \text{L.P.—2} \frac{1}{2} \text{ lbs.} \end{cases}$

Rule 60. Number of webs in piston, ... $\begin{cases} \text{H.P.—} \dfrac{D+20}{9}. \\ \text{M.P.—} \dfrac{D+20}{10}. \\ \text{L.P.—} \dfrac{D+20}{11}. \end{cases}$

This spacing of webs will ensure that Rule 32, page 59 is not violated. In calculating *local* strength of piston face, care must be taken to use the gauge pressure acting on it, and not the effective pressure on piston as in above rules. A good method of increasing local strength, and at the same time assisting the moulder, is to cast a solid cylindrical through tie at the centre of gravity of each sector of the piston; the diameter of these ties may be ·6 × L.P.æ. All holes provided for the removal of core material should be strengthened by an internal ring, the section of which may be, half thickness of face metal × the same.

Pistons of Compound and Triple engines are usually, for practical reasons, made all of the same depth (except in the case of tandem engines); but, if a piston is made of less depth than that given in Table XXXVI., the metal of the faces must be increased in proportion.

The proportions given for packing-rings and junk-ring bolts are equally applicable to cast-steel pistons.

All junk-ring bolts (or nuts, if studs are used), should be locked by means of a light wrought-iron ring, secured to junk-ring by studs having square bodies, which stand up through guard-ring, and through which stout split-pins are fitted: ⅝-inch studs and No. 1 L.S.G. split-pins are commonly used for large pistons.

When the ordinary packing-ring is used an excellent piston is produced by placing behind it,—one between every pair of junk-ring bolts,— a number of short spiral springs acting radially, and adjusted to give the pressures named above. For the pistons of auxiliary engines the Cameron spring (a corrugated steel ribbon) is most useful, and is also often employed for L.P. pistons of large size. The old fashioned coach spring should never be employed.

In horizontal engines solid packings should be used in lieu of springs for the lower one third circumference of piston; diagonal and oscillating pistons are also better for some solid packings.

The following Table is calculated by means of the above formulæ and shows at a glance the various proportions of H.P., M.P. & L.P. pistons, from 15 inches to 110 inches diameter, for Triple engines working at about 160 lbs., and Compound engines working at about 100 lbs. pressure:—

6

Table XXXVI.—Proportions

Diameter of piston in inches.	Values of x. $=·008D\sqrt{p}+\begin{cases}·3 \text{ H.P.}\\·45 \text{ M.P.}\\·6 \text{ L.P.}\end{cases}$			Depths at centre. $=3·2×x.$			Thickness of Metal $=·33×(x+·5)$ at centre and			
							H.P.		M.P.	
	H.P.	M.P.	L.P.	H.P.	M.P.	L.P.	Centre.	Edge.	Centre.	Edge.
15	1·44	1·37	1·25	4·60	4·38	4·00	·64	·58	·62	·56
20	1·82	1·68	1·46	5·82	5·37	4·67	·76	·69	·72	·65
25	2·20	1·99	1·68	7·04	6·37	5·36	·89	·81	·82	·75
30	2·58	2·30	1·89	8·25	7·35	6·05	1·01	·92	·92	·84
35	2·96	2·61	2·11	9·47	8·33	6·75	1·14	1·04	1·02	·93
40	3·34	2·91	2·33	10·69	9·31	7·45	1·27	1·15	1·12	1·02
45	3·72	3·22	2·55	11·91	10·30	8·14	1·40	1·27	1·23	1·11
50	4·10	3·53	2·76	13·12	11·29	8·83	1·52	1·38	1·33	1·20
55	4·48	3·84	2·98	14·34	12·27	9·52	1·65	1·49	1·43	1·30
60	4·86	4·14	3·19	15·55	13·25	10·21	1·77	1·60	1·53	1·39
65	...	4·45	3·41	...	14·24	10·90	...	...	1·63	1·48
70	...	4·76	3·62	...	15·23	11·58	...	...	1·73	1·57
75	...	5·07	3·84	...	16·22	12·27	...	...	1·84	1·67
80	...	5·38	4·05	...	17·21	12·96	...	...	1·94	1·76
85	...	...	4·27	...	...	13·66	...	...	...	...
90	...	...	4·49	...	...	14·37	...	...	...	...
95	...	...	4·71	...	...	15·06	...	...	...	...
100	...	...	4·92	...	...	15·74	...	...	...	...
105	...	...	5·14	...	...	16·43	...	...	...	...
110	...	...	5·35	...	...	17·12	...	...	...	...

TABLE XXXVI.—PROPORTIONS OF CAST-IRON PISTONS.   83

of Cast-Iron Pistons.

of Faces. ·3×(x+·5) at edge. L.P.		Thickness of boss round rod ·7×x		Number of Webs			Packing Ring.		Dia. of Junk-ring bolts (L.P.x×2)+·5
				H.P. D+20/9	M.P. D+20/10	L.P. D+20/11	Depth ·75×(L.P.x+1·5).	Thick. ·22×(L.P.x+1·5).	
Centre.	Edge.	H.P.	L.P.						
·58	·52	1·01	·37	4	...	...	2·06	·60	·75
·65	·59	1·27	1·02	5	4	4	2·22	·65	·79
·72	·66	1·54	1·17	5	5	5	2·38	·70	·84
·79	·72	1·81	1·32	6	5	5	2·54	·75	·88
·86	·79	2·07	1·48	6	6	5	2·71	·80	·92
·93	·85	2·34	1·63	7	6	6	2·87	·84	·96
1·01	·92	2·60	1·78	8	7	6	3·03	·89	1·01
1·08	·98	2·87	1·98	8	7	7	3·19	·94	1·05
1·15	1·05	3·13	2·08	9	8	7	3·36	·99	1·10
1·22	1·11	3·40	2·23	9	8	8	3·52	1·03	1·14
1·30	1·17	...	2·38	...	9	8	3·68	1·08	1·18
1·37	1·23	...	2·53	...	9	9	3·84	1·13	1·22
1·44	1·30	...	2·68	...	10	9	4·00	1·18	1·27
1·50	1·36	...	2·83	...	10	9	4·16	1·22	1·31
1·58	1·43	...	2·99	...	...	10	4·33	1·27	1·36
1·66	1·50	...	3·14	...	...	10	4·49	1·32	1·40
1·73	1·56	...	3·29	...	...	11	4·65	1·37	1·44
1·80	1·62	...	3·44	...	...	11	4·81	1·41	1·48
1·87	1·69	...	3·59	...	...	12	4·97	1·46	1·53
1·93	1·75	...	3·74	...	...	12	5·13	1·51	1·57

When using above Table for Compound engines simply omit M.P. columns. The dimensions are given in inches and decimals only, as the proportions for pistons of intermediate diameter are more readily seen with this notation.

PISTON RODS.

These should be made of steel having an ultimate tenacity of not less than 35 tons per square inch,—as softer steels do not wear satisfactorily. The permissible stress per square inch of material should diminish with the size of the rod, in the same manner as laid down for bolts, &c., (see Table XXXI. page 73), but,—on the principle that the more serious the accident that would result from the breakage of the part, and the more difficult of inspection it is, the more lightly it should be loaded,—**(Rule 61.)** the maximum permissible stress per square inch at bottom of thread, or through cotter-hole, should not exceed 8000 lbs. for steel, and 6000 for wrought-iron.

For small rods it is not advisable, in ordinary work, to exceed the limits given in the following Table, which is based on :—

Working stress per sq. in. = (Area at bottom of thread)$^{\frac{1}{3}}$ × C,
where **C** = 5000 for steel and 3800 for iron.

Table XXXVII.—Strengths of Piston Rods.

Diameter of piston-rod screw in inches.	35 Ton Steel.		Wrought Iron.	
	Working stress in lbs. per sq. inch.	Effective strength of rod in lbs.	Working stress in lbs. per sq. inch.	Effective strength of rod in lbs.
⅞	3700	1,560	2800	1,180
1	4100	2,270	3100	1,720
1⅛	4400	3,070	3300	2,270
1¼	4800	4,290	3600	3,220
1⅜	5100	5,400	3900	4,130
1½	5400	7,020	4100	5.330
1⅝	5700	8,380	4300	6,320
1¾	6000	10,500	4600	8,060
1⅞	6300	12,500	4800	9,530
2	6600	15,200	5000	11,500
2¼	7100	20,700	5400	15,800
2½	7700	28,700	5800	21,600
2¾ and upwards.	7700	...	5800	...

The body of the rod should, theoretically, be considered as a strut, or column in compression, but as the condition of the ends (whether "fixed" or "loose") cannot be very accurately determined, and as there is also a tolerably constant relation of diameter to length in ordinary marine engines, the safest and simplest way is to fix upon a special value of the working stress for each type of engine.

A convenient form of the formula is then,—

Rule 62. Diameter of piston-rod $= \dfrac{\text{Diameter of cylinder}}{F} \times \sqrt{p}$,

where $p =$ greatest difference of absolute pressures on the two sides of the piston ; and $F =$ co-efficient whose values are,—

Naval engines,—direct-acting,	.	.	.	.	56
Mercantile,—ordinary stroke, direct-acting,	.	.	52
,,	long ,, ,,	.	.	49
,,	very long ,, ,,	.	.	46
,,	medium stroke, oscillating,	.	.	47
,,	long ,, ,,	.	.	44

The stroke of Naval engines rarely exceeds dia. of L.P. cylinder × ·6 and is usually between ·5 and ·6 × L.P. diameter.

" Ordinary " stroke, for the engines of a merchant steamer, is,— diameter of L.P. cylinder × ·60 to ·66 ; whilst ·7 to ·8 would be " long," and ·8 to 1 " very long."

A " medium " stroke for an oscillating engine is,—diameter of cylinder × ·95 ; or, if a compound engine,—diameter of L.P. cylinder × ·8.

Compound diagonal engines would come under the heading " long," since they are usually given a stroke of L.P. diameter × ·75, though the practice of different makers varies between ·7 and 1·3 × L.P. diameter.

If it is more convenient to use *areas* of cylinders and of rods, and to proportion the rods to carry a certain stress per square inch of section, the above formula may be written,—

Rule 62a. Area of section of piston-rod $= \dfrac{\text{Area of cylinder} \times p}{F^2}$

where F^2 will then be the stress in lbs. per square inch,—in each case the square of the numerical co-efficient given above, viz.:—

Naval engines, direct acting,	.	.	.	.	3100
Mercantile, ordinary stroke, direct acting,	.	.	2700
,,	long ,, ,,	.	.	2400
,,	very long ,, ,,	.	.	2100
,,	medium stroke, oscillating,	.	.	2200
,,	long ,, ,,	.	.	1900

Attachment of Piston to Rod.—In small auxiliary engines it is a common practice simply to reduce the diameter of the piston-rod about ¼ inch, and fit the piston up against the shoulder ; but above 15 inches diameter of piston, it is better to leave only as much shoulder as may be desired for " trueing-up " the rod at some future time, and then give a taper of 1 in 4 on the diameters **(Rule 63.)** (*i.e.*, if taper is 4 inches long, the smaller diameter will be 1 inch less than the larger), and continue it until it dies away into the parallel part near screw, the diameter of which has been determined as directed above.

Where a piston-rod is fitted into a cross head, the same shoulder, taper, &c., should be used as for a piston.

The depth of piston-rod nut need not exceed the diameter of the screw; and it may be recessed as far into the piston as desired, as the strength of the piston is not materially affected thereby.

The piston-rods of oscillating engines are commonly made with a sort of cylindrical "bolt-head," which is recessed into the lower face of the piston, whilst the nut is similarly recessed (to its full depth) into the upper face.

The cast-steel pistons of large Naval engines are sometimes made without any central boss, and attached to a circular flange (resembling a shaft coupling) on the piston-rod by means of a number of comparatively small bolts. This method, which was first used by the late Dr Kirk, has many good points, but the chief is perhaps the ease with which the connection can be made and broken again.

Rule 64. The diameter of this flange is about, **diameter of piston-rod × 2·25,** and its thickness about **diameter o piston-rod × ·33.** The attaching bolts (usually between 2½ inches and 3½ inches diameter) should be of mild steel, but should not carry more than 6000 lbs. per square inch.

The cotter used for attaching the piston-rod head in the case of oscillating engines may have the following proportions:—

Rule 65.   Breadth of cotter = Dia. of rod where "shouldered down" × ·82.

Rule 65a.   Thickness of cotter = Breadth × ·33.

Rule 65b.   Taper of cotter = ⅜ inch per foot, or 1 in 32.

These proportions are adopted in order to save all possible height, and, with a properly proportioned rod, they give a shearing stress on the cotter (which should be of fairly hard steel) of about 9000 lbs. per square inch.

The stress on these cotters is in one direction only.

PISTON ROD GUIDES AND GUIDE BLOCKS.

Piston-rod guides should, wherever possible, be made of hard close-grained cast-iron, and the guide blocks, or crosshead shoes should be made of the same material, since no combination of metals gives better results than cast-iron rubbing on cast-iron.

White metal is often used for the faces of the shoes (either fitted or cast into recesses in the cast-iron or gun-metal shoes), and gives results which, though quite satisfactory, are no better than those obtained with cast-iron on cast-iron, when the surfaces are properly proportioned and well looked after during the first few hours after being put to work.

Where steam is cut off later than half stroke, the greatest pressure on the guide occurs when the crank is at right angles to the centre line through cylinder, and may be determined graphically as follows:—

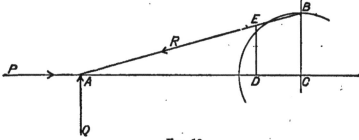

FIG. 13.

Let AC represent centre line through cylinder; AB connnecting rod; and BC crank; P effective load on piston; R re-action of connecting rod; and Q pressure on guide. Lay off P to some convenient scale on AC, and let it be represented by AD; then the perpendicular DE will be equal to the thrust Q, on the same scale.

Or, by calculation,—

$$P : Q :: AC : BC;$$

Rule 66. $$\therefore Q = P \times \frac{BC}{AC} = P \times \tan. BAC.$$

and, AB and BC being known, AC can be found from the relation

$$AC = \sqrt{\overline{AB^2} - \overline{BC^2}}.$$

Figs. 14, 15, 16, and 17 show types of piston-rod crosshead in common use for vertical engines,—the two former for small engines, and the two latter for large ones. The type shown in fig. 17 is specially suited for production as a steel casting, and is then strong, without unnecessary weight, and gives most excellent results.

The common method of fixing the sizes of crosshead shoes is to assume that the pistons, when at half stroke, have still the maximum effective pressures acting on them; to calculate the pressures on the guides from these loads in the manner described above; and then to fix the areas of the shoes so that the pressure per square inch may not exceed 80 lbs. **(Rule 67.)** For ordinary speeds of piston 70 lbs. is a fair value, but for the shoes of high-speed Naval engines 60 to 65 lbs. is a more usual pressure. Of course, where the indicator diagrams from similar engines are available, the actual effective pressures at half stroke may be ascertained and used, but the pressures per square

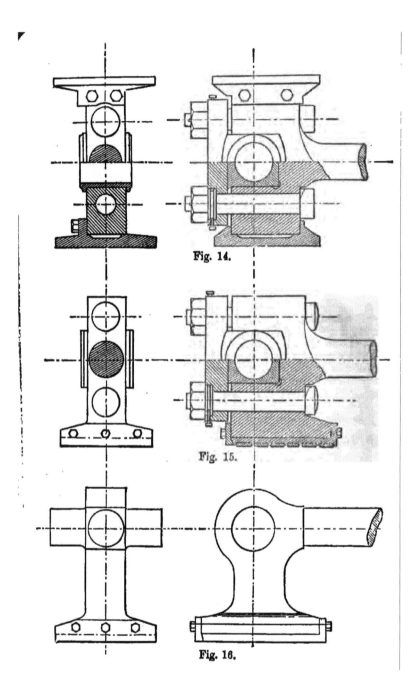

Fig. 14.

Fig. 15.

Fig. 16.

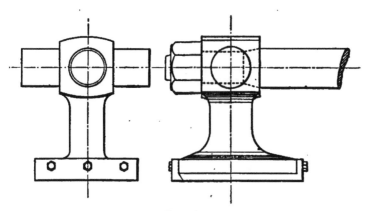

Fig. 17.

inch must then be limited to from 40 to 60 lbs., and should not exceed the latter figure.

It is very desirable to keep the crossheads and guides of compound and triple engines all of the same dimensions, and to effect this, the *mean* of the piston loads may be used to calculate from. The surfaces of the astern shoes may be from ·5 to ·7 of those of the ahead ones.

For gudgeons, and bolts and caps, see under **Connecting Rods.**

Where the gudgeon brasses are recessed into the piston-rod head, the inner or half-round brass may have a thickness,—

Rule 68. Thickness of inner half brass $=\dfrac{\text{Diam. of gudgeon}}{8}+ \cdot 2$,

whilst for the outer or flat half the thickness may be,—

Rule 68a. Thickness of outer half brass $=\dfrac{\text{Diam. of gudgeon}}{7}+ \cdot 2$.

See note with reference to white metal under "Connecting rod gudgeons," page 91 ; also Table XXXVIII., page 94.

CONNECTING RODS.

The length of the connecting rod, measured from centre of gudgeon to centre of crank pin, should not be less than twice the stroke of the piston.

So far as tensile stress alone is concerned, a rod of the same diameter as the piston-rod at the bottom of the thread would suffice, but there are also to be considered, compressive stress, bending stress due to the inertia of the rod itself, and bending stress due to the friction of the gudgeon and crank pin.

As marine connecting rods are usually between 10 and 16 diameters (measured at mid-length) long, they must, when compressive stresses are under consideration, be treated as struts jointed at the ends.

Hodgkinson & Gordon's formula for the *breaking* strength of such a strut, of circular section, is,—

$$P = \frac{fs}{1 + 4a\frac{l^2}{d^2}}$$

where P is breaking load in lbs.; l length from centre to centre, and d diameter, both in inches; s area of section in square inches; and f and a co-efficients, whose values for wrought-iron or mild steel are 36,000 and $\frac{1}{3000}$ respectively.

It should be noted that the greatest thrust on the connecting rod is to the effective load on the piston as AB (fig. 13) is to AC.

As the ratio of length to diameter rarely exceeds 16, however, it is probable that the following empirical rule (suggested by Grashof, and quoted by Unwin) will give more accurate results,—

Rule 69. $$P = \frac{KAI}{\frac{Al^2}{c} + I}$$

where P is greatest safe load in lbs.; A sectional area of rod in sq. ins.; I moment of inertia (equal to ·05d^4 for circular section); l length in inches from centre to centre; and K and c constants whose values are respectively 12,000 and 5000 for steel, and 10,000 and 5600 for wrought-iron. The maximum working load on a connecting rod should not exceed ·75P, and is commonly about ·6P.

The bending stress, due to the inertia, may be found from the formula,—

Rule 70. $$f = ·137 \frac{v^2 l^2}{g R d}$$

where v is velocity of crank-pin in feet per second; l length of rod from centre to centre, and d diameter of rod, both in inches; R radius of crank in feet; and $g = 32$.

The value of f will be found to lie between 800 and 1800 lbs. per sq. inch for the various types of engine and numbers of revolutions met with in ordinary practice, and this value must be added to the compressive stress in estimating the total stress on the material of the rod.

In a three-crank engine, when one of the crank-pins heats, the bending stress set up in the connecting-rod by the continued action of all the pistons may be very great (if a "seize" could really occur the rod would be destroyed instantly), but its magnitude cannot be even approximately calculated, and the case is, therefore, one that can only be met by an empirical formula. The above formulæ may, therefore, be used as *checks*, but the following empirical rule will be more readily applied, and the results given by it will be found to agree very closely with good modern practice :—

Rule 71.
$$D = \frac{\sqrt{L \times K}}{4}$$

where D is diameter of rod at middle, and L length from centre to centre, both in inches ; and

K = ·028 $\sqrt{\text{Effective load on piston in lbs.}}$—for Mercantile engines.

K = ·022 $\sqrt{\text{Effective load on piston in lbs.}}$—for Naval engines.

Connecting-rods are usually made tapered from the gudgeon end to the middle, and parallel from the middle to the crank end.

Rule 71a.—The gudgeon end may have a diameter equal to ·875 of the diameter at the middle.

Connecting Rod Gudgeons.—The rods of small engines may have the gudgeon shrunk into the jaws or sides of the double-eye, and working in brasses fitted into a recess in the piston rod head, and secured by a cap and bolts ; but larger engines should have the gudgeon shrunk into or formed with the piston-rod head or crosshead, and both jaws of the connecting-rod fitted with brasses, caps and bolts, &c. The great advantages of the latter plan are,—the ease with which the brasses can be overhauled and adjusted, and the fact that it does not require the piston-rod crosshead to be forged solid with the rod. It also braces or "trusses" the joint between the two rods, and, in a measure, compensates for the absence of side guides, which experience has shown are better omitted ; and, further, should unequal wear of the brasses take place, it occurs at a distance from the axis, and is of less relative importance than unequal wear of the two sides of a single central brass would be.

Rule 72.—The loads on gudgeon bearings, from maximum effective pressures on pistons, should never exceed 1200 lbs. per sq. inch, and, when double brasses are used, they can be kept down to about 1000 lbs. per sq. inch without using dimensions in any way abnormal, and, of course, work all the better for it.

Where the gudgeons cannot conveniently be made of hard steel, but are of mild steel, or wrought-iron, they must be case-hardened, and then carefully ground true.

White metal must not be used in the brasses for these bearings, as experience has shown that it is not suited for cases in which the pressures are great, and the angular movement small.

Rule 73.—The diameter of the gudgeon (when shrunk into the connecting-rod) may be 1·25 × Diameter of piston-rod, and the length is then given by,—

Length of Gudgeon = $\dfrac{\text{Max. effective load on piston}}{\text{Diameter} \times 1200}$.

Rule 74.—When the gudgeon is fixed into, or formed with, the piston-rod crosshead, the length of *each* end should be not less than ·75 × Diameter of piston-rod, and is commonly made about ·875,— with a diameter about the same as that of the piston-rod ; but in Naval engines the proportions rise to 1 or 1·2 for the lengths, and 1·25 for the diameters.

Connecting Rod Caps and Bolts.—The stresses on bolts of piston-rods and connecting-rods, per square inch area at bottom of thread, may be those given in Table XXXI., page 73,—the bolts being in all cases of mild steel ; but, where the load is carried by four bolts in place of two, the stress per square inch should be one-eighth less,— to allow for possible inequality in screwing up.

The bodies of the bolts should be turned down to the same diameter as the bottom of the thread, except where necessary for steadying the caps and brasses, and these remaining plain portions should be of slightly larger diameter than the screw thread,—say $\frac{1}{32}$ inch up to 2½ inch diameter, and $\frac{1}{16}$ inch for larger sizes.

The nuts should be of wrought-iron,—as steel nuts on steel bolts have a great tendency to "seize" and tear when imperfectly lubricated.

The necessary section of cap is most readily determined by means of the ordinary formula for a beam supported at the ends, and uniformly loaded, viz. :—

Rule 75.
$$w = \frac{8fz}{l^2}$$

where w is the working load per inch of length for any given value of f ; l the length in inches, from centre of bolt to centre of bolt ; and z the modulus of the section,—equal to $\dfrac{\text{Breadth} \times \text{Depth}^2}{6}$ for a rectangular section. For wrought-iron, or mild steel, the values of f may be,—

Flat-backed brasses and caps,—load carried by one cap,—$f = 9,000$.
Half-round   „   „   „ $f = 10,000$.
Flat-backed   „   „   two caps,—$f = 8,000$.
Half-round   „   „   „ $f = 9,000$.
In estimating the breadths of caps, care should be taken to deduct the diameters, or breadths, of all oil holes.

Connecting-Rod Brasses.—The overall thickness of these (*i.e.*, the thickness including any white metal) should be as follows :—

Rule 76.   Flat-backed brasses for gudgeon end,—
$$\text{Thickness} = \frac{D}{7} + \cdot2.$$

Rule 76a.   Flat-backed brasses for crank end,—
$$\text{Thickness} = \frac{D}{8} + \cdot2.$$

Rule 76b.   Round brasses for gudgeon end,—

$$\text{Thickness} = \frac{D}{8} + \cdot 2.$$

Rule 76c.   Round brasses for crank end,—

$$\text{Thickness} = \frac{D}{9} + \cdot 2.$$

Rule 76d.—Thickness of white metal $= \cdot 02D + \cdot 125.$

Rule 76e.—When no white metal is $\left\{ \begin{array}{l} \frac{D}{10} + \cdot 15 \text{ when flat backed.} \\ \frac{D}{11} + \cdot 12 \quad \text{,, round ,,} \end{array} \right.$ used at crank end the thicknesses may be

See Table XXXVIII., page 94, for proportions given by these rules.

The recesses in the brasses should be carefully tinned before the white metal is run in; and the practice of hammering the white metal to consolidate it should be avoided, as it tends to detach it from the brass.

Rule 77.—The width of the crank end of the connecting-rod should not be less than $\cdot 7 \times$ length of crank-pin,—as otherwise the brasses will not be properly supported.

The proportions of the double-eye at gudgeon end (fig. 18) should be as follows :—

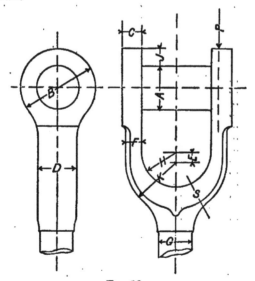

Fig. 18.

Table XXXVIII.—Thickness of Brasses.

Diameter of gudgeon or pin.	Connecting rod gudgeon when in piston rod head.		Connecting rod gudgeon when on forked end of connect. rod.		Crank-pins.			Main Bearings.		
	Inner half. $\frac{D}{8}+\cdot2$	Outer half. $\frac{D}{7}+\cdot2$	Round brass's. $\frac{D}{8}+\cdot2$	Flat backed bras.'z. $\frac{D}{7}+\cdot2$	Round brasses. $\frac{D}{9}+2$	Flat backed brasses. $\frac{D}{8}+\cdot2$	White metal ·02D+·125	Round brasses. $\frac{D}{10}+\cdot25$	Flat backed brasses. $\frac{D}{9}+\cdot3$	White metal ·02D+·125
2	·45	·48	·45	·48	·32	·37	...	·30	·35	...
2½	·51	·56	·51	·56	·37	·43	...	·34	·40	...
3	·57	·63	·57	·63	·42	·48	...	·39	·45	...
3½	·64	·70	·64	·70	·47	·54	...	·45	·50	...
4	·70	·77	·70	·77	·64	·70	·20	·65	·74	·20
4½	·76	·84	·76	·84	·70	·76	·21	·70	·80	·21
5	·82	·91	·82	·91	·75	·82	·22	·75	·85	·22
5½	·89	·99	·89	·99	·81	·89	·23	·80	·91	·23
6	·95	1·06	·95	1·06	·87	·95	·24	·85	·97	·24
6½	1·01	1·13	1·01	1·13	·93	1·01	·25	·90	1·03	·25
7	1·07	1·20	1·07	1·20	·98	1·07	·26	·95	1·08	·26
7½	1·14	1·27	1·14	1·27	1·04	1·14	·27	1·00	1·14	·27
8	1·20	1·34	1·20	1·34	1·09	1·20	·28	1·05	1·19	·28
8½	...	...	1·26	1·41	1·15	1·26	·29	1·10	1·25	·29
9	...	...	1·32	1·48	1·20	1·32	·30	1·15	1·30	·30
9½	...	...	1·39	1·56	1·26	1·39	·31	1·20	1·36	·31
10	...	...	1·45	1·63	1·31	1·45	·32	1·25	1·41	·32
10½	...	...	1·51	1·70	1·37	1·51	·33	1·30	1·47	·33
11	...	...	1·57	1·77	1·42	1·57	·34	1·35	1·52	·34
12	...	...	1·70	1·91	1·53	1·70	·36	1·45	1·63	·36
13	...	...	1·82	2·06	1·64	1·82	·38	1·55	1·74	·38
14	...	...	...	...	1·75	1·95	·40	1·65	1·85	·40
15	...	...	...	...	1·87	2·08	·42	1·75	1·97	·42
16	...	...	...	...	1·98	2·20	·44	1·85	2·08	·44
17	...	...	...	...	2·09	2·33	·46	1·95	2·19	·46
18	...	...	...	...	2·20	2·45	·48	2·05	2·30	·48
19	...	...	...	...	2·31	2·58	·50	2·15	2·41	·50
20	...	...	...	...	2·42	2·70	·52	2·25	2·52	·52

For rows 2½ to 3½ the crank-pin Round brasses follow $\frac{D}{10}+\cdot12$; the crank-pin Flat backed brasses follow $\frac{D}{9}+\cdot15$; the Main Bearing Round brasses follow $\frac{D}{11}+\cdot12$; the Main Bearing Flat backed brasses follow $\frac{D}{10}+\cdot15$.

Note.—Thicknesses given in above Table are *total* thicknesses, including white-metal, if any.

Rules 78 and 78a. $\quad \dfrac{B}{A} - 1\cdot82$, and $\dfrac{C}{A} - \cdot475$.

Rules 78b and 78c. $\quad \dfrac{D}{G} - 1\cdot2$, and $\dfrac{F}{G} - \cdot45$.

The sectional area at D, or F, is then about ·62 of that at G; and area C × J about ·39 of that at G.

Rule 78d. $\qquad\qquad$ Also $\dfrac{E}{H} - \cdot275$.

When double brasses are used the proportions differ slightly from the above, and are approximately,—

Rule 78e, 78f, and 78g. $\quad \dfrac{D}{G} - 1\cdot25$; $\dfrac{K}{H} - 1\cdot5$; and $\dfrac{E}{H} - \cdot47$.

But of course, in this latter case, much depends on the type of crosshead employed, and the side of the jaw should be considered as a cantilever, and the section at S made such that the moment of P does not impose a load of more than 9000 lbs. per square inch, when P is taken as five-eighths of the total effective load on the piston. The proportions given above are for the case where the crosshead consists of a central cubic block, into which the piston-rod end is fitted, and from which the gudgeons project, one on either side,—as shown in fig. 17.

SHAFTING.

The resistance of a plain cylindrical shaft to a simple twisting stress is given by,—

Rule 79. $\qquad\qquad T - \cdot196\ D^3 \times f,$

where T is the twisting moment in inch lbs.; D the diameter of the shaft in inches; and f the greatest shearing stress on the material,—in lbs. per square inch.

When the shaft is hollow, the formula becomes,—

Rule 79a. $\qquad\qquad T - \cdot196\dfrac{D^4 - d^4}{D} \times f,$

where D and d are respectively the external and internal diameters.

The values of f should not exceed the following:—

	f	
	Shafts below 10″ dia.	Shafts above 10″ dia.
Wrought-iron forging,	9,000	8,000
Scrap steel forging, .	10,500	9,000
Ingot steel forging, .	12,000	10,000

The torsional stiffness of shafts may be estimated by the following formulæ (Rankine):—

Rule 80.   For solid shafts $\theta = \dfrac{10\cdot2\ Tl}{CD^4}$ nearly.

Rule 80a.   For hollow shafts $\theta = \dfrac{10\cdot2\ Tl}{C(D^4 - d^4)}$ nearly.

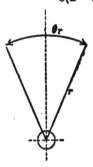

FIG. 18A.

where θr is length of arc moved through at radius r; l length of shaft in inches; C constant, values of which are given below; and other symbols, as defined above.

For cast-iron,   .   .   C = about 2,850,000.
  „ wrought-iron,   .   C = 8,500,000 to 10,000,000.
  „ steel,   .   .   C = 10,000,000 to 12,000,000.

Where a shaft is revolving *uniformly*, and transmitting power, the relation between the twisting moment and the horse-power applied is given by,—

Rule 81. $T = \dfrac{\text{I.H.P.}}{R} \times 63,000.$

The great majority of shafts, however, are turned by steam engines acting through cranks, and do not revolve uniformly,—because the tangential pressure on the crank-pin is constantly varying,—and must therefore be designed to resist the *maximum* twisting moment, instead of the mean given by the above equation.
The rule therefore takes the form,—

Rule 82. D^3, or $\dfrac{D^4 - d^4}{D}$, $= \dfrac{\text{I.H.P.}}{R} \times F.$

Where F is a co-efficient, the values of which depend on the number and relative positions of cranks, distribution of steam in the cylinders, &c., and are given in the following Tables :—

TABLE XXXIX.—SHAFTS FOR SCREW ENGINES. 97

Table XXXIX.—Shafts for Screw Engines.

Description of Engine.	Value of F for Crank-shafts.	Value of F for Tunnel shafts.
Single crank, single cylinder, cut-off ·5, .	150	130
Do. two cylinder compound, .	130	110
Two crank two cylinder compound, cranks at right-angles, ...	100	85
Three crank triple engine, cranks at 120°, .	85*	74*
Four crank quadruple engine, cranks at 90°,	95*	84*

Table XXXIXa.—Shafts for Paddle Engines.

Description of Engine.	Value of F Intermediate shaft journal.	Value of F Paddle shaft inner journal.	Value of F Paddle shaft outer journal.
Single crank, single cylinder, .	...	80	100
Two crank, two cylinder; cranks virtually at right angles, and connected by link, ...	...	58	65
Two crank, two cylinder, with intermediate shaft; cranks at right angles, . . .	58	50	65
Two crank, two cylinder; solid crankshaft; cranks at right-angles, ...	...	55	65

For paddle steamers working only in smooth water the above values of F may be reduced 20 per cent.

The above values of F will suffice for iron shafts, but it is better not to reduce them when employing mild steel, unless the forgings are from ingots of the highest quality.

Hollow Shafting.—Shafting for ships of war is now almost invariably made hollow, in order to get maximum strength and stiffness with minimum weight.

The internal diameter is usually half the external, so that the saving in weight is 25 per cent., whilst the reduction of strength is only about 6¼ per cent.

"Tail"-shafts are usually "set in," at the after end, after boring, so as to reduce the diameter of the hole within and near the propeller boss.

* *Note.*—For torpedo craft and cases where use of full power is occasional and of brief duration, values of F may be as low as half those here given.

7

CRANKSHAFTS IN GENERAL.

Crankshafts must be strong enough to bear, not only the maximum twisting stress, but also bending and shearing stresses, the magnitudes of which depend on the positions of the bearings. The conditions of loading and support are rather difficult to determine exactly,— depending as they do on the rigidity of the surrounding parts,—but it suffices for most purposes to assume that, in a simple case of overhang, they are represented by the formula,—

Rule 83. $$W = \frac{fz}{l},$$

where W is maximum load on pin, in lbs. ; f the greatest permissible stress on the material in lbs. per square inch (for which *see* page 95) ; l length in inches from mid-length of crank-pin to mid-length of bearing ; and z, for a circular section, is ·0982 D^3.

Where the crank has two arms, and is supported by a bearing on each side, the formula may be taken as,—

Rule 84. $$W = \frac{6fz}{l},$$

l being the distance, in inches, between the *centres* of the two bearings.

Equivalent twisting moment.—When a shaft is subject to simultaneous twisting and bending, the combined stress on any section of it may be measured by calculating what is called the

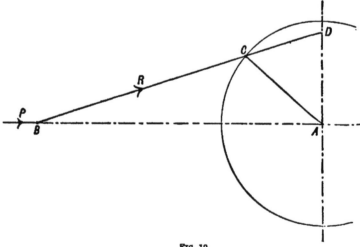

Fig. 19.

equivalent twisting moment; that is, the two stresses may be so combined as to be treated as one twisting stress only, and the size of the shaft calculated accordingly. The formula for combining the stresses is,—

Rule 85. $$T_1 = M + \sqrt{M^2 + T^2}$$

where T is the twisting moment; M the bending moment at any section ; and T_1 the equivalent twisting moment.

The shearing stresses on crank-shafts are relatively small, and are usually allowed for in the value assumed for f or F.

Curve of twisting moments.—The twisting moment, at any position of the crank, may be determined graphically as on Fig. 19.

Let AB (Fig. 19) be the centre line of the engine, through tne cylinder and shaft centres, AC the position of the crank, BC the connecting rod, and AD a line at right angles to AB. Produce BC to cut the line AD. Then, if P be the effective pressure on the piston, when the crank is in position AC, the twisting moment is P × AD.

Let the twisting moment be determined at intervals of say 10° of angular movement of the crank, so that there will be 18 values obtained for the half revolution. Draw a line AB (Fig. 20), and

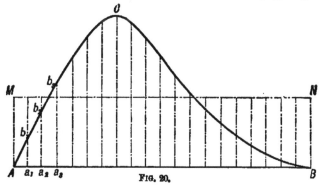

FIG. 20.

divide it into 18 equal parts, Aa_1, a_1a_2, &c. ; erect perpendiculars at these points, and cut off parts a_1b_1, a_2b_2, &c., representing to scale the 18 values obtained ; and through the points b_1, b_2, &c., draw a curve, which will be the curve of twisting stress on the shaft during one stroke of the piston. By prolonging AB, and going through a similar operation for the second half of the revolution, the curve of stress during the return stroke may be obtained.

If the area enclosed by the curve and the line AB be divided by the length of AB, the quotient (AM in Fig. 20) is the *mean* twisting moment. The value of AM may be calculated by taking a mean of the values of a_1b_1, a_2b_2, &c. Where there are two or more pistons acting on the same shaft, the curve of combined twisting moments is obtained by laying off the ordinates of the second piston, or crank, above or beyond the first curve, which is used as the base line, and so on, care being taken to step, or displace, the various curves in the direction AB, by an amount due to the angles between the cranks.

The polar form of this diagram, which is preferred by some engineers, is obtained by laying off the ordinates of the various curves radially from a circle which represents the crank path, instead of from a straight line representing the stroke of piston.

The effective pressures on the pistons at each point are obtained from the indicator diagrams in the manner explained on page 8.

The bending moment on a section of the shaft will vary exactly with the pressure on the crank-pin, and to find the maximum equivalent twisting moment on a section, it is only necessary to construct a secondary curve from the formula $T_1 = M + \sqrt{M^2 + T^2}$, between the point of maximum twisting and that at which the pressure on the piston is greatest.

When steam is not cut off in the cylinder before ·4 of the stroke, the maximum load on the piston may be used to calculate the bending moment which is to be combined with the maximum twisting moment.

Effect of Inertia of Reciprocating parts.—The attention of practical engineers was first drawn to this very interesting subject by Mr Arthur Rigg, in his "Treatise on the Steam Engine," and it has since been treated by various other writers. So far as ordinary marine machinery is concerned (and this term may be taken to include the high-speed machinery of Torpedo Gun-Boats), the general effect of this inertia is to equalise the action of the crank in the same manner as a fly-wheel,—the energy expended in accelerating the reciprocating parts during the early part of the stroke being given out again during the latter part, when the steam pressure is lower. When a curve of twisting moments is constructed as described above, and corrected for the effects of inertia, it will generally be found to show that by far the greater part of the steam pressure at the commencement of the stroke is absorbed in producing acceleration, and that the remaining *effective* pressure is but a very small fraction of the whole.

It is not necessary, however, to take these effects into consideration when proportioning the various parts of an engine, since calculations based on the statical stresses will quite cover all that is necessary to provide for dynamical stresses.

Surfaces of Crank-pins and Main Bearings.—If the effective bearing surfaces of pins and journals be considered as equal to diameter multiplied by length, then, the pressures per square inch should not xceed those given in the following Table :—

e

TABLE XL.—PRESSURE ON CRANK-PINS, ETC. 101

Table XL.—Pressures on Crank-pins and Main Bearings.

Crank-pin Screw engine may carry	400 to 500 lbs. per square inch.		
Main bearing do.	do.	300 to 400	do.
Crank-pin Paddle engine	do.	800 to 900	do.
Main bearing do.	do.	500 to 600	do.

The pressures referred to in the above Table are those resulting from the maximum effective loads on the pistons.

The main bearings of modern Naval engines (surfaces calculated as described above) are not usually loaded to more than about 275 lbs. per square inch (**Rule 86**), but the surface is reduced by recessing the brasses at the sides, so as to keep them clear of the shaft for one-third of its diameter at each side. The crank-pins, on the other hand, are loaded rather more heavily than Table allows, viz.,—to about 550 lbs. per square inch (**Rule 87**), and this is further increased by recessing the brasses similarly to those of the main bearings.

Crankshafts of Screw Engines.

Where a shaft has two, three, or more cranks in it, the after cranks have not only to resist the stresses imposed by their own cylinders, but have also to transmit the twisting stresses produced by all cylinders forward of them, so that,—were it not for practical reasons,—the sizes of crank-arms, journals, and pins, might be progressively reduced from after to forward end of shaft.

On the forward journal and forward arm of forward crank, there is little more twisting stress than is necessary to drive the forward eccentrics (if any), and the bending stresses are those due to half the load on the pin; whilst the after arm of the same crank, and the journal adjoining it, are subject to similar bending stress, and to nearly the whole of the twisting stress due to the load.

If the bending moment in either journal be expressed by $\frac{M}{2}$, then, the equivalent twisting moment in after journal will be,—according to formula given on page 99 (Rule 85),—

$$T_1 = \frac{M}{2} + \sqrt{\left(\frac{M}{2}\right)^2 + T^2}.$$

Coming now to the aftermost crank,—if T_2 be the maximum total twisting moment produced by all cylinders forward of it; T_3 the maximum twisting moment from its own cylinder; and M_3 the corresponding bending moment; then, for the forward journal, the twisting moment is T_2, and the bending moment $\frac{M_3}{2}$, so that the equivalent twisting moment is,—

$$T_4 = \frac{M_3}{2} + \sqrt{\left(\frac{M_3}{2}\right)^2 + T_2^2},$$

whilst, on the after journal, the twisting moment is $T_2 + T_3$, and the bending moment $\frac{M_3}{2}$, so that the equivalent twisting moment is,—

$$T_5 = \frac{M_3}{2} + \sqrt{\left(\frac{M_3}{2}\right)^2 + (T_2 + T_3)^2}.$$

The maximum bending moment, acting in a plane perpendicular to the axis of the shaft, at any section of the aftermost crank-arm, distant x from the crank-pin centre is,—

$$M_5 = \frac{T_2 + T_3}{L} \times x,$$

and for the forward arm of the same crank,—

$$M_4 = \frac{T_2}{L} \times x,$$

where L is the length of the crank in inches.

The bending moment at the mid-length of the pin of a solid or rigidly built-up shaft is,—

$$M_6 = \frac{R \times L}{8},$$

where R is the load on pin, and L the distance between the centres of the main bearings on either side,—the shaft being considered as a continuous girder supported at the centres of the bearings.

In marine crankshafts, where the crank-arms are not less than ·7 of the diameter in thickness, and where the bearings are close to the crank-arms, there is really little or no bending action at the journals.

At and near the ends of the stroke the crank-arms are subjected to a bending stress which acts in the plane of the axis of the shaft and has a magnitude,—

$$M_7 = \frac{R \times L_1}{8},$$

where L_1 is the distance from the mid-length of pin to the fore-and-aft centre of the arm (Fig 21). Hence the thickness (a) of the forward arm of forward crank must never be less than,—

$$a = \sqrt{\frac{3R \times L_1}{4bf}}.$$

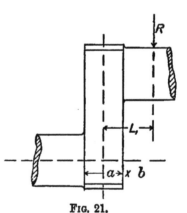

FIG. 21.

In ordinary practice, however, to secure interchangeability, the crankshafts of merchant steamers are made with all journals of the same diameter, and all crank-arms of the same dimensions ; but in light machinery for Naval purposes the forward crank-arms are some-times reduced in fore-and-aft thickness.

Crank-arms, when forged solid with the shaft, will be found to satisfactorily resist all stresses to which they are subject, if made of the following proportions :—

Rule 88.   Breadth of arm — 1·1 × Diameter of shaft.

Rule 88a.   Thickness of arm — ·75 × Diameter of shaft.

When for any reason it is not possible to make the arms so thick as ·75 × diameter of shaft, they may be made as thin as ·7, or even, in special cases, ·65 ; but when this is done, care must be taken to make the breadth such that, —

$$\frac{a \times b^2}{6} = \cdot 196 \, D^3.$$

The most destructive of all stresses to crankshafts are, however, those resulting from the main bearings getting out of line, and the levels of these should therefore be verified every few months, and the shaft lifted and re-bedded whenever any unequal wear is apparent.

When crank-pins and shaft journals are designed for the pressures given in Table XL., page 101, they will be found to have about the following proportions :—

Rule 89.

Diameter of crank-pin .   .   .   .   . = 1 × diam. of shaft.

Rule 89a.

Length of crank-pin (two crank compound) . = 1 to 1·5 ×   do.

Rule 89b.

Length of crank-pin (three crank triple)   . = 1 to 1·25 ×   do.

Rule 89c.

Length of main bearing (two crank compound) = 1·1 to 1·3 ×   do.

Rule 89d.

Length of main bearing (three crank triple)   = ·8 to 1·1 ×   do.

In high-speed Naval engines the bearings and crank-pins are often made longer than this, but the power transmitted through a given crankshaft, when forced draught is in use, is frequently 50 per cent. more than is usual in the merchant marine ; in other words,—what is taken from the diameter of the journal is added to the length.

To ensure equal wear in the main bearings, they should, theoretically, increase in length from forward aft, in proportion to the twisting moment at each journal,—as the turning force at the crank-pin tends to cause the shaft to move round eccentrically within the bearing, and so cause the wear ; and, as a matter of fact, the after bearing *is* found, in practice, to wear more rapidly than the others, and should not be cut away at the sides.

Shaft Couplings and Coupling Bolts.—(**Rule 90.**) When couplings are forged solid with the shaft they should have a thickness of ·28 × diameter of shaft.

Coupling bolts should have little beyond a shearing stress to resist, and their size is therefore given by,—

Rule 91. Area of one bolt $= \dfrac{\cdot196D^3}{N \times r}$ or $\dfrac{\cdot196\left(\dfrac{D^4 - d^4}{D}\right)}{N \times r}$

if the material of the bolts and that of the shaft offer the same resistance to shearing ; D being diameter of shaft, N number of coupling bolts, and r radius at which they are placed.

By way of making some allowance for screwing-up stresses on small bolts, coupling bolts below 1¼-inch diameter may be made 20 per cent. stronger than given by above rule.

Where the bolts in any coupling may require to be removed at times, they should be made with a taper of about ¼-inch per foot on the diameters.

Rule 92.—It is by no means necessary, in large bolts, to make the screwed part equal in diameter to the body of the bolt, and bolts between 2½ and 3½ inches diameter may be reduced ¼-inch, and those above 3½ inches, ½-inch in diameter. The nuts may have a thickness of ·75 to ·8 × diameter of screw.

In two, three, and four throw crankshafts, above 10 inches diameter, each crank, with its adjacent journals, should be made as a separate forging, and the number and position of coupling bolts should be such that the shafts can be interchanged or reversed as may be desired.

When large cranks are coupled together in this way, a spigot on each, fitting into a recess in its neighbour, materially assists adjustment, and ensures the "truth" of the shafts when bolted together ; such spigots may have a diameter of ·5 to ·7 × diameter of shaft, and a length of say ⅜-inch for the projection, and ⁷⁄₁₆-inch for the recess.

Built Crankshafts.—With a view of obtaining sounder and more reliable work, crankshafts over 12 inches diameter are frequently built up of separate pieces, as shown in fig. 22 ; but great care is required in the construction of such shafts to ensure perfect truth when finished, and freedom from internal strain.

Rule 93.—The thickness of the crank-arms is the same as for solid cranks, viz.,—·75 × diameter of shaft ; and the external radius of end of arm should be 1·8 × radius of hole (into which pin or shaft body fits). The crank-arm is frequently stepped back so as to have a thickness of say ·65 at the pin end whilst remaining ·75 at the shaft end in order to get a longer pin. Of course such shafts are considerably heavier than solid ones, and are therefore not used in Naval work.

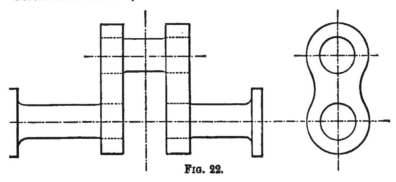

FIG. 22.

Crankshafts of Paddle Engines.

These may usually be classed under one of the following three heads :—

(a) Intermediate shaft type,—in which the outer crank-arms are keyed upon the paddle shafts, while the inner ones are similarly fixed to an intermediate shaft, the crank-pins being fixed in the inner crank-arms, and, to a certain extent, free in the outer ones, as shown in fig 23.

(b) The type in which the intermediate shaft and its crank-arms are absent, and the crank-pins are fixed in the paddle-shaft crank-arms, and left long enough to allow of the attachment of a link, or sling, which holds the crank-arms at the required angle with one another.

(c) The solid shaft,—in which the cranks are forged in one with the shaft,—as in the case of a screw engine,—and the paddle-shafts connected with it by solid flange couplings, a little elasticity being obtained by the interposition of a thickness of leather, by helical springs on the coupling bolts, or by other similar devices.

The effect of the first arrangement is to communicate very equable stresses to the paddle-shafts, as the pressures from the crank-pins are always at right-angles to the cranks on the paddle-shafts ; and, in smooth water, the power of each engine is very nearly equally divided between the two wheels, and the bending action on each paddle-shaft never exceeds half that due to its own cylinder, for, when near the dead-points, the bending moment is at its maximum, and is wholly taken by the inner crank-arm, to which the pin is secured.   For these reasons the intermediate shaft must be stronger than the paddle-shafts, when the ship is intended to work in rough water, as it may have to transmit the whole twisting force of one engine, and always takes, during certain parts of the revolution, the whole bending force from that engine.

Hence, if T be the maximum twisting moment from one piston of a double cylinder paddle engine, and M the maximum bending moment from the same piston, then,—as in the general case (page 99),—the maximum equivalent twisting moment T_1 will be,—

$$T_1 \text{ (on intermediate shaft)} = M + \sqrt{M^2 + T^2}$$

$$\text{and } T_1 \text{ (on paddle shafts)} = \frac{M}{2} + \sqrt{\left(\frac{M}{2}\right)^2 + T^2}.$$

Rule 94.—With the second type of shaft (b), the axes of the cylinders are necessarily at an angle with one another, but the cranks are usually so placed that the arrangement is equivalent to one in which the cylinders are in the same plane and the cranks at right-angles.

Each shaft takes the whole of the bending moment from its own cylinder ; and,—while usually transmitting half the combined twisting moment—may, in rough water, have to transmit the whole of the twisting moment of one engine. The inner journals of these shafts are therefore subject to precisely the same stresses as the intermediate shaft journals in type (a), and should be made of the same size, as indicated in Table XXXIXa., page 97.

The maximum pull in the coupling link may be taken as being equal to that in one connecting-rod, but its ordinary value will only be one half of this.

In the third case (that of the solid crank-shaft), the cranks will generally be at right-angles, and the twisting stresses will be similar to those in the previous cases ; the central part transmitting ordinarily half the stress from one piston, but occasionally the whole. The bending stresses in each crank-arm and journal will be those due to half the load on one piston, and may be determined by means of the formulæ previously given.

Paddle shafts.—The outer end of each paddle shaft is subjected to a bending stress, which is the resultant of the stresses due to the weight of the wheel, and to the re-action of the water on the floats, and the moment of which (M_1) may be taken as,—

$$M_1 = \sqrt{(M \text{ of weight})^2 + (M \text{ of re-action})^2}.$$

The twisting stress at the outer end of a paddle shaft is the same as at the inner journal. The outer bearing may have a length of from $1\cdot5$ to $2 \times$ diameter of journal, according to service for which boat is intended, and to weight of wheels.

Overhung cranks. Fig. 23 shows a pair of overhung cranks as commonly fitted in paddle wheel engines. The crank-pin is subject to bending and shearing stresses due to the thrust on the connecting-rod. The maximum bending stress in the pin is close to the face of the crank and is,—

$$M = R \times l$$

where R is the thrust on the connecting-rod, and l the length on pin from centre of connecting-rod to face of crank.

The diameter of the pin is given by,—

Rule 95. Diameter of crank-pin $= \sqrt[3]{\dfrac{R \times l}{f} \times 10 \cdot 2}$

where f is the greatest permissible stress on the material, in lbs. per square inch. This rule gives the diameter requisite for strength, but it may be necessary to make the pin larger in order to get sufficient surface to comply with Table XL., page 101.

The crank-arm is to be treated as a lever, so that,—if a is thickness measured parallel to axis of shaft, and b breadth, at a section x inches from the crank-pin,—the bending moment at that section is,—

$$M = R \times x$$

and, $\qquad \dfrac{a \times b^2}{6} = \dfrac{M}{f}$

or $a = \dfrac{6M}{b^2 \times f};$

Rule 96. The bending moment decreases with the distance from the crank-pin, while the shearing stress is the same throughout the crank-arm ; consequently the section near the crank-pin should have an extra square inch for each 8000 lbs. of thrust on the connecting-rod, beyond the area necessary to resist the bending moment.

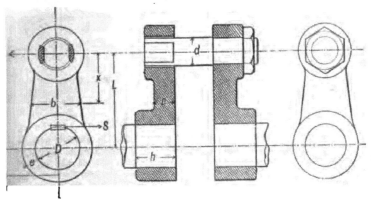

FIG. 23.

The dimensions of the boss to receive the shaft may be determined from,—

Rule 97.
$$T = 2R \frac{L}{D+e}$$

where T is the total stress on the section $h \times e$, and the other letters have the meanings shown in Fig. 23. The length h is usually ·75 to 1 × Diameter of shaft, (**Rule 98.**) and,—when crank and shaft are of the same material,—h and e may have the following relative values :—

> When $h =$ D then $e = $ ·35D.
> ,, $h = $ ·9D ,, $e = $ ·38D.
> ,, $h = $ ·8D ,, $e = $ ·40D.
> ,, $h = $ ·7D ,, $e = $ ·41D.

The crank-eye, or boss into which the crank-pin is fitted, should bear the same relation to the pin that the shaft boss does to the shaft.

The diameter of the shaft end, on which the crank is fitted, should be 1·1 × diameter of journal.

Keys for Cranks, etc. These should be made of steel several grades harder than the shaft and crank, and should be of the following proportions :—

Rule 99.   Breadth of Key $= \dfrac{D}{4} + $ ·125.

Rule 99a.   Thickness of Key $= $ ·5 × Breadth.

Depth of key-way in shaft, at edge, should be ·17 × Breadth. Some engineers use two (smaller) keys placed 90° apart, in order to obtain at least three bearing points on shaft, and so avoid all risk of "rocking"; and large shafts are sometimes fitted with three keys. Where a lever is placed at some distance from the end of a shaft, so that the key cannot be fitted and driven from the end, it is a good plan to fit a small driving key alongside of the sunk key, as shown in Fig. 24.

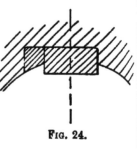

FIG. 24.

Tail Shafts of Screw Engines.

These must be strong enough to resist the bending stresses arising from the weight of the propeller, and from its action when partially immersed, in addition to the twisting stress. The former can of course be readily determined by the rules already given for bending strains on crankshafts, but the calculation of the stresses resulting from "racing" of the propeller is too complex a subject to enter on here. Experience has shown, however, that the tail shaft should have a diameter not less than that of the crank-shaft; and it is not at all uncommon now to make them 10 per cent. stronger than the crank-shaft, since breakage of one may involve serious consequences. Much also depends on the manner in which tail-shaft is supported.

Rule 100.—The taper of the end that fits into the propeller boss should not be less than ¾-inch per foot, and may, with advantage, be a little steeper,—say 1 inch per foot, *i.e.*, for every foot of length in the tapered portion the diameter will alter 1 inch.

The thread of the large nut that holds the propeller on the shaft is commonly made 2½ threads per inch, regardless of the diameter ; it should be left-handed when the propeller is right-handed, and *vice-versâ*. The nut should be securely locked, preferably by a plate fixed to the after end of the boss by tap screws.

The propeller should be secured by one feather or key extending the whole length of the boss, the proportions of which may be,—

Rule 101.—Breadth of key = ·22 × largest diameter of shaft + ·25.

Rule 101a.—Thickness of key = ·55 × breadth.

The diameter of the screwed end of the shaft should be sufficiently reduced to allow of the key being fitted in, from the after end, clear of the thread.

The diameters over the brass casings at the bearings should differ by ⅛-inch, to enable the shaft to be got in and out of the stern-tube more easily. The thickness of brass casing should not be less than,—

Rule 102.—Thickness of brass casing at bearings = ·055D_1 + ·15 where D_1 is the diameter of the shaft at the bearing, under the casing.

See Table XLIV., page 118.

Board of Trade Regulations as regards Shafting.

Main, tunnel, propeller, and paddle shafts should not be passed if less in diameter than that found by the following formulæ, without previously submitting the whole case to the Board of Trade for their consideration. It will be found that first-class makers generally put in larger shafts than those obtained by the formulæ.

When shafts are proposed to be fitted whose diameters are less than 5 inches, the case should be submitted for consideration.

For compound condensing engines with two or more cylinders, when the cranks are not overhung :—

$$S = \sqrt[3]{\frac{C \times P \times D^2}{f\left(2 + \frac{D^2}{d^2}\right)}}$$

$$P = \frac{f \times S^3}{C \times D^2}\left(2 + \frac{D^2}{d^2}\right)$$

Where S = diameter of shaft in inches.

> d^2 = square of diameter of H.P. cylinder, in inches ; or sum of squares of diameters when there are two or more H.P. cylinders.
>
> D^2 = square of diameter of L.P. cylinder, in inches ; or sum of squares of diameters when there are two or more L.P. cylinders.
>
> P = *absolute* pressure in lbs. per square inch,—that is, boiler pressure plus 15 lbs.
>
> C = length of crank in inches.
>
> f = constant from following table.

Note.—Intermediate pressure cylinders do not appear in the formulæ.

For ordinary condensing engines, with one, two, or more cylinders when the cranks are not overhung :—

$$S = \sqrt[3]{\frac{C \times P \times D^2}{3 \times f}}$$

$$P = \frac{3 \times f \times S^3}{C \times D^2}$$

Where D^2 is square of diameter of cylinder in inches ; or sum of squares of diameters when there are two or more cylinders ; and other letters have same meanings as above.

Table XLI.—Constants for Shafts.

For two Cranks. Angle between Cranks.	For Crank and Propeller (tail) Shafts. f		For Tunnel Shafts. f
90°	For paddle engines	1,047	1,221
100°	of the direct act-	966	1,128
110°	ing type multiply	904	1,055
120°		855	997
130°	constant in this	817	953
140°		788	919
150°	column suitable	766	894
160°	for angle of cranks	751	877
170°		743	867
180°	by 1·4.	740	864
For three Cranks. 120°		1,110	1,295

Note.—When there is only one crank, the constants applicable are those in the Table, opposite 180°.

Lloyd's Rules with regard to Shafting.

The diameters of crank and straight shafts are to be not less than those given by the following formulæ :—

Table XLII.—Diameters of Shafts.

Description of Engine.	Diameter of Crank Shaft in inches.
Compound—Two cranks at right angles . .	$(\cdot04A + \cdot006D + \cdot02S) \times \sqrt[3]{P}$
Triple — Three cranks at equal angles . .	$(\cdot038A + \cdot009B + \cdot002D + \cdot016S) \times \sqrt[3]{P}$
Quadruple—Two cranks at right angles . .	$(\cdot034A + \cdot011B + \cdot004C + \cdot0014D + \cdot016S) \times \sqrt[3]{P}$
Do. —Three cranks . .	$(\cdot028A + \cdot014B + \cdot006C + \cdot0017D + \cdot015S) \times \sqrt[3]{P}$
Do. —Four cranks . .	$(\cdot033A + \cdot01B + \cdot004C + \cdot0013D + \cdot0155S) \times \sqrt[3]{P}$

Where A is diameter of H.P. cylinder in inches.
  B   do.   first I.P. cylinder in inches.
  C   do.   second I.P.   do.
  D   do.   L.P.   do.
  S   stroke of pistons in inches,
  P   boiler pressure above atmosphere in lbs. per square inch.

The screw (tail)-shaft to be the same diameter as is required for the crank shaft.

Intermediate (tunnel) shafting should be at least $\frac{15}{16}$ths of the diameter required for the crank shaft.

Where engines are of extreme proportions the case should be specially submitted to be dealt with on its merits.

36. All shafts to be examined when rough turned and when finished.

37. Gauges of an improved description for testing the truth of crank shafts are to be supplied with all new engines, and adjusted in the presence of a Surveyor.

THRUST BLOCKS.

To find the thrust along the shaft of a screw engine, it is necessary to know the speed of the ship, and the effective horse power.

The effective horse power is the power actually employed in propelling the ship, and its relation to the indicated horse-power depends on the combined efficiency of the engines and propeller.

For the purpose of calculating the surface of thrust collars it is sufficient to assume that E.H.P. is equal to two-thirds I.H.P.

Then,—

Rule 103.　Indicated Thrust $= \dfrac{\text{I.H.P.} \times 33,000}{S \times 101\cdot3} = \dfrac{\text{I.H.P.} \times 326}{S}$,

where S is the speed of the ship in knots ; and,—

Rule 104. Mean Normal Thrust $= \dfrac{\text{I.H.P.} \times 22,000}{S \times 101\cdot3} = \dfrac{\text{I.H.P.} \times 217}{S}$.

As the thrust varies with the I.H.P., and inversely with the speed (*see* Table XLIII.), it may rise considerably above the "mean normal" value, if, from any cause, the speed is reduced without a corresponding decrease in the power, as, for instance, when "thrashing" against a head wind and sea, or when towing, and allowance must be made for such temporary increases.

Rule 105.　The pressure per square inch, due to the mean normal thrust, should never exceed 70 lbs.; and, for tug boats and vessels specially exposed to severe weather, it should not exceed 50 lbs.

In ordinary practice the pressure is about 50 lbs. per square inch in Naval work, and about 40 lbs. per square inch in mercantile steamers, but when white metal is used these loads may be safely increased by 25 per cent. For shafts of "mild" steel, as now generally use, gun-metal does not make a satisfactory rubbing surface, and white metal should always be employed.

The relation between depth and thickness of collars on shaft is given by,—

Rule 106.　Thickness $= \begin{cases} \cdot25\,(D - d) + \cdot4 \text{ inch (Types 1 and 3).} \\ \cdot17\,(D - d) + \cdot4 \text{ inch (Type 2).} \end{cases}$

where D is diameter over collars, and d diameter between collars,—both in inches.

The spaces between collars on shaft should be as follows :—

Rule 107.　When rings in block are solid gun metal, as shown in fig. 25.

Space $= \cdot4\,(D - d)$.

Rule 107a.　When block is fitted with cast-iron or cast steel "horse-shoes" faced with brass or white metal, as shown in fig. 26.

Cast-iron,—

Space $= 1\cdot8\sqrt{(D - d) - 1\cdot5}$.

Cast Steel,—

Space $= 1\cdot5\sqrt{(D - d) - 1\cdot5}$.

Rule 107b.　When rings are cut out of solid metal of block and cap, and entire corrugated surface is covered with uniform layer of white-metal, as shown in fig. 27.

Space $= \cdot4\,(D - d) + \cdot6$ in.

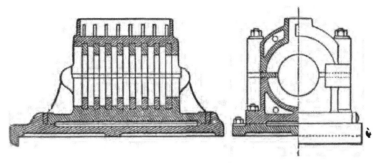

FIG. 25.—Thrust Block (Type 1).

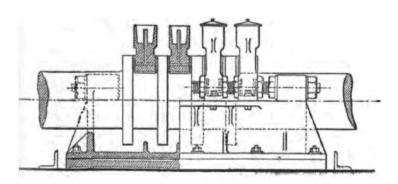

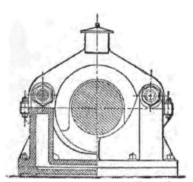

FIG. 26.—Thrust Block (Type 2).

8

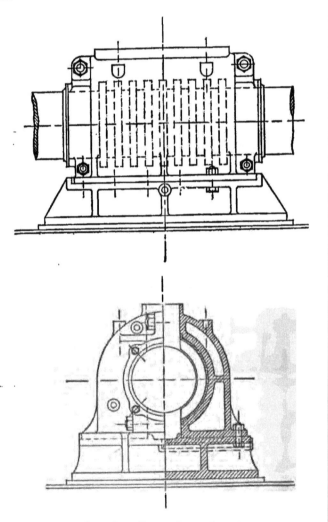

FIG. 27.—Thrust Block (Type 3).

Rule 108. The number of collars employed is, to some extent, optional with the designer, but is commonly ·8 to ·9 × diameter of shaft in inches, for types 1 and 3, and a good approximate rule for type [2] is,—one collar for shafts up to 6 inches diameter, and an additional one for every 2 inches of diameter beyond.

As in all other bearings, good working depends almost entirely on good lubrication, and the satisfactory results obtained with blocks of type (2) are undoubtedly mainly due to the oil-bath in which the collars run ; but this type of block has the great disadvantage of requiring very deep collars, to form which it is necessary to cut very deeply into the forging (as much as 6 to 7 inches a side in the shaft of a large liner), thus leaving only what is, in very many cases, decidedly the inferior portion of the forging to take all the twisting stress.

To meet this objection, Mr Rounthwaite has designed the block shown in fig. 27, which has the following important advantages over those hitherto used :—

(a) The most reliable portion of the shaft forging is not cut away as in type (2).

(b) The shaft may be run entirely submerged in oil, if so wished.

(c) The shaft can be stripped bare for examination in a few minutes, and the block, if injured in any way, can be removed and replaced without disconnecting and lifting the thrust-shaft.

(d) There is no risk of injury to the block or shaft by meddlesome or careless interference with the adjustment of the " horse-shoes" as in type (2).

(e) Satisfactory results can be obtained with less weight and bulk of block than with any other type.

There should always be an ordinary plummer block close to the thrust block, to take the weight of the shaft, and prevent vibration, and so ensure an even and steady pressure over the whole surface of the collars. In Naval work this plummer-block and the thrust-block are frequently parts of the same casting.

In order to facilitate the examination of the thrust-blocks of large engines, when type (3) is not adopted, the thrust collars should be placed upon a special short length of shafting which can be got out with *comparative* ease.

Large blocks of types (1) and (3) are much improved by having hollow caps and bodies, and arrangements for keeping up a continuous circulation of water through them.

Thrust blocks of all types should be carefully scraped and bedded to a perfect bearing on the shaft before leaving the workshops.

Rule 109. The number and diameter of cap bolts for type (1) should be such that the shearing stress per square inch, at the bottom of the thread, due to the mean normal thrust, may be between 5000 and 6000 lbs. when the bolts are assisted by spigot pieces of usual proportions.

The sizes of bolts usually employed vary from 1 inch to $1\frac{3}{4}$ inch.

Rule 110. The number and diameter of holding down bolts should be such that the shearing stress per square inch, at the bottom of thread due to the mean normal thrust, may be between 1800 and 2200 lbs., when the bolts are unassisted by stops or angle bars riveted to the top plate of the seating, and between 3200 and 3800 lbs. when such stops are fitted. Stops should always be fitted where possible, as, when they are absent, the stress on the bolts is largely a bending stress, owing to the unavoidable presence of a certain thickness of packing.

The sizes of bolts usually employed vary from ⅞ inch to 1½ inch. Packing should be of cast-iron, *not* of wood.

Table XLIII.—Thrust Surfaces per E.H.P. for Various Speeds of Vessel.

Speed in knots.	Surface necessary to keep pressure 50 lbs. per sq. inch.	Speed in knots.	Surface necessary to keep pressure 50 lbs. per sq. inch.	Speed in knots.	Surface necessary to keep pressure 50 lbs. per sq. inch.	Speed in knots.	Surface necessary to keep pressure 50 lbs. per sq. inch.
	Sq. inch.		Sq. inch.		Sq. inch.		Sq. inch.
8	·54	12	·36	16	·27	20	·21
9	·48	13	·33	17	·26	21	·21
10	·43	14	·31	18	·24	22	·21
11	·39	15	·29	19	·23	23	·19

STERN-TUBES.

In merchant vessels the stern-tube is almost invariably of cast-iron, fitted with a brass bush, or bushes, to carry the lignum-vitæ strips which form the actual bearing surface for the shaft.

In Naval vessels, on the contrary, the stern-tube is almost always of gun-metal, and the bushes are not always fitted, as the lignum-vitæ strips may be fitted into grooves in the tube itself.

In twin-screw Naval vessels the shaft brackets are fitted with gun-metal bushes, carrying strips of lignum-vitæ,—very similar to those used for the after ends of stern-tubes in merchant vessels ; and the forward ends of the stern-tubes are also fitted with long lignum-vitæ bearings, for the purpose of reducing vibration in the unsupported outboard shafts. Some foreign warships have been fitted with a second or intermediate shaft bracket to more effectually perform the same office.

The stern-tubes of merchant vessels should be secured in place by a ring nut on the after end, made to screw up against the after face of the stern-post and draw the tube aft until a collar, formed on it, is close up to forward side of post; the forward end of tube being, of course, jointed to the collision bulkhead in the usual manner.

Naval stern-tubes do not require any fixing at the after end, as they are always surrounded by steel tubes, ½ to ⅝-inch thick, which tie the collision bulkhead to the post, or, in the case of twin-screw ships, to the special stern-tube brackets ; they are simply drawn tightly into the post, or brackets, and bolted to the bulkhead at the forward end. When, as occasionally happens, they are put in from the after end, the flange at the stuffing box end is made separately and screwed on.

Cast-iron stern-tubes should be of the thickness given by,—

Rule 111.
$$T = \frac{D}{12} + \cdot 5,$$

and gun-metal tubes and bushes for shaft-brackets by,—

Rule 111a.
$$T = \frac{D}{22} + \cdot 35,$$

where **D** is the greatest diameter of shaft, over casings. (*See* Table XLIV., page 118.)

The stern-bush, for either shaft-bracket or stern-tube, should be of such length that the pressure per square inch (taking surface as length × diameter) does not exceed 50 lbs., and a good rule is,—

Rule 112. Length of bush = 3 to 3·5 × **diameter of shaft,**

where "diameter of shaft" is again greatest diameter over casings.

This rule will be found to limit the pressure to about 30 lbs. per square inch.

In Naval twin-screw vessels the bush at after end of stern-tube is generally three diameters in length.

Rule 113.—The bush at forward end of stern-tube should not be shorter than one diameter for single screw vessels, and, for twin-screw vessels, where there is a length of unsupported shaft outside the stern-tube, the length of this bush should not be less than 1·5 to 1·75 × diameter of shaft.

These bushes in stern-tubes should have a thickness of metal at the back of the lignum-vitæ not less than,—

Rule 114.
$$T_1 = \frac{D}{28} + \cdot 25,$$

where **D** is diameter of shaft over casing.

The ribs between the lignum-vitæ strips may have a thickness of ⅜-inch for a shaft 6 to 8 inches in diameter, increasing to ¾-inch for a shaft of 18 inches diameter.

The lignum-vitæ strips themselves may vary from ½-inch × 2 inches to 1 or 1⅛-inch × 4 inches, for shafts of the diameters just mentioned.

The stuffing-box and gland at the inner end of stern-tube may have the following proportions :—

Rule 115.

Depth of stuffing-box $= \dfrac{D}{2} + 1 \cdot 5.$

Rule 115a.

Amount gland can enter $= \dfrac{D}{5}+1$.

Rule 115b.

Diameter of stuffing-box $= (1\cdot1 \times D)+1\cdot5$ for bushed glands.

Rule 115c.

Diameter of stuffing-box $= (1\cdot06 \times D)+1\cdot5$ for solid G.M. glands.

The gland studs may vary from four, 1 inch in diameter, for a shaft of 6 to 8 inches diameter, up to eight, 1¼ inch diameter, for a 16 or 18 inch shaft. If pinion nuts and toothed rings are used, the gland studs must, of course, be made stouter to stand the bending stresses, and should then vary from 1¼-inch to 1⅝-inch.

The following Table shows at a glance the thicknesses of shaft casings, stern-tubes, and stern bushes, and the dimensions of stern-tube stuffing-boxes, as calculated by Rules 102, 111, 114, and 115 :—

Table XLIV.—Stern-Tubes and Bushes, etc.

Dia. of shaft under casing. (D_1)	Dia. of shaft over casing. (D) (casing $= \cdot055D_1 + \cdot15$)	Thickness of cast-iron tube. $\dfrac{D}{12}+\cdot5$	Thickness of gun-metal tube or shaft-bracket bush. $\dfrac{D}{22}+\cdot35$	Thickness of tube bushes at back of lignum-vitæ. $\dfrac{D}{28}+\cdot25$	Depth of stuffing box. $\dfrac{D}{2}+1\cdot5$	Amount gland can enter. $\dfrac{D}{5}+1$	Dia. of stuffing box for bushed cast-iron gland. $1\cdot1\,D+1\cdot5$	Dia. of Stuffing box for solid gun-metal gland. $1\cdot06\,D+1\cdot5$
6	7	1 1/16	1 1/16	1/2	5	2 3/8	9 1/4	9
7	8	1 5/16	3/4	1/2	5 1/2	2 5/8	10 1/4	10
8	9 1/8	1 1/4	3/4	9/16	6	2 7/8	11 1/2	11 1/8
9	10 1/4	1 3/8	1 3/16	5/8	6 5/8	3	12 3/4	12 3/8
10	11 3/8	1 7/16	7/8	11/16	7 1/4	3 1/4	14	13 1/2
11	12 1/2	1 ⁹/16	15/16	11/16	7 3/4	3 1/2	15 1/4	14 3/4
12	13 5/8	1 5/8	15/16	3/4	8 1/4	3 3/4	16 1/2	16
13	14 3/4	1 3/4	1	3/4	8 3/8	4	17 3/4	17 1/8
14	15 7/8	1 13/16	1 1/16	13/16	9 3/8	4 1/4	19	18 3/8
15	17	1 15/16	1 1/8	7/8	10	4 3/8	20 1/4	19 1/2
16	18	2	1 3/16	7/8	10 1/2	4 5/8	21 3/8	20 1/2
17	19 1/8	2 1/8	1 3/16	15/16	11	4 7/8	22 1/2	21 1/4
18	20 1/4	2 5/16	1 1/4	1	11 ½	5	23 3/4	23
19	21 3/8	2 1/4	1 5/16	1	12 1/8	5 1/4	25	24 1/8
20	22 1/2	2 3/8	1 3/8	1 1/16	12 3/4	5 1/2	26 1/4	25 5/8

In designing gun-metal stern-tubes, shaft casings, &c., great care must be taken to avoid, as far as possible, all shoulders or steps in the longitudinal section, as the castings set so rapidly, and consequently rip the core so quickly, that everything tending to hinder the longitudinal contraction is a source of danger; want of attention to this will probably cause a long tube to be torn into two or three separate pieces in cooling. The middle portions of Naval stern-tubes are commonly made about 30 per cent. thinner than the ends or portions containing the bearings. In torpedo boats, &c., the tube-shafts seldom have any casings, and run in white metal bearings.

MAIN BEARINGS.

The surfaces and lengths of main bearings have already been dealt with on pages 101 and 103.

Main Bearing Bolts.—These should be of "mild" steel, with iron nuts, and, to allow for variations in adjustment, &c., should be proportioned as follows :—

Rule 116. When there are two main bearings to each crank, and two bolts to each cap,—assume that each bolt must carry one third of maximum effective load on piston, and select the proper diameter from Table XXXI., page 73.

Rule 116a.—When there are two bearings to each crank, but four bolts to each cap,—assume that each bolt must carry one fifth of maximum effective load on piston, and select diameter from same Table.

Rule 116b.—Where there is only one long bearing with four cap bolts, between two cranks of a triple engine (cranks at 120°),—assume that each bolt must carry one fourth of maximum effective load on piston, and select diameter as before.

Rule 116c.—When the crank is overhung, as in various types of paddle-engine, and the whole thrust of the piston-rod, at the ends of the stroke, is resisted by two cap bolts,—call the thrust P ; the distance from centre of piston-rod to centre of main bearing bolts pw ; and the distance from centre of piston-rod to centre of outer bearing (in case of paddle-shaft), or of bearing at other end (in case of intermediate shaft) pf. Then (fig. 28),—

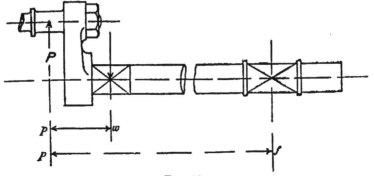

FIG. 28.

$$\text{Stress on main bearing bolts} = P \times \frac{pf}{wf}$$

and each bolt must carry one half of this.

Rule 116d. When the conditions are similar to (c), but there are four bolts in each cap,—assume that each bolt must carry one third of total load on cap, and select suitable diameter from Table XXXI., page 73.

In cases (c) and (d) the stresses on the main bearing bolts due to the weight and re-action of the wheels should be calculated on the principle indicated above, and allowed for if necessary ; their magnitude of course depends very greatly on the length of the paddle shaft.

Main Bearing Caps or Keeps.—In order to allow for variations in adjustment, either unnoticed or such as occur when a bearing warms up and has to be slacked, the loads on main bearing caps may be assumed to be as follows :—

Rule 117.

Two bearings to each crank (cases 116 and 116a) .   .   .

.   .   .   . $\frac{3}{4}$ × max. effect. load on piston.

Rule 117a.

One long bearing between two cranks (case 116b) .   .   .

.   .   .   . $\frac{3}{4}$ × max. effect. load on piston.

Rule 117b.

One bearing to each crank (cases 116c and d) .   .   .   .

.   .   .   . $\frac{pf}{wf}$ × max. effect. load on piston.

The load on a main bearing cap is neither a single central load, nor is it a uniformly distributed load, though probably nearer the latter than the former. If suitable values for the working stresses are employed, the most generally convenient formula is that for a beam supported at ends, and uniformly loaded, viz.:—

Rule 118. $w = \dfrac{8fz}{l^2}$,

where w = load in lbs. per inch of span (l) ;

$\quad\quad$ l = span, or distance from centre of cap bolt to centre of cap bolt ;

$\quad\quad$ z = Modulus of section $\left(= \dfrac{\text{breadth} \times \text{depth}^2}{6} \right.$ for a rectangular section ; for other sections see Table LXXXVI., p. 302. $\left. \right)$;

and f has the following values :—

Flat-backed brasses and caps $\begin{cases} \text{Cast-iron} & . & . & . & . & 2,500 \\ \text{Wrought iron or ``mild'' steel} & 9,000 \\ \text{Cast steel} & . & . & . & 10,000 \end{cases}$

Half round brasses and caps $\begin{cases} \text{Cast-iron} \quad . \qquad . \qquad . \qquad . \quad 3,000 \\ \text{Wrought iron or ``mild'' steel } 10,000 \\ \text{Cast-steel} \quad . \qquad . \qquad . \quad 11,000 \end{cases}$

When the caps are made of cast steel the section at the centre may, with advantage, be made to approximate to that of a "channel" bar by employing two deep external ribs; but in this case the value of f must not exceed 6000 (**Rule 119**). The values given above for cast steel are no doubt rather low, but there is,—in the present state of the manufacture,—always some risk of unsoundness, blow-holes, &c.

In estimating the breadths of caps care should be taken to deduct the breadths, or diameters, of all hand holes, oil holes, &c.

Main Bearing Brasses.—For small engines, auxiliary engines, &c., the inner or under brass may be of a semi-octagonal form, and the outer or upper one flat on the back; but in larger engines the under brass at least should be of the "half-round" type, and, when the framing can be arranged to allow of it, it is both cheaper and better to make the upper one "half-round" also.

Although brasses of good gun-metal will give excellent results, when properly designed and looked after, less risk and trouble will be incurred if white metal is used for the rubbing surfaces.

The overall thickness of the brasses (*i.e.* the thickness including white metal) should be as follows :—

Rule 120. Flat-backed brasses Thickness $= \dfrac{D}{9} + \cdot 3$-inch.

Rule 120a. Round brasses . . Thickness $= \dfrac{D}{10} + \cdot 25$-inch.

Rule 120b. White metal . . . Thickness $= \cdot 02\,D + \cdot 125$ inch.

where D is diameter of journal in inches.

The recesses in the brasses should be carefully tinned before the white metal is run in, and the practice of hammering the white metal to consolidate it should be avoided.

When no white metal is used the thickness of brass at crown should be,—

Rule 121.

Thickness of brass (when no white metal) $= \begin{cases} \dfrac{D}{10} + \cdot 15 \text{ when flat-backed.} \\ \dfrac{D}{11} + \cdot 12 \quad \text{when round.} \end{cases}$

Rule 122. Brasses should never overhang the frames or caps by more than "thickness of brass," as given by above rules, at each end.

See Table XXXVIII., page 94, for the proportions given by the above rules.

The tendency of brasses to close, after being warm, and grip the shaft, should be provided against either by securing them to the frames by screws or bolts, or, a simpler and better way, by **H** section liner strips fitted as shown in fig. 29.

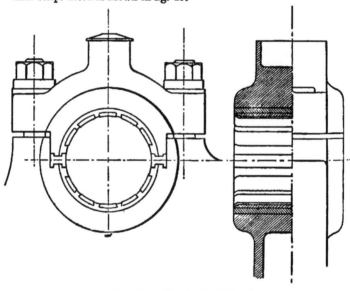

FIG. 29.—Crank Shaft Bearing.

In large main bearings it is very useful to have the lock-rings of the nuts graduated, in order that, after being slacked back, the nuts may be returned exactly to the old positions.

FRAMINGS.

The section of the girder that carries the main bearing is usually either of the "**H**" or "box" type. For small engines either the **H** or open bottom box is the best type, and for large engines the box, —if the frame is of cast-iron,—or the **H**,—if cast-steel is used.

In the case of vertical screw engines for Naval purposes this girder is sometimes cut down to the very slenderest proportions, so far as the engine builder is concerned, and is built of steel plates and angles as a part of the engine seating. In oscillating paddle engines, on the other hand, the girder receives no support from any seating, and must be strong enough to deal with all stresses communicated to it from the bearing. Between these two extremes almost every possible inter-mediate case is found in practice, and the designer must meet them by using one or other of the general formulæ for girders, and must use his judgment in deciding what co-efficients best express the conditions of loading, supporting or fixing, &c.,—not forgetting that the material is subject to alternate tensile and compressive strains.

Rule 123. In the worst case (that of the oscillating paddle engine) the girder should be taken as *supported* only, and the load considered as a single central one, and the values of f may be 3000 for cast-iron, and 8000 for cast-steel. In this particular example there are frequently other stresses,—due to the unsymmetrical positions of the columns or pillars with regard to the bearings,—which must also be allowed for.

In vertical screw engines there are stresses on these girders communicated from the columns, and due to the thrusts on the guides, and also to the weight of the cylinders, &c., when the ship is rolling.

In paddle-wheel vessels there are also (as previously mentioned) stresses due to the weights and re-actions of the wheels, &c., &c.; but it is quite impossible to investigate each case separately here, and these few hints must suffice.

Very valuable additional strength and stiffness may sometimes be obtained by using a forged or cast steel cap with lips at the ends which prevent the springing open of the gap containing the brasses, or, spigots or projections may be used which will prevent it either opening or closing

COLUMNS.

The tensile stresses on the material of columns, and also on the bolts connecting them to the cylinders have been dealt with under the heading "Column feet and Bolts," page 67 ; but in vertical engines there are also bending stresses due to the thrusts on the guides, and to the weight of the cylinders, pistons, valves, rods, &c., when the ship is rolling. These may all be dealt with by means of the formula for a beam fixed at one end and loaded at the other, viz.:—

$$W = \frac{fz}{l},$$

and need not therefore be considered in detail here.

Where the guide is on one column only (port or starboard) the other may be considered as taking half the thrust on the guide, because the two are so tied together by the cylinder as to constitute one frame.

The bolts attaching the column feet to the bedplate or frame must of course be strong enough to carry the direct tensile stress, plus the stress caused by the tendency of the column to cant over on one edge of its base ; and it must be noticed that only about one-fourth of the bolts in each column foot are, on on average, able to offer resistance to the latter stress.

When weight is of great importance, as in Naval work, and framings are made of the lightest possible sections, special stresses,—such as those resulting from the action of the reversing engine, &c.,—must be carefully considered and allowed for.

CONDENSERS.

1. Jet Condensers.

The Capacity of a jet condenser should not be less than one-fourth that of the cylinder, or cylinders, exhausting into it, and need not be more than one-half,—unless the engine is a very quick running one : one-third the capacity of cylinder is generally sufficient.

The Form of a jet condenser is not a matter of much consequence, and depends, to some extent, on the type and design of engine to which it is fitted. The inlet for steam should be high enough to prevent water getting into the cylinders, and the lower portion should be so shaped that the water will all drain to the air-pump.

The Position of the Jet Pipe or nozzle, and the form of the delivery openings in it depend so much on the form of the condenser that no definite rule can be given ; the nozzle pipe may have a number of small holes drilled it it, or the water may issue from transverse or longitudinal saw-cuts or slots, or may all issue in a sheet, from the end of the pipe, through a nozzle resembling a section of pipe which has been flattened and nearly closed.

The Quantity of Injection Water depends on the weight of steam to be condensed, and on its temperature, and,—to ascertain the exact quantity,—the temperature of the injection water, and the required temperature of feed water must also be known.

The vacuum with this type of condenser rarely exceeds 25 inches, and is more commonly 24 inches,—which corresponds to a temperature of about 140°; the temperature of the hot-well varies, in practice, from 110° to 130°.

The number of pounds of injection water (Q) necessary to condense one pound of steam is given by,—

$$Q = \frac{1114 + (\cdot 3 \times T_S) - T_H}{T_H - T_W},$$

where T_S = temperature of steam at exhaust. *
$T_H =$,,   of hot-well.
$T_W =$,,   of cooling water.

Rule 124. It is usual to make an allowance of injection water of from 27 to 30 times the weight of steam to be condensed, for vessels running in temperate climates, to 30 to 35 times for those running in the tropics.

The relation of weight of steam to volume, to temperature, and to release pressure is given in the following Table :—

* Strictly, *absolute* temperatures (Fahrenheit temperatures+461°) should be ed in this and all similar calculations.

Table XLV.—Weight, Pressure, and Temperature of Steam.

Absolute pressure in lbs. per sq. inch.	Number of cubic feet to weigh 1 lb.	Tempera-ture (Fahren-heit).	Absolute pressure in lbs. per sq. inch.	Number of cubic feet to weigh 1 lb.	Tempera-ture (Fahren-heit).
2·8	122	140°	10·1	37	194°
3·6	98	149	12·2	31	203
4·5	80	158	14·7	26	212
5·5	65	167	17·5	22	221
6·8	54	176	20·8	19	230
8·3	44	185	24·5	16	239

See also complete Table of Pressures, Temperatures, Volumes, Total Heats, &c., pp. 333 to 339.

The Area of Injection Orifice, and size of pipes, is governed by the head of water, the vacuum, and the length of piping and number of bends, &c., or, in other words, by the equivalent head at the condenser.

Neglecting the resistance to flow at the orifice, and in the pipes and passages, the velocity at the condenser may be found as follows :—

Let h be the head, in feet, above the valve on the condenser ; p, the pressure in the condenser in lbs. per sq. inch ; h_1 the equivalent head ; and g, gravity. Then,—

$$h_1 = h + (15 - p)\ 2\cdot3,$$

and velocity in feet per second is $\sqrt{2\,gh_1}$ or $8\sqrt{h_1}$.

Rule 125. In practice, owing to the loss of head resulting from resistances at valves, and in pipes, &c., the actual velocity is only about half that given by the above rule ; hence, in designing, it is usual to calculate on a velocity of only 25 feet per second for shallow draught steamers, and 30 feet per second for deeper ones.

The following more concise expressions are derived from the above rules,—

Rule 126.

$$\text{Area of orifice in sq. ins.} = \frac{\text{No. of c. ft. of injection water per minute}}{10\cdot4 \text{ to } 12\cdot5 \text{ according to circumstances}}.$$

Rule 126a.

$$\text{or Area of orifice in sq. ins.} = \frac{\text{Weight of injection water in lbs. per min.}}{650 \text{ to } 780 \text{ according to circumstances}}.$$

The handle or lever for working the injection valve should be placed close to the starting and reversing handles, in order that water may be promptly shut off when engine is stopped.

A snifting or overflow valve, held on its seat by atmospheric pressure only, should be provided, to prevent undue accumulation of water in the condenser : its diameter should be the same as that of the injection valve.

A Bilge Injection Valve should also be fitted, having an area of about two-thirds that of the sea injection, and suitable means should be adopted to prevent its becoming choked by dirt from the bilges.

2. Surface Condensers.

These are now fitted to all classes of vessels excepting some swift low-pressure paddle steamers, employed in the cross-channel passenger services, in which the saving of weight is the first consideration. The extra coal consumption due to the necessary "blowing off" quite prohibits the use of the jet condenser on steamers making long voyages.

The net saving of fuel by the use of the surface condenser averages 15 per cent., and may, with care, reach 20 per cent.; and, in addition, the life of the boilers is very much prolonged.

On the other hand, the surface condenser is heavier and more costly, and occupies more space than the jet, and also requires an additional (circulating) pump.

The Cooling Surface per I.H.P. should be as follows :—

Rule 127. For modern Mercantile triple engines . . 1·3 sq. ft.

Rule 127a.   ,,   Naval   ,,   . . 1·1

The proportion of 1·1 sq. ft. per I.H.P. is found to give very satisfactory results in Naval vessels, although the pressure at release, when the engines are exerting maximum power, is not generally below 10 lbs. absolute ; and a vacuum within 2½ inches of the weather barometer is usually obtained.

The following Table shows the increased surfaces necessary in cases where the terminal or release pressures are higher than in those referred to above :—

Terminal pressure	20 lbs. absolute	. . . .	1·70 sq. ft. per I.H.P.
,,	15 lbs. ,,	. . . .	1·57 ,, ,,
,,	12½ lbs. ,,	. . . .	1·50 ,, ,,
,,	10 lbs. ,,	. . . .	1·43 ,, ,,
,,	8 lbs. ,,	. . . .	1·37 ,, ,,
,,	6 lbs. ,,	. . . .	1·30 ,, ,,

As the efficiency of the condenser depends not only on the extent of the surface, but also on the mean difference of temperature between steam and water, if the surface be made less than above directed, the difference of temperature must be proportionately greater, and, to effect this, the speed of water must be greater, the general result being the discharge of a larger quantity of cooling water at a lower temperature. If the quantity of water be not increased the temperature of condenser, and corresponding back pressure, will of course increase.

Correspondingly, if the surface be made greater, the action may be less intense, and the quantity of water, and the pumping power less, the general result being the discharge of a smaller quantity of cooling water at a higher temperature. If the quantity of water be not diminished the temperature of condenser will fall and the result be cold feed water.

It would therefore appear that to keep the temperature of condenser constant, while condensing a given quantity of steam with sea water at a given temperature, the quantity of water should be (approximately) inversely proportional to the surface. The fact that the first foot of tube through which the water passes is the most efficient condensing surface, and each succeeding foot less and less efficient is a complicating element in these calculations.

The circulating pumps must of course be proportioned to suit the highest sea temperature that the vessel will meet with in ordinary work.

Condenser Tubes.—These are, as a rule, made of brass, solid drawn, the composition being 68 per cent. of best selected copper, and 32 per cent. of best Silesian spelter ; the Admiralty, however, specify 70 per cent. copper, 29 per cent. spelter, and 1 per cent. tin.

Condenser tubes vary in diameter from ⅛ inch to 1 inch, but ¾ inch tube, with a thickness of 18 L.S.G., is by far the most frequently used size, both in Merchant and Naval services ; the Admiralty specifications usually state the thickness as ·05 inch however, whilst 18 L.S.G. corresponds with ·048 inch.

Length unsupported.—(Rule 128) When tubes are secured at the tube-plates by screw ferrules and packing, the unsupported length should not exceed 100 diameters ; or, when they are held by tightly fitting ferrules, 120 diameters. If the tubes are longer than this, they should be supported by intermediate diaphragms of rolled brass, from ⅜-inch to ¾-inch in thickness, according to size ; but care must be taken, in arranging the design, that these diaphragm plates do not interfere with the free flow of the steam to all parts of the condenser.

Tube-plates.—These should be of rolled brass of the following thicknesses :—

Rule 129.

When wood ferrules are used . . 1·5 × diameter of tube.

    ,,   screw glands   ,,   . . 1·1 ×   ,,   ,,

" Diameter of tube " signifies, in all cases, *external* diameter.

Fig. 30 shows the method of securing a tube by a wood ferrule, and Fig. 31 by means of a screw gland and packing ; the dimensions on the figures give the proportions ordinarily adopted for ¾-inch tubes.

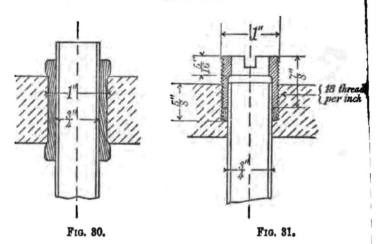

FIG. 80.   FIG. 81.

Spacing of Tubes.—The tubes should be spaced zigzag, and equilaterally (*i.e.*, the centre of any tube should be at the same distance from the centre of each of the immediately surrounding tubes). The pitch may be as follows :—

Rule 130.

When wood ferrules are used—Dia. of hole for ferrule + ¼ in.

When screw glands are used—Extreme dia. of gland + ⅛ to ³⁄₁₆ in.

The following Table will be of assistance in making rough calculations for size of condenser :—

Table XLVI.—Size of Surface Condenser.

External diameter of tube—¾ inch.					
Pitch of tubes, . .	1⅛″	1⁵⁄₃₂″	1³⁄₁₆″	1⁷⁄₃₂″	1¼″
Number of tubes per sq. ft. of tube plate, ...	131	124	118	112	106
Cooling surface per cubic ft. of tubes,	25·7 sq. ft.	24·3 sq. ft.	23·1 sq. ft.	21·9 sq. ft.	20·8 sq. ft.

The surface of 1 foot of ¾ inch tube is ·1963 sq. ft.

See also Table of Surfaces of Tubes, page 323.

Quantity of Cooling Water.—The quantity of water necessary for condensation is given by the formula,—

Rule 131.
$$Q = \frac{1114 + (\cdot 3 \times T_S) - T_H}{T_W - t_W},$$

where T_S = temperature of steam at release,

T_H =   „   hot-well,

t_W and T_W = initial and final temperatures of cooling water.

With temperature of sea 60°, and ordinary engine temperatures, this quantity will be found to be about 27 times the weight of the steam condensed; and with sea at 75°, about 42 times.

If the weight of steam discharged into the condenser by a modern triple engine be taken at ·2 lb. per I.H.P., and the cooling surface provided be at the rate given above,—viz., 1·3 sq. ft. per I.H.P.,—the rate of condensation will be say ·16 lb. per sq. ft. per minute, and the speeds of cooling water through tubes necessary to effect this are found to be,—

When sea temperature is 60° . . about 110 ft. per minute.

„   „   75° . . „ 170 „ „

Rule 132. At these velocities the water requires to pass through about 20 ft. of tube in order to reach a temperature, at discharge, of 100°.

Where the tubes are very short this may be effected by causing the water to travel three times through the condenser; if this is not possible the speed of the water must be reduced in proportion, so as to give it time to rise to the discharge temperature. In Naval con-densers the average run of the water is not more than 15 feet, and the speeds 80 feet and 125 feet per minute respectively.

Whenever possible the tubes should be arranged horizontally, as experiment seems to show that vertical tubes have an efficiency of rather less than ·7 of that of horizontal ones.

Application of Cooling Water.—The simplest method of applying the cooling water is to allow it to run direct from the sea into the condenser, and then to pump it overboard from the condenser; where this cannot be conveniently arranged, and the water is forced through the condenser by a reciprocating pump, a large air-vessel should be provided as near to the pump as possible, to guard against shock from "racing" or sudden starting.

The centrifugal pumps, invariably employed for circulating purposes on ships of war, suck from the sea and drive the water through the condensers, but they are of course incapable of communicating any shock to the structure.

In order to reduce the shock from reciprocating circulating pumps some engineers fit a screw-down by-pass valve which is opened more or less whenever "racing" occurs.

In merchant steamers the condenser commonly forms a part of the

9

framing, and is made of cast-iron ; its strength of course depends largely on its form and position, and has little relation to the stresses resulting from its action as condenser.

In war vessels, on the other hand, the condenser rarely, if ever, forms any part of the framing ; it is commonly cylindrical in form, made of naval brass or gun-metal, and strong enough to bear an internal test pressure of 30 lbs. per sq. inch, which is applied to test the sound-ness of the casting, and the tightness of joints and tube packings.

Condensers for torpedo gun-boats, &c., up to about 5 feet diameter, usually have the barrels made of sheet brass (8 to 6 L.S.G.) with flanges and stiffening rings riveted and soldered on.

Manholes, sight-holes, mud-holes, and air-cocks should always be provided ; in Naval condensers fittings for introducing soda, and for boiling out, are always fitted in addition.

AIR-PUMPS.

Wherever practicable these should be vertical, and single-acting, since this type is in every way the most efficient and satisfactory.

In vertical screw engines the air-pump should be driven by levers and links from the L.P. crosshead or gudgeon ; when driven direct from the piston or crosshead the speed is objectionably high, and the pump is not usually convenient of access.

When the pump must be placed horizontally, or in an inclined position, its design,—whether it.be single or double acting,—is a more difficult problem, and each detail and point should have the careful attention of a skilled designer.

In horizontal screw engines the air-pump is usually placed hori-zontally beneath the condenser, and driven direct from the L.P. piston ; in twin-screw vessels the condenser and pump are generally behind the cylinders, and in single-screw ships in front of the cylinders, and on the opposite side of the shaft.

In paddle engines the air-pumps are placed in different positions, and driven in different ways according to the type of engine. In oscillating engines they are usually inclined, and driven by connecting rods from a crank formed in the middle of the intermediate shaft, or by an eccentric fixed in the same position ; but are sometimes placed vertically, and driven through a lever or beam. In almost all other types of paddle engine they can be placed vertically without difficulty, and may be driven as found most convenient.

Size of Air-Pump.—Whether for jet or surface condenser, the quantity of water to be removed by the air-pump can be easily determined by the rules given under "Condenser," but the quantity of air is variable and cannot be calculated ; and it is therefore necessary, especially in the case of surface condensing engines, to rely more on experience than on calculation.

The most convenient formula is probably the following :—
Rule 133.

Capacity of Pump, in cubic inches, $= \dfrac{\text{I.H.P.}}{\text{Revolutions per minute}} \times C,$

where C has the following values :—

For single-acting vertical pump and jet condenser . . C = 7 0
For single-acting ·vertical pump and surface condenser
   (mercantile) . . . . C = 300
  ,,   ,,   ,,   (naval) C = 250
  ,, double-acting ,, horizontal pump   ,, . C = 470

The proportions given above for the air-pumps of surface condensing
engines are based on the assumption that no jet injection apparatus is
fitted,—experience having shown that it is quite unnecessary.

As some engineers still prefer, when fixing the sizes of air-pumps
for ordinary merchant steamers, to make the pump capacity bear a
direct relation to the capacity of the L.P. cylinder, it may be of
service to give the following approximate equivalent of the above rule
(Rule 133) for this particular case :—

Rule 133a.

Capacity of single-acting vertical air-pump ⎫
for surface condensing triple engines of ⎬ $= \dfrac{\text{Capacity of L.P. cyl.}}{23}$
ordinary merchant steamer, . . . ⎭

Air-Pump Rods.—These may be made either of iron cased in
gun-metal, or of one of the rolled bronzes. Where a guide is fitted, and
the stroke of the pump does not exceed 2¼ times its diameter, the rod
(without regard to its material) may have a diameter given by,—

Rule 134. Diameter of rod $= \dfrac{\text{Diameter of pump}}{9} + \cdot 6 \text{ inch.}$

See Table XLVII. (page 134) for sizes given by this rule.

Where air, circulating, feed, and bilge pumps are grouped together,
and the links are long, the guide is sometimes omitted, but it is not
a plan that can be recommended, except for pumps of very short

stroke $\left(\text{say less than } \dfrac{\text{Diameter}}{2}\right)$; if it is followed however, the

diameter of the rods should be increased by 15 per cent., and the stuffing
box should be kept as far above the end of the barrel as possible.

In the case of horizontal pumps, placed as described above, for
twin-screw vessels, the length of unsupported rod between the stuffing
boxes is not usually so great as to prevent Rule 134 from applying ; in
single-screw vessels a guide should be fitted if the pump is placed and
driven as described above.

The Size of the Screw on the air-pump rod, by which it is
attached to the air-pump crosshead, is given by,—

Rule 135. Area at bottom of thread $= \dfrac{\text{Area of bucket}}{20 \times \sqrt{\text{diameter of bucket}}}.$

See Table XLVII. (page 134) for sizes given by this rule

This allows a stress of about 2000 lbs. per sq. inch of material in the rod for a 12-inch pump, increasing to 3800 lbs. per sq. inch in the rod for a 40-inch pump, the load on the bucket being taken as 30 lbs. per sq. inch. The rule may be taken as giving the size of screw whether the rod is of bronze or of wrought-iron or steel, as the bodies of the rods will be of the same diameter for both materials, and nothing is gained by reducing the ends of iron or steel rods.

When the number of reciprocations is high (say 200 to 350 per minute) the rods should be made 10 per cent. larger in diameter than given by the rule, although the stroke may be short, or even very short, since there is always a liability to sudden strain, owing to irregular or intermittent action of the pump, and the intensity of these strains will vary as the squares of the bucket speeds,—other things being equal.

Very long rods must be specially considered,—as struts,—fixed or jointed at the ends as the case may be.

Connecting Rods for Trunk Air-Pumps.—When of the type commonly employed for driving the pumps of oscillating paddle-engines these may have the following proportions :—

Rule 136.   Area of section of rod = ·01 × area of bucket ;

or, in case of a round rod,—

Rule 136a.   Diameter of rod = ·1 × diameter of bucket ;

when two bolts are used to connect the brasses at the end,—

Rule 136b.

Diameter of each bolt (in body) = ·056 × diameter of bucket.

Air-pump buckets are usually packed with manilla rope, from ½ to ¾-inch diameter; in the Navy an external ring gland is sometimes fitted to compress the packing, but it is unnecessary in ordinary cases, and may be a source of danger in careless hands.

When the air-pump valves are of india-rubber care should be taken to fit a quality that is capable of resisting the heat, &c.; the Admiralty specification is as follows :—

"To be made of the best Para caoutchouc, with no other materials whatever than sulphur and white oxide of zinc; the sulphur is not to exceed 1¾ per cent., and the oxide of zinc not to exceed 70 per cent. The india-rubber is to be made of the best materials, not re-manufactured, and to be of a homogeneous character throughout, thoroughly compressed, free from air-holes, pores, and all other imperfections. "Samples of the india-rubber will be tested, and must be capable of enduring a dry heat test of 270° Fahrenheit for one hour, and a moist heat of 320° Fahrenheit for three hours, without impairing its quality."

Air-Pump Barrels.—For pumps below 20 inches in diameter the gun-metal barrel is usually surrounded by a cast-iron one, to which it forms a "liner"; above 20 inches the more common plan is to omit the outer barrel and secure the gun-metal one to cast-iron head and foot boxes by means of ribbed flanges,—or, as in Naval work, to make barrel, and head and foot boxes all one gun-metal casting. Whichever type of construction is used, the working barrel should be of thickness given by,—

Rule 137.

$$\text{Thickness of air-pump barrel} = \frac{\text{Diameter of pump}}{60} + \cdot 25\text{-inch}.$$

See Table XLVII. (page 134) for thicknesses given by this rule.

When the pump is of the Naval type the portions above and below the working barrel may be from $\frac{1}{16}$ to $\frac{1}{8}$ inch thinner than the barrel, according to size and design.

Where the gun-metal barrel is not enclosed in an outer casing, it is very desirable to have a manhole in the side, through which access may be had to the bucket and foot valves, without disturbing the cover or guides. Similarly, when the size of the pump renders it possible, hand or sight holes should be provided in the sides of the head box, for getting at the head valves.

The Areas through the Valve Gratings of foot, bucket, and head valves should be kept as large as possible ; on an average, those of the foot and bucket valves run about one-third of the gross area of bucket, and those of the head valves a little larger ; they can of course be increased by increasing the diameter of the pump, and decreasing the stroke.

Speeds of Bucket.—In ordinary mercantile engines the speed of bucket employed varies from 200 feet to 350 feet per minute ; but in Naval engines it commonly ranges from 300 feet to 450 feet, and with pumps driven direct from the pistons is of course higher still. When the pump works in connection with a jet condenser the speed should not much exceed 200 feet per minute.

Sizes of Suction and Discharge Pipes.—For the pumps of surface condensing engines (where no jet apparatus is fitted) the diameter of suction pipe from condenser, in inches, is given by,—

Rule 138. $\quad$ Diameter of air-pump suction pipe $= \frac{D}{50} \times \sqrt{S}.$

or, Equivalent mean speed in pipe must not exceed 2500 ft. per min. where D is diameter of bucket in inches, and S its mean speed in feet per minute.

If there is a supplementary hot-well or feed-tank, the pipe connecting it with the air-pump should be of diameter given by,—

Rule 139. $\quad$ Diameter of air-pump delivery pipe $= \frac{D}{62} \times \sqrt{S}.$

or, Equivalent mean speed in pipe must not exceed 3850 feet per minute.

In addition, an air-pipe should be fitted as close to the pump as possible; its diameter may be,—

Rule 140. Diameter of air-pipe $= \frac{D}{80} \times \sqrt{S}$.

or, **Equivalent mean speed in pipe must not exceed 6400 feet per minute.**

If the pipe to feed-tank is long, or has many or sharp bends in it, these points must, of course, be considered and allowed for in fixing the size.

When the pump works in connection with a jet condenser, the diameter of suction and delivery pipes may be,—

Rule 141. Diameter of Suction and delivery pipes $= \frac{D}{25} \times \sqrt{S}$.

or, **Equivalent mean speed in pipe must not exceed 625 feet per minute.**

The delivery pipes from air-pumps of jet condensing engines should have good large air-vessels (say capacity not less than capacity of pump) placed as near to the pumps as possible.

Table XLVII.—Air and Circulating Pumps.

Diameter of pump (single-acting).	Thickness of barrel (single-acting).	Thickness of barrel +10 per cent. (double-acting).	Diameter of rod.	Diameter of screw on outer end of rod.	Diameter of double acting pump of same capacity (circulating only).
1	2	3	4	5	6
inches.	inch.	inch.	inches.	inches.	inches.
8	3/8	...	1 1/2	1 1/4	...
10	7/16	...	1 3/4	1 1/2	...
12	7/16	...	2	1 3/4	...
14	1/2	...	2 1/4	2	...
16	1/2	1/2	2 1/2	2 1/4	11 1/2
18	9/16	1/2	2 1/2	2 1/4	13
20	9/16	9/16	2 3/4	2 1/2	14
22	5/8	9/16	3	2 3/4	15 1/2
24	5/8	9/16	3 1/4	2 3/4	17
26	11/16	5/8	3 1/2	3	18 1/2
28	11/16	5/8	3 3/4	3	20
30	3/4	11/16	4	3 1/4	21
32	3/4	11/16	4 1/4	3 1/2	22 1/2
34	13/16	11/16	4 1/2	3 1/2	24
36	7/8	3/4	4 1/2	3 3/4	25 1/2
38	7/8	3/4	4 3/4	4	27
40	7/8	13/16	5	4	28 1/2

Column 2 is calculated in accordance with Rule 137.
„ 4 „ „ „ 134.
„ 5 „ „ „ 135.
Column 2 is only to be read with column 1, and column 3 only with column 6; columns 4 and 5 may be read with either column 1 or column 6. Column 3 gives thickness in cases where barrel is fixed or carried by a flange at the mid-length only.

RECIPROCATING CIRCULATING PUMPS.

These may be either single or double acting, and, as their efficiency is practically unaffected by position, they may be either vertical, horizontal, or inclined. In general design the various parts may resemble the corresponding parts of the air-pump (in practice the same patterns are sometimes used for both pumps), but circulating pumps require, in addition, large air vessels, non-return air or pet cocks, and, in the case of large double-acting pumps, a by-pass valve.

The air-vessel for a single-acting pump should have a capacity equal to twice that of the pump, when possible, but never less than once and a half. The delivery of the pump should be limited, when necessary, by opening the air-valve, *not* by partially closing the suction valve. Air or pet valves may vary from 1 inch for a 10 inch pump, to 2 inch for a 20 inch pump,—extra area being given if the speed of bucket is exceptionally high.

The by-pass valve may have an area of about one-tenth the area of the bucket.

When the size of pump required would, if made single-acting, exceed 20 inches in diameter, it is better to fit a double-acting pump of half the bucket area, and thus obtain a steadier delivery, whilst at the sam time reducing the magnitude of the stresses and the sizes of the various parts.

Size of Circulating Pump.—This depends on quantity of cooling water required, and number of strokes per minute.

Let Q = cubic feet of cooling water required per minute.
n = number of strokes per minute.
S = length of stroke in feet.

Then,—

$$\text{Capacity of Circulating Pump} = \frac{Q}{n} \text{ cubic feet.}$$

and,—

$$\text{Diameter of Circulating Pump} = 13.55 \sqrt{\frac{Q}{n \times S}} \text{ inches.}$$

In determining Q, the temperature of cooling water should be taken at the highest ordinary temperature encountered on the routes the vessel is intended to steam over.

If the efficiency of the pump be taken as unity, the capacity necessary to deliver water equal in weight to 42 times the steam condensed is,—

Capacity of Circulating Pump (cubic ins.) $= \dfrac{\text{I.H.P.}}{\text{Revs. per minute}} \times 227.$

allowing, however, an ordinary efficiency of pump, and for possible reduction in efficiency due to the use of air-valve and by-pass valve, the capacity should be,—

Rule 142.

Capacity of Circulating Pump (cubic ins.) $= \dfrac{\text{I.H.P.}}{\text{Revs. per minute}} \times 300.$

i.e., in the case of single-acting pumps working with surface condensers, the *same capacity* as for a single-acting air-pump.

Circulating Pump Rods.—For single-acting pumps, these should be of the same materials as the air-pump rods, and the same formula may be used to obtain size. See Table XLVII.

Wherever possible the air and circulating pump rods should be made of exactly the same dimensions, in order that one spare rod may serve for either pump. With this object, when the circulating pump is made double-acting, the rod should be of the size suitable for a single-acting pump of the same capacity per double stroke ; this will give a rather stiffer rod than is absolutely necessary to resist the compressive stress, but will be no disadvantage.

The rule for rods of double-acting circulating pumps will then be,—

Rule 143.   Diameter of Rod $= \dfrac{D}{6\cdot4} + \cdot6$ inch (See Table XLVII.).

Although manilla rope and lignum-vitæ packings are both occasionally used, the buckets of circulating pumps really do not require any packing, and are better without it, since the unnecessary friction is thereby saved.

The thicknesses of gun-metal barrels or liners for circulating pumps may be determined by the rule previously given for air-pump barrels (Rule 137), when the barrels are of similar type and similarly supported. When the barrel is fixed in place by means of a single flange placed at mid-length (a common method for double-acting circulating pumps), the thickness should be 10 per cent. above that given by the rule (see Table XLVII.).

The areas through the foot, bucket, and delivery valves should be kept as large as possible, but, as in the case of air-pumps, they will usually average one-third of the gross bucket area.

The bucket speeds in common use are from 200 to 400 feet per minute.

The sizes of suction and delivery valves may be as follows :—

Rule 144.   Diameter of Pipe $= \dfrac{D}{F} \times \sqrt{S.}$

Where D is the diameter of bucket in inches, S its mean speed in feet per minute, and F a co-efficient, the values of which are as follows :—

Double-acting { Suction— F = 25 (mean speed in pipe, 625 feet).
   pumps   { Delivery—F = 24 (,,   ,,   576 ,,).

Single-acting { Suction — F = 27 (,,   ,,   729 ,,).
   pumps   { Delivery—F = 24 (,,   ,,   576 ,,).

When any pipe is under 10 inches in diameter the velocity of flow through it should be kept 10 per cent. below that allowed by the above rule. Specially long or specially tortuous pipes must also have special consideration,—the velocity of flow being reduced in accordance with the circumstances of the case.

Lloyd's rules require that a bilge injection or a bilge suction to the circulating pump shall be fitted, the diameter of which shall be at least two-thirds that of the sea-inlet.

CENTRIFUGAL CIRCULATING PUMPS.

In Naval work this type of pump is almost invariably used, and it is also not infrequently fitted in all classes of mercantile steamers. It is, no doubt, preferable to the reciprocating type where large bodies of water are to be moved against a merely nominal head, and especially where the main engines run at a high number of revolutions per minute, since there is an entire absence of shock, and, when the pipes are properly arranged, the work is done with a less expenditure of power. On the other hand, the centrifugal pump is more costly than the reciprocating, occupies more space, requires more attention, and necessitates rather larger pipes and valves.

Size of Pipes.—The size of pipes should be such that the speed of water does not exceed 500 feet per minute, even when the sea temperature is 75°, or, in other words, when the sea temperature is 60° the speed should be about 330 feet per minute.

When the pipes are under 10 inches diameter these speeds should not exceed 450 and 300 feet respectively, and if the pipes are not short and direct, with easy bends, the efficiency of the pump will be improved by keeping the speeds 50 feet per min. lower again.

A convenient form of the rule for size of pipe will then be as follows :—

Rule 145.   Diameter of pipe in inches = $\cdot 64\sqrt{Q}$

or,—   Diameter of pipe in inches = $\sqrt{\dfrac{\text{I.H.P.}}{20}}$

where Q is the maximum number of cubic feet of cooling water per minute. This corresponds with a speed of 450 feet per minute in the pipes.

Size of Fan or Wheel.—A good proportion for the diameter of pump wheel is,—

Rule 146.

Diameter of pump wheel in inches = 2·8 × Diameter of pipes,

or,—

Rule 146a.   Diameter of pump wheel in inches = 1·8√Q

where Q is maximum number of cubic feet of cooling water per minute
or, reversing the equation,—

Rule 146b. $Q = \dfrac{D^2}{3\cdot24}$ cubic feet.

Width of Wheel at Periphery.—The width of wheel at the
periphery, or width of vane at the tip may be,—

Rule 147.

Width of Wheel at Periphery $= \dfrac{\text{Diameter of Pipes.}}{4}$

Stroke of Piston.—The stroke of the centrifugal pump engine
may be conveniently fixed as follows :—

Rule 148.   Stroke of piston = ·3 × Diameter of Fan.

Size of Steam Cylinder.—As centrifugal circulating pumps are
invariably fitted with bilge suctions so as to be available as bilge
pumps in case of need, and as the work done when pumping from the
bilge is far in excess of that done when merely circulating water
through the condenser, the size of steam cylinder must evidently be
fixed with regard to the former duty.

The efficiency of the pump depends very greatly on the pipe
arrangement ; with direct pipes of good size, and easy bends, it may,
when pumping from the bilge, approach 30 per cent., but a bad
arrangement, with sharp bends, will very seriously affect it, and may
easily reduce it below 20 per cent. ; in ordinary practice, where
reasonable care is exercised in scheming the pipes, it may be assumed
to be 25 per cent. The efficiency also tends to diminish as the lift
increases, so that a rather lower efficiency should be assumed in the
case of large deep ships, and *vice versa*.

Then $\dfrac{\text{Indicated Horse-power}}{\text{Water Horse-power}} = 4$

where "water horse-power" is $\dfrac{\text{lbs. water pumped per min.} \times \text{lift in ft.}}{33,000}$

or, putting it into another shape,—

Rule 149.

Tons water pumped per hour × lift in feet × 150 = P_m × A × S

where P_m = mean pressure in steam cylinder ;
          A = area of piston in square inches ;
          S = piston speed in feet per minute.

Speed of Periphery of Pump Wheel.—The speed of periphery of pump wheel necessary to discharge water at the velocities mentioned above against a head (h), under average conditions, is given by,—

Rule 150. $$V = 11 \sqrt{h}$$

where V is velocity of periphery in feet per second, and h the actual lift in feet. With a very good arrangement of pipes the co-efficient may be as low as 9, in place of 11, but a bad arrangement will send it up to 13 or over.

In ordinary circulating work when inlet and outlet are both submerged, as is usual in Navy, V requires to be about 24 feet per second (1440 feet per minute), and the resistances may therefore be assumed to be equivalent to a head of nearly 5 feet.

When pumping from the bilge, V varies from 30 feet to 50 feet per second (1800 feet to 3000 feet per minute) in ordinary Naval work, but, as shown by the above equation, it increases as the square root of the head.

Size of Bilge Suction Pipe.—A convenient rule for area of this pipe is,—

Rule 151.

One square inch area for each 7 tons per hour to be pumped.

This limits the speed of water in the suction pipe to 580 feet per minute.

It is not desirable to have the area of bilge suction more than about ·6 of the area of the circulating pipes, as the size of steam cylinder required soon becomes cumbersome for the ordinary circulating work.

The pump should be capable of performing the specified bilge duty with a steam pressure not exceeding two-thirds of the ordinary boiler pressure.

FEED PUMPS, &c.

Capacity of Feed Pumps in jet-condensing engines.—This depends mainly on the degree of saltness at which the water in the boilers is to be maintained, or, in other words, on the proportion of the *gross* feed-water that is "blown off."

Let Q represent the *net* feed-water, or the quantity required as steam by the engines, and say water in boilers is to be kept down to n times the density (or saltness) of sea-water, then the amount of sea-water that must be pumped into boilers is,—

$$\text{Gross feed-water} = \frac{n}{n-1} \times Q.$$

To reduce the time required to bring the water in the boilers up to the working level again after "blowing off," it is usual to make *each* feed-pump capable of pumping *twice* this quantity; therefore,—

Rule 152.   **Each pump should supply** $\dfrac{2n}{n-1} \times Q$.

The **net** feed-water (Q) required by a jet condensing engine, working with steam of about 30 lbs. pressure, may be taken as 26 lbs. per I.H.P. per hour.

Of course, when pumps of this capacity are fitted, the question of the efficiency of the pump need not be raised.

The amount of scale or salt deposited in the boilers does not depend on the density at which the water is kept, but only on the quantity of sea-water pumped into the boilers ; 35°, or 3½ times the density of sea-water, is a good density at which to work the boilers of jet condensing engines, and 40° to 45°, or 4 to 4½ times, for boilers of surface condensing engines.

The Hydrometer, or Salinometer, by means of which the density of the water is ascertained, is, in the Navy, graduated in degrees, so that, when floating in pure water, the zero point is at the surface, and when in clean sea-water it marks 10°, when in water of twice the density of sea-water, 20°, and so on. In the Merchant Service, engineers either express the density in ounces per gallon (sea-water containing about 5 oz. per gallon), or in "thirty-twos,"—sea-water containing about $\frac{1}{32}$ of its weight of solid matter.

Capacity of Feed-pumps for Surface Condensing Engines.— It will be sufficient, with this type of engine, if the total capacity of each pump be made equal to **twice** the net feed water required ; this allowance also covering any want of efficiency in the pumps.

If Q be the net feed water required in lbs. per hour, l length of stroke of feed pump in inches, and n number of strokes per minute ; then,—

Rule 153.

Diameter of each feed-pump plunger, in inches $= \sqrt{\dfrac{1 \cdot 15 \times Q}{l \times n}}$.

For ordinary compound engines the net feed water (Q) required may be taken at 18 lbs. per I.H.P. per hour, for triple engines 15 lbs. per I.H.P. per hour, and for quadruple engines 13·5 lbs. per I.H.P. per hour.

Or, if the formula be written,—

Rule 154.   Capacity of feed-pump, in cubic inches $= \dfrac{\text{I.H.P.}}{\text{R}} \times C$,

the values of C will be as follows :—

> For Compound engines,   C = 16·5.
> „   Triple   engines,   C = 13·7.
> „   Quadruple engines,   C = 12·4.

In determining the sizes of feed pumps for Naval boilers, it will ~nerally be sufficient to take the natural draught power as the basis of the calculation; they are always fitted as independent auxiliary engines.

Where the feed pumps are driven by the main engines, there should be, except in very small engines, two pumps, each capable of supplying the boilers at full power, and so arranged that either may be worked independently of the other, and be easily put out of gear when not required.

Relief Valves.—When the pumps are driven by the main engines, or when they are independent auxiliary engines, but are not of a self-regulating type, each pump should be fitted with a relief valve, of a diameter equal to two-thirds that of the delivery pipe from that pump, and loaded to 1½ times the boiler pressure. Provision should be made for leading all water from relief valves back into suctions.

Feed-pump valve-boxes and valves.—These should be of best gun-metal; and since the seats, as well as the valves, wear out rapidly, loose seats should be fitted. The faces of the valve seats should be made flat, and the seating area, or area of faces in contact, should not be less than 20 per cent. of the gross area through valve; that is,—

Rule 155.

Width of faces in contact = ·0475 × Diameter of valve.

The diameter of valve, measured inside the seat, should be equal to that of the delivery pipe.

The noise made by gun-metal valves on gun-metal seats may be much reduced by loading them with light spiral springs, made of plated steel or of hard brass wire.

Each pump should have an air vessel of twice its own capacity. Non-return air, or "pet" valves, should be fitted between the suction and delivery valves; and sometimes the construction of the pump is such that a small cock is necessary to let out air that accumulates in the upper part of the chamber or barrel, but this type of construction should be avoided where possible.

Feed pipes.—The pipes leading to and from the feed pumps should be of such size that the velocity of flow, when working steadily at full power, does not exceed 400 feet per minute. The velocity should also be less as the pipes get smaller, to allow for the increased percentage of friction; and, if 400 feet be adopted as the limit in a 6-inch pipe, the limit in a 1½-inch pipe should be 200 feet; or, more briefly, velocity for any diameter of pipe should not exceed that given by,—

Rule 156. $\qquad V = 164 \sqrt{\text{Diameter}},$

where V is in feet per minute and diameter is in inches.

The maximum deliveries of pipes, according to this formula, may then be shown in tabular form as follows :—

Table XLVIII.—Delivery of Feed Pipes.

Diameter of pipe.	Delivery in cubic feet per minute. (C)	Delivery in lbs. per hour.	Diameter of pipe.	Delivery in cubic feet per minute. (C)	Delivery in lbs. per hour.
1½	2·45	9,260	3½	20·43	77,200
1¾	3·60	13,600	3¾	24·30	91,800
2	5·04	19,050	4	28·63	108,200
2¼	6·79	25,660	4¼	33·30	125,800
2½	8·81	33,300	4½	38·42	145,200
2¾	11·22	42,400	5	50·00	189,000
3	13·94	52,700	5½	63·36	239,500
3¼	17·00	64,200	6	78·92	298,300

Then, to find diameter of feed pipe, determine value of C in equation,—

Rule 157.

$$\frac{D^2 \times S}{184} = C,$$

where D is diameter of plunger in inches, and S its mean speed in feet per minute ; look for the corresponding figure in column 2 or 5 of Table ; and opposite, in column 1 or 4, will be found the appropriate diameter of pipe. If the exact figure does not appear in column 2 or 5, take the next higher. Of course, if the pump is single-acting, the deliveries will be one-half those given in above Table.

Although pumps of the plunger or single-acting type only propel the water during the " in " stroke, the mean speed should be calculated as though pump were of the solid bucket, or double-acting type, since the water, when in motion, travels at very nearly the same speed in each case,—air vessels producing little effect at the high pressures now used.

Feed pump plungers.—These should be of good gun-metal, and when of the single-acting ram type may be of thickness given by,—

Rule 158.

$$\text{Thickness of metal of plunger} = \sqrt{\frac{P \times D}{4660}} + \cdot 15 \text{ inch,}$$

where P is working pressure in lbs. per square inch, and D is diameter of ram in inches.

The greater thickness at the end should merge gradually into the thickness of body, and the corner of the inner end should have a radius of about ·4 × diameter of ram, when the end is not made quite hemispherical.

Feed Tanks.—To avoid waste of water through the occasional overflow of the hot-well, it is the rule in the Navy to fit a feed-tank, or supplementary hot-well, into which the air pumps deliver, and from which the feed pumps draw ; and, where fresh water only is used, such an arrangement is quite necessary.

Rule 159.—The size of tank is usually about 1 cubic foot for every 40 I.H.P., but where space is obtainable, and weight is not very much restricted, a greater capacity is very desirable.

The tanks are commonly of $\frac{5}{16}$-inch steel plates, flanged over and single riveted, stayed for a pressure of 10 to 15 lbs. per square inch, and galvanised after completion. Each tank should have a manhole, an air pipe, an overflow pipe, three or four zinc blocks to prevent corrosion, and a water gauge giving a visible range of at least 3 feet ; and where there are two or more engine rooms, with a tank in each, the various tanks should be connected one with another.

The feed-tanks should be placed as close to the air-pumps, and as low down as possible ; where they are under the platforms a float and index is more convenient than a water-gauge.

Reserve Tanks, &c.—**Rule 160.**—Where the boilers are fed with fresh water only, a sufficient reserve supply should be carried to provide about 21 cubic feet, or say 1300 lbs. " make up," or auxiliary feed, per 100 I.H.P. per 24 hours.

This supply may either be carried in reserve tanks, in the double bottom or elsewhere ; or may be produced as required by evaporators delivering into the L.P. valve-casing, or into the condenser ; or, a combination of these two methods may be employed.

Feed Heaters. — Various types of feed heater are now in common use ; and, where space and weight allow of their adoption they are, without doubt, a very desirable adjunct, since they keep a great deal of deposit out of the boilers, and, —when the exhaust steam from the auxiliary engines can be used in them,—save a considerable amount of heat that would otherwise be wasted.

Board of Trade Rules relating to Feed Pumps, Pipes, &c.

Paragraph 136. — Each boiler of a passenger vessel, whether old or new, should be fitted with suitable check valves between it and the feed pipes, and the boilers of all *new* passenger vessels, passenger tugs, and other small passenger vessels, should be fitted with separate feeding arrangements in addition to, but unconnected with, the main feed pipes and valves. It is desirable that the main feed check valve chest on each boiler be separate and distinct from that of the auxiliary feed, and that a stop cock or stop valve be fitted in each chest or between each chest and the boiler, so that the latter may be shut off, and either of the check valves examined while the other feed is at work. In very small vessels an efficient hand-pump, instead of the donkey pump, may be passed if the Surveyor has satisfied himself as to its efficiency when steam is up, and provided there are separate feed pipes and valves as directed above. This is to apply also in the case of old vessels of the above description, when being fitted with new boilers.

The Surveyor should discourage the practice of using the same pump for the bilges and for feeding boilers.

Lloyd's Rules relating to Feed Pumps, Pipes, &c.

26. The engines are to be fitted with two feed pumps, each capable of supplying the boilers ; the pumps, &c., to be so arranged that either can be overhauled whilst the other is at work.

28. In engines of 70 H.P. and under, one feed pump will be deemed sufficient, provided it is of adequate capacity.

The main feed pumps may be worked by independent engines, provided they are fitted with automatic regulators for controlling their speed. If only one such pump is fitted for the main feed, the auxiliary feed pump required by paragraph 31 should also be fitted with an automatic speed regulator.

31. A steam pump is to be provided capable of supplying the boilers with water, and is to have suctions from hot-well and from sea. A steam pump is to be so fitted as to pump from each compartment, to deliver water on deck, and, if no hand pump is fitted in engine-room, it must be fitted to be worked by hand. In small vessels in which only one steam pump is fitted, it must comply with all these requirements.

BILGE PUMPS.

There is no basis of calculation for the size of these pumps, and no generally recognised rule. For jet condensing engines they are generally of the same size as the feed pumps, and for surface condensing engines the following rule may be used :—

Rule 161.   Capacity of Bilge pump $= \dfrac{\text{Capacity of Cylinder}}{350}$.

In the case of a compound engine, capacity of L.P. cylinder is to be understood.

Where the pumps are driven by the main engines there should be,—except in the case of very small engines,—two pumps of the size given by the above rule, arranged so that either may be worked quite independently of the other, and be readily put out of gear when not required.

The hinged clack valve is the best for this class of pump, as it is less likely to choke than other types.

Covers or doors of pump valve-boxes and mud-boxes should be secured by hinged bolts and some form of wing nut, in order that they may be easily and quickly removed, and replaced whenever the choking or stoppage of the pump renders it necessary.

All strainer, or mud, boxes should be placed in easily accessible positions *above* the floor-plates, as terminal rose-boxes are always troublesome, and may be a cause of serious danger.

The "directing," or "distribution" valve boxes should also be above the floor plates, and easily accessible ; and each cover, or hand-wheel should bear a name plate indicating the compartment with which it opens communication ; a very good, and also very cheap plan, is to cast the required letters or words on the upper face of the hand-wheel rim.

Where one bilge pump must also act as wash-deck or fire pump (to comply with Board of Trade regulations), a three-way open-bottom cock, with one port in plug, should be fitted, to prevent the possibility of sea water entering the ship; when the pipes are over three inches in diameter, however, it is not always convenient to fit a cock, and non-return valves may be fitted,—though they are less safe than the cock.

Board of Trade Regulation referring to Bilge Pipes, &c.

Paragraph 139.—In all cases where pipes are so led or placed that water can run from the boiler or the sea into the bilge, either by accidentally or intentionally leaving a cock or valve open, they should be fitted with a non-return valve and a screw not attached, but which will set the valve down in its seat when necessary ; the only exception to this is the fireman's ash cock, which must have a cock or valve on the ship's side, and be above the stoke-hole plates.

Lloyd's Rules relating to Bilge Pumps, Pipes, &c.

27. The engines are to be fitted with two bilge pumps, which are to be so arranged that either can be overhauled whilst the other is at work.

28. In engines of 70 H.P. and under, one bilge pump will be deemed sufficient, if of adequate capacity.

29. A bilge injection, or bilge suction to the circulating pump, is to be fitted.

30. The engine bilge pumps are to be fitted to pump from each compartment of the vessel. The mud-boxes and roses in engine-room are to be placed where they are easily accessible, and to the satisfaction of the Surveyor.

31. A steam pump is to be provided capable of supplying the boilers with water, and is to have suctions from hot-well and sea. A steam pump is to be so fitted as to pump from each compartment, to deliver water on deck, and, if no hand pump is fitted in engine-room, it must be fitted to be worked by hand. In small vessels in which only one steam pump is fitted, it must comply with all these requirements.

35. Bilge suction pipes to be arranged to pump direct from each compartment,—the roses to be fixed in easily accessible places ; for numbers and sizes of suctions required, see Rules relating to hulls of ships, p. 276.

10

43. Cocks and valves connecting all suction pipes to be fixed above the stoke-hole and engine-room platforms.

44. The arrangement of pumps, bilge injections, suction and delivery pipes, is to be such as will not permit of water being run from the sea into the vessel by an act of carelessness or neglect. Any defective arrangement to be reported to the Committee.

PUMP LEVERS AND LINKS, &c.

In order to obtain quantitative results for guidance in proportioning pump gear, it is necessary to assume some maximum load per square inch of bucket, and the load here assumed is 30 lbs. for both air and circulating pumps.

Pump levers.—The strength of these may be determined by means of the ordinary rule for beams,—

Rule 162. $$W = \frac{fz}{l},$$

and,—when the central hole cut in the plate to admit the gudgeon boss does not exceed ·4 × depth of beam, and the attachment of the central boss or gudgeon flange is made with rivets in the usual way,—the values of f may be as follows:—

For thin plates, $f = 4500$ for screw engines, and 5500 for paddle engines.

For solid levers (forged) $f = 5500$ for screw engines, and 6500 for paddle engines.

In determining z $\left(= \frac{bd^2}{6}\right)$, b is four times the thickness of the plate, for ordinary plate beams, and d is the total depth or width of the plate at the gudgeon.

A lower value of f is adopted for screw engines because, owing to their higher speeds, there is a much greater liability to severe strains from sudden starting and from irregular action of the air-pump. When air and circulating pump are both driven by the same pair of levers, the values of f should be reduced 15 per cent.,—as one lever of the pair may have to carry considerably more than half the load at times.

The outline of the lever is often determined by drawing lines tangent to the previously determined radii at centre and ends, but this gives rather a clumsy-looking form, and is, of course, heavier than one in which the modulus of section varies more nearly as its distance from the centre or fulcrum.

Pump links.—For the various reasons stated above, the bolts in pump links require to be made very heavy.

Rule 163.—The link bolts of paddle engines, when two in number, may carry loads equal to those given in Table XXXI.; and when four in number, 10 per cent. less, or 20 per cent. less according as air-pump only, or both air and circulating pump are driven by one pair of levers. The link bolts of screw engines (usually four in number) may, in similar circumstances, carry loads 30 per cent., and 40 per cent. below those given in Table XXXI., page 73.

The loads that link bolts may carry under the various conditions mentioned will then be as follows :—

Table XLIX.—Strengths of Pump Links.

Diameter of link bolt in inches.	Total load in lbs. on one bolt.				
	Paddle Engine.			Screw Engine.	
	Two bolts.	Four bolts (air-pump).	Four bolts (air and circulating pumps).	Four bolts (air-pump).	Four bolts (air and circulating pumps).
1	2,150	1,930	1,720	1,500	1,300
1⅛	3,000	2,700	2,400	2,100	1,800
1¼	4,200	3,800	3,400	2,900	2,500
1⅜	5,400	4,800	4,300	3,800	3,200
1½	7,100	6,400	5,700	5,000	4,200
1⅝	8,500	7,600	6,800	6,000	5,100
1¾	11,000	10,900	9,300	7,700	6,600
1⅞	13,100	11,800	10,500	9,100	7,800
2	16,100	14,500	12,800	11,200	9,600
2¼	20,400	18,300	16,200	14,200	12,200

When the links are composed of pairs of parallel bolts, on to the ends of which the brasses are threaded, it will generally suffice to make the diameters of the middle portions of the bolts, between the brasses, the same as those of the end or bolt portions,—since the compressive stresses are usually much less than the tensile.

When the length of the plain or middle portion of the bolt reaches 20 diameters, however, it is advisable to substitute one central rod with ⊤ ends for the two parallel ones.

If, in any case, the links are so placed that the compressive stresses exceed the tensile, and if the length of link be more than 10 diameters of bolt or pillar, but less than 24 diameters, a careful estimate of the compressive stress should be made (with due allowance for sudden starting, unequal loading, &c.), and the diameter be determined by means of Rule 69 ; in the case of a "parallel bolt" link the length may be taken as the length of the middle portions of the bolts, between the brasses, but for a "central pillar" link the length should be taken from centre of pin to centre of pin.

As the stroke of the pumps is commonly less than the stroke of the piston, the load on the links at the piston-rod end of the levers is less than the load on those at the pump end in inverse proportion to their distances from the fulcrum of the lever.

Surfaces of pins.—Rule 164.—The surfaces of the various pins (diameter × length) to which the pump links are attached, should be such that the load does not exceed 800 lbs. per square inch in large engines, and 600 lbs. per square inch in engines indicating less than 500 horse power.

Pump crossheads.—These should be calculated as beams loaded in accordance with the circumstances of the case.

Rule 165.—Considering the sections usually employed, and the nature of the load, the value of f for wrought-iron or steel may be taken at 7000. For general formulæ for beams, *see* pp. 298–305.

Pump lever gudgeons.—The load on the pump lever gudgeon, and on its bearings, cap bolts, &c., is the sum of the loads on the two ends. The cap bolts should not carry greater loads than link bolts of corresponding diameters. (*See* Table XLIX., page 147.)

Rule 166.—The bearings may carry a load not exceeding 600 lbs. per square inch. The diameter of gudgeon will be most conveniently determined by considering the stresses to which it is subject as shearing stresses simply; then,—

Rule 167.   Diameter of Gudgeon $= \sqrt{\dfrac{S}{K}}$

where S = the sum of the total loads on the two ends of the pair of levers;

    K = 1200,—when there are two bearings, one close to each lever, as is usual in screw engines;

    K = 1600,—when there is only one lever in place of a pair, with very short gudgeon and bearing at each side close to lever, as frequently fitted in paddle engines.

When bell-crank levers are used, the magnitude and direction of the stress on the gudgeon bearing will still be,—as in the case of straight levers,—the resultant of the stresses on the links at the ends.

SLIDE-VALVES.

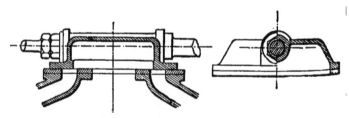

FIG. 32.—Common Locomotive Slide-Valve.

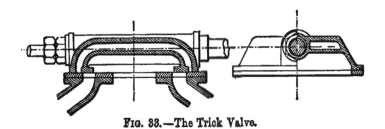

FIG. 33.—The Trick Valve.

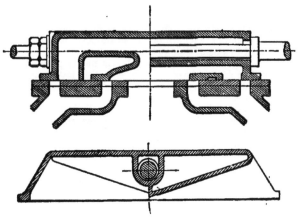

FIG. 34.—Common Double-Ported Valve.

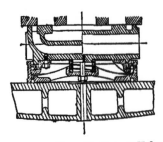

FIG. 35.—Church's Patent Valve.

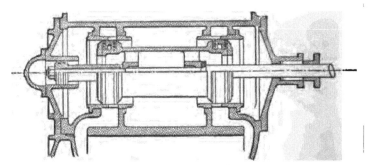

FIG. 36.—The Piston Slide-Valve.

Travel of Valve.—(Rule 168) The travel of a slide-valve should be from 2½ to 3 times the length of the steam-port in cylinder face, measured parallel to axis of slide-rod.

As the work done in moving the valve is nearly proportional to the length of travel, it is important to keep this latter as short as possible, and the best way of effecting this is to use double or triple-ported valves.

Single-ported valves should not be used for cylinders of more than 25 inches diameter; whilst triple-ported valves can only be conveniently employed for large engines of fairly long stroke, and are therefore not usually applicable in Naval work.

Since the introduction of the triple engine, piston valves have been re-introduced, and are at present the least objectionable form of valve that can be used for H.P. cylinders; their worst feature is their inability to leave the face when any dirt or grit is deposited on it, and the consequent severe strains that may be thrown on the valve-gear; also, as ordinarily fitted, they are not so steam-tight as flat valves.

Surface of Valve.—Flat slide-valves often give unsatisfactory results owing to want of sufficient bearing surface,—that is to say, the pressure per square inch of rubbing surface is too great, and undue wear, cutting, &c., result.

Where the ports are more than 20 inches wide, the surface should be increased by the insertion of a central bar, and in large faces there may be three or four bars, dividing the ports into widths of 16 to 20 inches. Care must be taken to provide proper grooves or recesses in these bars, and also in the side bars, to admit of the entry of steam for lubricating purposes.

Area of passages in valves.—In comparatively slow-moving engines, where weight is not of great importance, the areas of steam and exhaust passages in double and triple-ported valves may be 10 or 15 per cent. greater than the port areas in the cylinder faces, but in other cases the areas should not exceed those of the cylinder ports.

Relief-rings.—Flat slide-valves should be fitted with some arrangement for relieving the pressure on the back, and so reducing wear and tear, and the driving power required ; and they also require a spring of some sort to replace them against the face, when they have been caused to leave it by the presence of water in the cylinder, or by other causes.

A very simple and cheap method of effecting both of these objects is to use a circular gun-metal "relief-ring," of angle-bar section, held out by about six spiral springs, against the valve-box door or other flat surface, and packed steam-tight by means of an india-rubber or asbestos ring.

Rule 169.—The pressure per square inch of rubbing surface, imposed by the springs, should not exceed 2½ lbs.

Small valves may have one relief ring, and very wide ones two rings side by side ; the areas may be made as large as can be got on the valve backs, when the relief pipe from M.P. valve is led to L.P. receiver, and that from L.P. valve to condenser in the usual manner ; a greater difference of pressure than given by this arrangement is not desirable, as it would then be difficult to prevent leakage.

Port openings.—With given leads and cuts-off, both reckoned as percentages of the stroke of piston, the openings to steam are proportional to the travels, and may therefore be increased or diminished by increasing or diminishing the travel.

Lead of valve.—The lead given to the valve is generally decided arbitrarily, and varies from a bare $\frac{1}{32}$ inch in small auxiliary engines to 1¼ inch in large L.P. valves ; but the piston speed and momentum of moving parts should be taken into consideration, as in M.P. and L.P. cylinders the compression is rarely sufficient in itself to absorb all the momentum, and a certain amount of lead is necessary to prevent shock.

The extent to which the engine will probably be run "notched" or "linked" up should also be considered.

Ordinarily, when the valves are properly set, the lead at the back or top end will be about half that at the front or bottom end, and the cut-off at the crank end will be earlier than at the other, but this will be compensated for by the larger opening and fuller diagram obtained.

Inside, or exhaust, lap.—As regards inside lap, the common plan of making length of exhaust port in valve exactly equal to (exhaust port in cylinder face + the two bars),—and then setting the valve so that its mid-position is slightly further away from the crank than the mid-length of the exhaust port in cylinder face,— tends to equalise the compressions, and gives very satisfactory results.

For a quick-running compound or triple engine, exhaust from H.P. cylinder should commence at ·85 of the stroke, and for a slow-working paddle engine, not later than ·95 of the stroke.

Proportions of slide-valves.—Figures 37 and 38 show some of the elementary proportions of the common, and of the Trick valve.

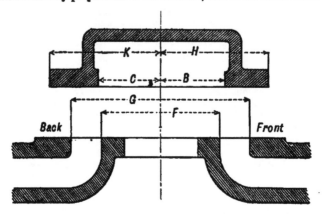

FIG. 37.—Proportions of a Common Valve.

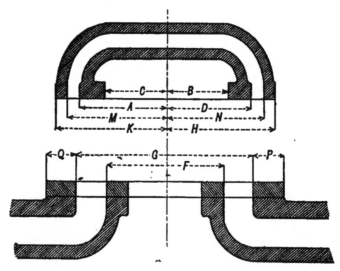

FIG. 38.—Proportions of a Trick Valve.

Let x be the *outside* lap of the valve at the front end, and y that at the back end. Then (Fig. 37),—

$$H = \frac{G}{2} + x; \text{ and } K = \frac{G}{2} + y.$$

Also, let z be the *inside* lap at the front, and w that at the back. Then (Fig. 37),—

$$B = \frac{F}{2} - z; \text{ and } C = \frac{F}{2} - w.$$

Referring now to Fig. 38,—let x, y, z, and w be the laps as before. Then,—

$$H = \frac{G}{2} + x; \text{ and } K = \frac{G}{2} + y.$$

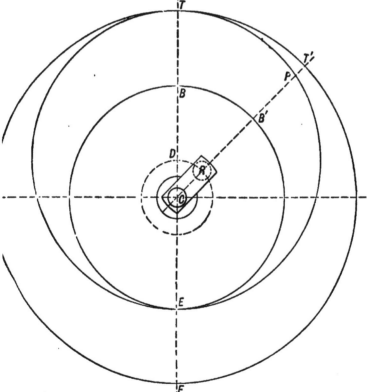

FIG. 39.—Diagram of the Piston Path.

Also,—

$$B = \frac{F}{2} - z; \text{ and } C = \frac{F}{2} - w.$$

$$A = \frac{G}{2} + \tfrac{1}{8} \text{ inch}; \text{ and } D = \frac{G}{2} + \tfrac{1}{8} \text{ inch.}$$

The openings through the valve laps or covers must be as large as possible, but need not exceed the ordinary opening of the valve to steam at the outer edge ; then,—

$$G + P = K + N; \text{ and } G + Q = H + M.$$

Valve diagrams.—Figures 39 and 40, when combined and used as described below, show at a glance the complete cycle of the operation of any proposed valve, and also the effect of varying any one of the elements,—travel, laps, leads, or openings.

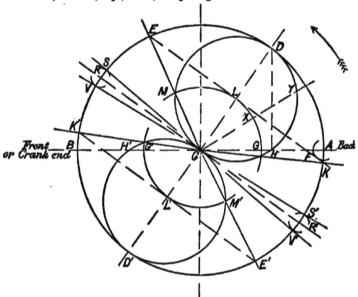

FIG. 40.—Zeuner's Diagram for the Common Valve Motion.

In constructing the diagram for a proposed valve, the diagram Fig. 40, the extreme diameter of which is equal to the travel of the valve, should be placed in the centre of the diagram Fig. 39,—the former being drawn full size, and the latter ¾ inch or 1 inch to the foot, as found most convenient.

Referring to Fig. 39, the outer circle TT'F represents the top or back end of the cylinder ; the inner circle BB'E represents the bottom, fiont, or crank end ; and the eccentric circle gives the position of the piston corresponding to any angular position of crank. The diagram is drawn by setting off CD equal to crank, and DT equal to connecting rod, and then swinging DT round on D as a centre. T is the position of piston on top or back centre, and E its position on bottom or front centre ; and the position corresponding to any other position of crank,—such as CR,—is found by producing CR to cut the circles, when PT' is the distance of the piston from one end of cylinder, and PB' its distance from the other.

Referring now to Fig. 40, the problem most frequently met with is, —Given the travel, leads, and cuts-off, to determine the laps and position of eccentric,—and it is solved as follows :—

Draw CE so that, when produced, it will cut the eccentric circle at a point corresponding to the given point of cut-off (i.e., if cut-off is to be at x inches from the back end of the stroke, the point in the eccentric circle must be x inches from the outer circle) ; then ACE is the angle through which the crank must move to arrive at the position of cut-off. With A as centre, and radius AF equal to the lead, draw part of a circle, and then draw EK. Next, draw CD perpendicular to, and bisecting EK in L. Then CL is the lap required, and BCD is the angle between the crank and eccentric ; and since CD is the half-travel, LD is the maximum opening of port.

To extend the usefulness of the diagram,—on CD as diameter, describe a circle ; and from centre C with radius CL, strike the arc GLM, which is called the lap circle. The part GH is then equal to AF, and represents the lead, or opening of valve when the crank is "on the centre," or in the position CA. XY likewise represents the port opening when the crank is in position CY.

To determine the operation of the valve at the other end, produce DC to D', and describe a circle on CD'. Then let H'G' be the lead at this end ; from C, with radius CG', draw the lap circle G'L'M', and through C and M' draw CE'. Then CE' is the position of the crank at cut-off, and CG' is the lap.

The positions of the crank, when the port commences to open, are CK and CK'.

If the valve has no inside lap (i.e., if the ports are both just closed to exhaust when the valve is in mid-position), release will occur when the crank is at CR, a position at right angles to CD, and compression will commence at CR', also at right angles to CD.

If, however, the valve has positive inside lap at the front end, and negative inside lap at the back end, the positions of release and of commencement of compression will be altered as indicated, release taking place at S and S', and compression commencing at V and V'.

The points S, S' and V, V' are obtained by striking the small arcs indicated with the inside lap as radius ; of course, if the lap is positive, the valve will open for exhaust later and close earlier ; but if the lap is negative, the exhaust will be earlier and the compression commence later.

To obtain the position of piston corresponding with the various

crank positions, each radial line indicating a crank position must be
continued until it meets the outer circle TT'F, and the position of
piston can then be scaled off.

Effect of linking up.—The effect of "linking up" on the operation
of the valve is most readily exhibited by the application to the above
diagrams of the following very closely approximate construction,
suggested by Mr Macfarlane Gray :—

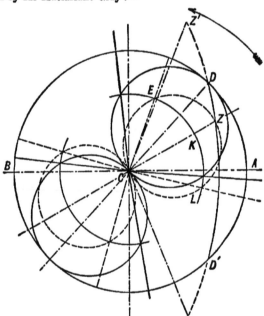

FIG. 41.—Diagram showing the effect of " Notching up."

Suppose the link (Fig. 42) to be notched up so that the link block
M is distant MT from the point at which the eccentric rod is attached
to the link.

FIG. 42.—Link Motion " Notched up."

Draw the valve diagram (Fig. 41) due to the position and throw of the eccentrics in Fig. 42, and through D and D' draw the arc of a circle with a radius found as follows :—

$$\text{Radius} = \frac{DT \times DD'}{2\,TN} \quad \text{(see Fig. 42).}$$

Then divide this arc at Z so that DZ is to DD' as TM is to TN ; join CZ, and on it as diameter describe a circle cutting the lap circle in L and E ; and draw the lines CL and CE, the former indicating position of crank when valve opens, and the latter the position when valve closes.

It will thus be seen that the effect on the valve of notching the link up to the point M is the same as though it were driven by an eccentric having the angular position and eccentricity CZ ; the lead is earlier, the opening is reduced to KZ, the cut-off is earlier, and a further examination would show that release occurs earlier, and compression commences earlier than when in full gear.

Open and crossed rods.—When the eccentric rods are arranged as shown in Fig. 42, they are said to be "open," but when the gear is so arranged that D is joined to N, and D' to T, whilst the crank remains turned away from the link TN, the rods are said to be "crossed" ; in the diagram for this latter case the arc DD' must be drawn convex towards C, or its centre must be on side A.

When it is intended to work the engines linked-up to any considerable extent, the rods should be of the "open" type, as a greater range of expansion can be obtained with less reduction of port openings than is possible with "crossed" rods.

Overhung gear.—Where the valve-gear is of the "overhung" type (*i.e.*, the type in which T is always outside of M, even in full gear), the diagrams are constructed on the same principle as Fig. 41, and the position and eccentricity of the eccentrics are determined by producing the arc DD', and taking a point Z' *beyond* D, so that Z'D is to (DD'+2Z'D) as TM is to TN, and joining CZ'. CZ' is the required eccentricity of the sheaves, and BCZ' is the angle between them and the crank.

Obliquity of eccentric rods. — The valve diagram, as above described, takes no account of the effect produced by the obliquity of the eccentric rods ; when the rods are so short as to make this disturbing effect noticeable, the diagram may be corrected as follows :—

From A and B (Fig. 43) drop perpendiculars AN and BN' upon RR', and through N and N', with length of eccentric rod as radius, strike an arc ; also, with same radius, strike similar arcs through O and P and O' and P'. Then, if the points where these arcs cut the travel circle be joined with C, the corrected crank positions, at which the various events of the cycle occur, will be obtained ; the radial

lines indicating these new crank positions may then be produced
outwards so as to cut the eccentric circle and indicate the piston

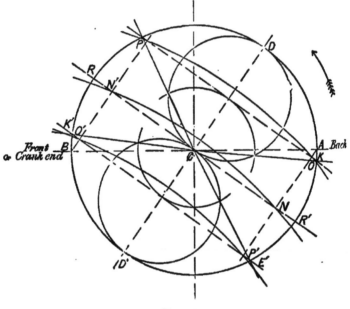

positions as before. The correcting arcs for the exhaust laps have
been omitted as tending to complicate the diagram, but they will be
drawn on either side of the arc NN', at distances from it equal to
the laps, and will give new points on the travel circle in a precisely
similar way.

When the eccentric rods are very short, and the locomotive or
"slot" link is used, as in some types of horizontal engine, no diagram
will give more than approximately accurate results, and the final
adjustments should be made on a model ; one that will serve every
purpose may be rigged up on a drawing board with a few pieces of lath
and some stout pins, in little more than half-an-hour.

For the solutions of other problems connected with the diagram for
the ordinary slide valve, and for the methods of constructing diagrams
for various types of expansion valve, see Mr Seaton's "Manual of
Marine Engineering."

Diagram for oscillating engine.—When the diagram Fig. 39.
is applied to the case of an oscillating engine, the co·centric circle
deviates slightly from a true circle, owing to the fact that there is

virtually a connecting rod of varying length ; the necessary correction
in the eccentric circle may be made as follows :—

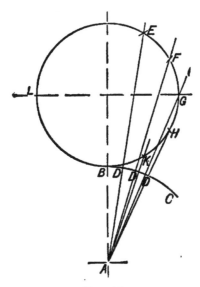

FIG. 44.

Let A be the trunnion centre, and BLE the crank path ; divide the
half crank path into any number of equal parts (say six), at K, H, G,
&c. ; join each point with A ; and from A, with radius AB, strike the
arc BDC. Then, referring to Fig. 45, set off the corresponding points
E, F, G, &c., and transfer the lengths ED, FD, GD, &c., from Fig.
44 to the positions indicated by the corresponding letters on Fig. 45,
and through the points D, D, D, &c., draw a curve ; this curve will
then be the true curve of successive positions of piston corresponding
to the crank positions indicated by any radial lines.

GL in Fig. 45 (page 160) must, of course, represent GL in Fig. 44,
either to the same or some other convenient scale.

VALVE GEAR.

It must be understood that the various proportions given in the
following sections on "Valve Gear" and "Eccentrics" are such as
are suitable for *ordinary* speeds of engines of the two categories,
Naval and Mercantile, and that both increased strengths and increased

surfaces will be required if the engines are to run at exceptionally high numbers of revolutions.

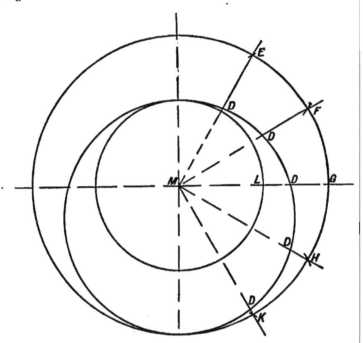

FIG. 45.

Valve or slide rods.—The power requisite to move a slide valve on its face is a constantly varying quantity, and care must be taken to use its highest ordinary value in all calculations for dimensions of valve gear.

To obtain this, the gain from using any relief arrangement, and also that due to the pressures on some portions being in equilibrium, should be neglected, and the total pressure on the valve taken as the maximum difference of absolute pressures between which it works, acting on its gross area,—these pressures being those of the steam-chest or receiver on one side, and those of receiver or condenser on the other.

If L be the length and B the breadth of a valve, both in inches; d the maximum pressure per square inch, obtained as described above, and ·2 the co-efficient of friction for metallic surfaces rubbing together dry ; then,—

Stress on valve rod $= ·2 (L \times B \times p)$ lbs. ;

and,—Rule 170,—

Smallest diameter of valve rod $= \sqrt{\dfrac{L \times B \times p}{12500}} + ·25$;

and,—Rule 171,—

Diameter of valve rod in gland $= \sqrt{\dfrac{L \times B \times p}{6500}} + ·25.$

The value of p should be taken as 30 lbs. for L.P. valves, and, as it is desirable to have the rods all of the same dimensions, the highest value of the above expression for any one cylinder should be used for all.

When piston-valves are used for the H.P. cylinders, the valve-rod is usually made of the same size as those for the flat slides of the M.P. and L.P. cylinders.

Theoretically, the power required to move a piston-valve should be little more than is necessary to overcome the friction at the stuffing-box, but practically,—as mentioned under "Slide-valves,"—the power required is greatly in excess of this, and may have almost any value.

The above-mentioned diameter at the gland $\left(\text{which is on an average equivalent to } \dfrac{\text{Extreme diameter of valve}}{7}\right)$, is, no doubt, greater than necessary, but is convenient, and gives ample strength,—a consideration of perhaps greater importance in marine than in any other class of work. Some makers have used rods of diameter at the gland equal to $\dfrac{\text{Extreme diameter of valve}}{13}$.

Rule 172. The cap-bolts for valve-rod end or eye must each carry a load equal to $\dfrac{·2 (L \times B \times p)}{2}$, and the size suitable for this load will be found from Table XXXI., page 73. Their diameter will be found to very closely approach $·5D + ·2$ inch, where D is diameter of valve-rod found by Rule 170 above.

Valve-rod guides.—All valve-rods should be fitted with guides, placed as close to the link as possible ; and, if they are not fitted with balance pistons, they should have guides at the tail ends as well.

The most satisfactory form of guide is one resembling a piston-rod slipper guide of the "single" type (*i.e.* one shoe and one guide for both ahead and astern) in miniature,—the shoe being cast with a small tubular crosshead which is bored to fit the valve-rod, and secured on it by a through cotter.

When this type is not adopted, and the guide is made to embrace

11

the valve-rod just below, or outside of, the gland, a square or two flats should (whenever space is available) be formed on the rod to take the guide,—as the various joints of the gear will then be to some extent relieved of the twisting stresses, due to the eccentrics and rods not being in exactly the same plane as the axis of the valve-rod, and will in consequence run longer without overhauling.

Reversing quadrants, or "links." — The old slot-link still retains its place, as the most convenient of application in oscillating paddle engines, and in some types of horizontal screw engines.

Its principal proportions should be as follows —the unit D being the diameter of valve-rod as found by Rule 170 :—

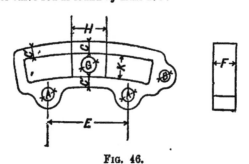

FIG. 46.

Rule 173, a to k.

(*a*) Centres of eccentric-rod pins (E) 2·75 to 3 × throw of eccentrics.

(*b*) Breadth of link (F)	1·1 D + ·4	inch.
(*c*) Thickness of bar (C) { with bridge . .	D + ·25	,,
{ without ,, . .	1·1 D + ·25	,,
(*d*) ,, ,, (C') { with ,, .	·9 D + ·25	,,
{ without ,, .	D + ·25	,,
(*e*) Length of block (H)	2·5 D + 1	,,
(*f*) Width of slot (K)	1·4 G + ·35	,,
(*g*) Diam. of eccentric-rod pin (A) { solid bush .	D + ·25	,,
{ adjustable brasses	1·1 D + ·25	,,
(*h*) ,, drag-rod pin (B) { as in fig. .	·55 D + ·25	,,
{ when on bridge .	·8 D + ·25	,,
(*j*) ,, block-pin (G) { when overhung .	1·4 D + ·25	,,
{ both ends secured	1·3 D + ·25	,,
(*k*) ,, drag-rod at ends { if one . .	·7 D + ·25	,,
{ if two . .	·48 D + ·25	,,

Theoretically, surfaces may be a little smaller for paddle engines, but it is better to give the extra surface and make them same as for screw engines.

For all vertical engines, and in all other cases where possible excepting where L.P. cylinder is below say 25 inches in diameter), the double-bar link is preferable, as large working surfaces and pins are more easily obtained, and they are more accessible and more easily adjusted.

The principal proportions of the double-bar link should be as follows,—the unit being again D, or diameter of valve-rod as found by Rule 170 :—

Rule 174, a to n.

(a) Centres of eccentric rod pins . . . 8 × throw of eccentrics, or roughly 7·8 D.

(b) Depth of bars . . . . . 1·4 D + ·75 inch.

(c) Thickness of bars . . . . . ·5 D + ·15 "

(d) Width between bars and thickness of valve-rod eye when gun-metal or bronze block } 1·2 D + ·25 "

(e) Width between bars when cast-steel block and gun-metal liners . . . . } 1·65 D + ·25 "

(f) Dia. of valve-rod eye (G.M. or bronze block) 1·55 × depth of link-bar.

(g) " " (steel block and G.M. liners) 1 × " "

(h) Length of sliding block . . . . 3 D to 3·75 D.

(j) Diameter of eccentric-rod pins . . ·85 D + ·5 inch.

(k) Length " " . . ·7 D + ·4 "

(l) Diameter of drag-rod pins . . . ·7 × diameter of eccentric rod pin.

(m) Length " " . . . 1 × length of eccentric rod pin.

(n) Diameter of bolt through ends of link bars . . . . } ·35 × depth of link bar.

For proportions given by above rules, *see* Table L. (page 164).

Whether the links are forged from iron or steel, all the pins and working surfaces should be thoroughly case-hardened before completion.

Position of suspension pins.—It is sometimes convenient to place the suspension, or drag-rod pins at the end, in the case of the slot link, but, when possible, they should always be placed at mid-length of the link,—so as to coincide in position with the block-pin when the link is in mid-gear,—as there will then be the least "slotting" motion, or "wash" of link.

When the block-pin is at the half-travel position, a line drawn from its centre, at right angles to the axis of the valve-rod, should bisect the versed sine of the arc described by the centre of the drag-rod pin in the end of the reversing lever.

The reversing lever, when in mid-position, commonly lies parallel to the valve-rod, but it is better to incline it slightly away from the valve-rod, and so reduce the "wash" of link when in ahead gear, whilst increasing it a little when in astern gear. The amount of inclination, measured on the arc, may be about one-eighth the length of the lever. *See* Fig. 47 (page 165).

Table L.—Proportions of Double-bar Links.

Diameter of valve-rod (in valve).	Centres of eccentric rod pins, (about) (a)	Depth of bars. (b)	Thickness of bars. (c)	Width between bars, G.M. block. (d)	Width between bars, steel block. (e)	Diameter of valve-rod eye, G.M. block. (f)	Diameter of valve-rod eye, steel block. (g)	Length of block on bars, (3·5 D) (h)	Diameter of eccentric rod pins. (j)	Length of eccentric rod pins. (k)	Diameter of drag-rod pins. (l)	Length of drag-rod pins. (m)	Diameter of end bolt. (n)
inches.	inches.	inches.	inches.	inches.	inches.	inches.	inches.	inches.	inches.	inches.	inches.	inches.	inches.
1½	12	2⅞	⅞	2	...	4½	...	5¼	1¾	1½	1	1½	1
1¾	14	3¼	1	2⅜	...	5	...	6	2	1⅝	1⅛	1⅝	1⅛
2	16	3½	1⅛	2⅝	...	5½	...	7	2¼	1¾	1¼	1¾	1¼
2¼	18	4	1¼	3	...	6¼	...	7¾	2⅜	2	1⅜	2	1⅜
2½	20	4¼	1⅜	3¼	...	6¾	...	8¾	2⅝	2⅛	1½	2⅛	1½
2¾	21	4⅝	1½	3½	4¾	7¼	4⅝	9½	2¾	2⅜	1⅝	2⅜	1⅝
3	23	5	1⅝	3⅞	5¼	7¾	5	10½	3	2½	1¾	2½	1¾
3¼	25	5¼	1¾	...	5⅝	...	5¼	11¼	3¼	2⅝	1⅞	2⅝	1⅞
3½	27	5⅝	1⅞	...	6	...	5⅝	12¼	3½	2⅞	2	2⅞	2
3¾	29	6	2	...	6½	...	6	13	3¾	3	2⅛	3	2⅛
4	31	6⅜	2⅛	...	6⅞	...	6⅜	14	4	3¼	2¼	3¼	2¼
4¼	33	6¾	2¼	...	7¼	...	6¾	14¾	4⅛	3⅜	2⅜	3⅜	2⅜
4½	35	7	2⅜	...	7⅝	...	7	15¾	4¼	3½	2½	3½	2⅜
4¾	37	7⅜	2½	...	8	...	7⅜	16½	4½	3¾	2⅝	3¾	2½

Note.—In Naval work the proportions and surfaces are usually somewhat greater than in Mercantile work; the above Table gives good average proportions. For paddle engines the surfaces may be same as for screw engines, since little is saved by reducing them.

The **drag**-rods should, of course, be as long as they can conveniently be made, and care should be taken that the angle they make with the reversing lever is not too obtuse when at its maximum.

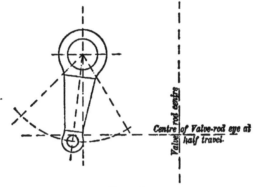

Centre of Valve-rod eye at half travel.

FIG. 47.

When the most convenient length of drag-rod has been fixed upon, and the reversing lever arc has been drawn in what is apparently the best position, the arrangement should be checked by drawing down the paths described by the ahead and astern eccentric-rod pins on the link, when respectively in gear. These paths should resemble attenuated figures of 8 (rather broader for the astern pin), and the arc described by the end of the drag-rod, should divide them fairly down the centre; if the division is nearer one side than the other, the length of drag-rod should be altered so as to set it right.

Trial should also be made with higher and lower positions of suspension at the lever end of the drag-rod, to see whether the width of the 8's cannot be reduced, or whether they cannot be twisted more nearly into line with the valve-rod axis, and so virtually reduced in width.

ECCENTRICS.

Eccentric sheaves.—These should be made of hard and tough cast-iron, and, when made in two pieces, the small piece should be of cold blast or other specially strong iron. The line of division should always be at right angles to the line through centres of shaft and of eccentric, and the key-way should be placed in the centre of the large piece. The joint separating the two portions should also have a small step in it, on each side of the shaft, to prevent motion of either portion in a plane perpendicular to the shaft axis.

The bolts holding the two portions together may either have

rectangular nuts fitted into recesses in the large portion, or ordinary nuts, or cotters, placed where the sheave is lightened out, or perforated. The recesses in which the bolt heads lie may be filled up with white-metal if desired.

A good method of construction for large sheaves is to lighten them out only,—leaving a continuous web or plate,—instead of completely perforating them; but the web must be placed close to one side to give room for the set screws for fixing sheave on shaft, and holes may be cut through it as required for getting at nuts or cotters. When sheaves are constructed in this way they can be very conveniently secured back to back by means of a stout through bolt, which has the effect of causing the astern key to always share the work of the ahead one, and *vice versâ.*

Eccentric straps.—These may be of good gun-metal for sheaves on shafts up to 10 inches in diameter, but for larger eccentrics they should be of cast-steel, with gun-metal liners.

For small high-speed engines the straps should be of gun-metal, lined with white-metal between the side lips.

The straps should always have side lips, as shown in Figs. 48 and 49; and when gun-metal liners are used they should have external projecting rings fitting into grooves in the cast-steel straps, as shown in Fig. 49.

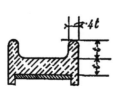 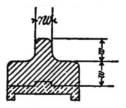

Gun-metal Strap lined with White-metal.   *Cast-steel Strap with Gun-metal liner.*
        FIG. 48.   FIG. 49.

When gun-metal straps are used, each half should be made thicker at the middle, to allow for wear; and whether gun-metal or steel is used, the strap should be stiffened by ribs as indicated in Figs. 48, 49, and 50,—one central rib being used for steel, and two side ones for gun-metal.

The general proportions of sheave and strap should be as follows,— the unit D being still the diameter of valve-rod, as found by Rule 170 :—

Rule 175, a to m.

(a) External diameter of sheave . $1 \cdot 2 \times$ (diameter of shaft + throw of eccentric), or roughly $7 \cdot 8$ D.

(*b*) Breadth of sheave at strap . {Mercantile   D+   ·6 in.
                                  Naval   .   D+1·25 "

(*c*) Thickness of metal round shaft {pierced sheave   ·6 D+   ·5 "
                                      lightened " . ·45D+ ·25 "

(*d*)   "   "   at circumference {pierced   " . ·4 D+   ·5 "
                                      lightened " . ·4 D+ ·25 "

(*e*) Breadth of key .   .   .   .   .   .   ·55D+   ·4 "

(*f*) Thickness "   .   .   .   .   .   ·27D+   ·3 "

(*g*) Diameter of bolt connecting portions of sheave . ·45D+   ·25 "

(*h*) Thickness of G.M. straps at middle (ex. lips) (*t*) . ·4 D+   ·5 "

(*j*)   "   "   "   sides   "   . ·33D+   ·5 "

(*k*)   "   cast-steel straps (*w*)   .   .   ·3 D+   ·5 "

(*l*)   "   G.M. liners for cast-} $\dfrac{\text{Dia. of sheave}}{70}$ + ·125 inch.
                steel straps .   }

(*m*) Diameter of bolts connecting parts of strap .   . ·5 D+   ·2 ,,

For proportions given by above rules, *see* Table LI., page 168.

As a check on the diameter of bolts in strap, it should be seen that the loads on them are not in excess of those given in Table XXXI. See page 73.

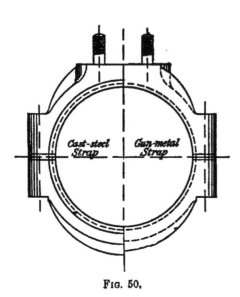

FIG. 50.

Table LI.—Proportions of Eccentric Sheaves and Straps.

Diam. of valve-rod (in valve).	External diam. of sheave (about)	Breadth of sheave at strap. (b)		Thickness of metal round shaft. (c)		Thickness of metal at circumference. (d)		Dimensions of Key. (e)	(f)	Diam. of bolt connecting sheave.	Thickness of gun-metal straps. (h)	(j)	Thickness of cast-steel straps.	Thickness of liners for steel straps.	Diam. of bolt connecting strap.
		Mercantile.	Naval.	Pierced	Lightened.	Pierced	Lightened.	Br'dth	Thickness.	(g)	Middle.	Sides.	(k)	(l)	(m)
ins.	ins.	ins.	ins.	ins.	ins.	ins.	ins.	ins.	in.	ins.	ins.	ins.	in.	in.	ins.
1½	12	2⅛	2¾	1½	1	1⅛	⅞	1¼	¾	1	1⅛	1	1	5/16	1
1¾	14	2⅜	3	1⅝	1⅛	1¼	1	1⅜	13/16	1⅛	1¼	1⅛	1 1/16	5/16	1⅛
2	16	2⅝	3¼	1¾	1⅛	1⅜	1⅛	1½	⅞	1⅛	1⅜	1 3/16	1⅛	⅜	1¼
2¼	18	2⅞	3½	1⅞	1¼	1⅜	1⅛	1⅝	15/16	1¼	1⅜	1¼	1 3/16	⅜	1⅜
2½	20	3⅛	3¾	2	1⅜	1½	1¼	1¾	1	1⅜	1½	1⅜	1¼	7/16	1½
2¾	21	3⅜	4	2⅛	1½	1⅝	1⅜	1⅞	1 1/16	1½	1⅝	1 7/16	1 5/16	7/16	1⅝
3	23	3⅝	4¼	2⅜	1⅝	1¾	1½	2	1⅛	1⅝	1¾	1½	1 7/16	7/16	1¾
3¼	25	3⅞	4½	2½	1¾	1⅞	1⅝	2¼	1 3/16	1¾	1⅞	1⅝	1½	½	1⅞
3½	27	4⅛	4¾	2⅝	1⅞	2	1¾	2⅜	1¼	1⅞	2	1 11/16	1 9/16	½	2
3¾	29	4⅜	5	2¾	1⅞	2	1¾	2½	1⅜	1⅞	2	1¾	1⅝	9/16	2⅛
4	31	4⅝	5¼	3	2	2⅛	1⅞	2⅝	1 7/16	2	2⅛	1⅞	1¾	9/16	2¼
4¼	33	4⅞	5½	3⅛	2⅛	2¼	2	2¾	1½	2⅛	2¼	1 15/16	1 13/16	⅝	2⅜
4½	35	5⅛	5¾	3¼	2¼	2⅜	2⅛	2⅞	1½	2¼	2⅜	2	1⅞	⅝	2½
4¾	37	5⅜	6	3⅜	2⅜	2⅜	2⅛	3	1 9/16	2⅜	2⅜	2⅛	1 15/16	⅝	2⅝

Eccentric rods.—These should be sound forgings of good wrought-iron,—the double-eye at the link end being cut out of the solid forging. Though much may be said in favour of the rectangular section for the body of the rod, experience has decided that, on the whole, the circular section is the best. The diameter at the link end should be that given by,—

Rule 176.

Diameter of eccentric rod at link end = ·8 D + ·2 inch.

The diameter at mid-length of rod may be calculated in the same manner as a connecting-rod (Rule 69), but for direct-acting engines of ordinary construction,—(where the length of connecting-rod is not more than 2 × stroke for Naval engines, and 2¼ × stroke for Mercantile engines),—the rods may be made with a straight taper from end to end, and may have the following diameters at the ends next eccentrics:—

Rule 177.

Diameter of eccentric rod at eccentric $\begin{cases} \text{Naval}..........\text{D} \\ \text{Mercantile}...\text{D}+ \cdot 4 \text{ inch,} \end{cases}$

where D has the same value as before.

The studs by which the T end of the eccentric-rod is secured to the strap should be of the same strength as the bolts for valve-rod caps (Rule 172), and for connecting the two parts of the strap (Rule 175 *m*), and their diameter may be calculated in the same manner.

The cap-bolts at link end of rod should also, collectively, have the same strength as the valve-rod cap-bolts.; but, as they will be considerably smaller, they must not carry so great a load per square inch of section at bottom of thread, and their diameter will be easily fixed by reference to Table XXXI., page 73.

Single eccentric gear.—This type of gear is well suited for use on small oscillating paddle-engines, where the power required to move the valves is not more than one man can easily exert.

The sheave should be cast with a balance weight, and with a "snug" or projection on the side, to engage with the stop on the shaft, and should be bored to be a nice working fit on the shaft.

The stop should be cut from a turned and bored ring of wrought steel, of rectangular section, and should be fixed to the shaft by two or three steel tap screws. The length of the stop may be such that it will embrace half the circumference of the shaft, and it must be fixed centrally *opposite* to the crank, as the action of the valve is reversed by the interposition of the valve-lever.

The "snug" on the sheave will then be placed centrally on the line connecting the shaft and eccentric centres, and will have a length, measured round the circumference of the shaft, given by,—

Rule 178.

Length of " snug " on sheave = $\left\{\begin{array}{l}\text{(Half circumference of shaft)} \\ \quad -\text{(part of circumference in-} \\ \quad \text{tercepted between centre} \\ \quad \text{lines of ahead and astern} \\ \quad \text{eccentric positions).}\end{array}\right.$

The arrangement will be readily understood on reference to Fig. 51.

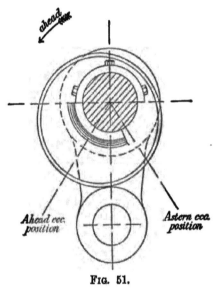

Ahead ecc. position

Astern ecc. position

Fɪɢ. 51.

JOY'S VALVE GEAR.

The general arrangement of one form of this gear is shown in Fig. 52. The method of laying down the centre lines is as follows :—

Take a point d (Fig. 53, page 172) on or near the axis of the connecting-rod, such that its extreme transverse vibration is about twice the full stroke of the valve (better rather more than less) ; and through its extreme positions d_1 d_2 draw zz. Then mark off e and e_1,—the extreme longitudinal positions of the point d,—and from these points draw two lines to meet in a point f on zz, the position of which is such that the angle efe is not more than 90° (if there is room a less angle is better).

The point f should be controlled so as to move as nearly as possible on the line zz; this may be effected either by an "anchor" link 2,—which may be placed on either side of zz,—or by a sliding guide.

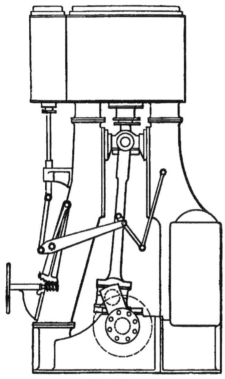

Fig. 52.

Next, on the valve rod centre 5, mark off g_1 and g_2, so that gg_1 is equal to lap + lead for the top or back end of cylinder, and gg_2 equal to same for crank end of cylinder. Then take a point j on the link $e_1 f$ so that $e_1 j$ is about $4/3 \times dd_1$, and draw jg_1, cutting zz in m; m will be the fulcrum of the lever 3, and will also be the centre of the curved reversing block in which the fulcrum slides. When swing links are used (as in Fig. 52), in lieu of a curved block, the arcs described by the link ends, when in ahead and astern positions, must intersect in M.

The position assumed for j in the first instance is approximate only, and may not be quite correct; its position must be such that the point *m* in the lever jmg_1 moves to an equal distance on either side of the centre of the reversing block,—also marked *m*. If not at first quite correct, it will generally be obtained on a second trial.

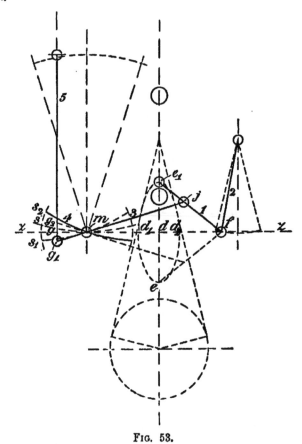

Fig. 53.

When the gear is correctly set out, the act of moving it from ahead to astern gear, or to any intermediate position, should leave the lead unaltered.

The valve-rod link, which is attached at *g*, may have any convenient length, but the length fixed on determines the radius of the curved guides in the reversing block, or the length of the swing links, which-ever construction is adopted.

It will be noticed that the lap and lead depend on the ratio of jm to mg_1, whilst the port openings (over and above the leads), depend only on the angle to which the reversing block is canted, and will be about 4/5ths of the distances ss_1 and ss_2 respectively.

Considerable deviations from the above-described positions and proportions may be made without materially affecting the nature of the results.

Where the distance from centre of piston rod to centre of valve rod is different for the different cylinders (H.P., M.P., L.P.), the valve-rod link may be inclined as much as 1 in 12, but beyond that the point m should be moved, and the lengths jm and mg_1 altered.

For complete investigation of Bremme's or Marshall's gear see Mr Seaton's "Manual of Marine Engineering."

REVERSING GEAR.

For all ordinary types of engine, when size renders hand-gear too slow in its action, the simplest and most efficient form of reversing engine is the direct-acting steam cylinder, combined with a brake cylinder containing oil or water. A very perfect form of this gear has long been associated with the name of Messrs Brown of Edinburgh ; by the adoption of valve-gear of the "hunting" type they have produced a machine which is handled with the greatest ease and certainty, and which is, at the same time, quite automatic in its action. The action of the gear is as follows :—On moving the hand-lever through any fraction of its possible travel, the steam valve is dis-placed, or opened, a corresponding amount, and the piston begins to move ; but, so soon as the gear begins to move, the steam valve begins to return to its central or closed position, and when the piston has travelled through the fraction of its stroke corresponding with the original movement of the hand-lever, the valve is again central, and the gear at rest.

Direct steam gears should never be fitted without the brake cylinder, as, in careless hands, the steam-power is apt to "take charge," overpower the controlling hand-gear, and "carry away" some of the brackets, &c.

Another very excellent type of reversing gear, which is also adaptable to almost any type of engine, is that known as the "all-round" gear. In this gear a connecting-rod from the main lever on the weigh-shaft takes hold of a crank-pin on the side of a worm-wheel, and the worm gearing with this wheel is revolved by a small steam engine.

The great advantage of this gear is that,—as implied in its name,—it has no "stops," but can travel round continuously, without risk of carrying anything away, and is therefore perfectly safe in even the most careless hands. It has also the further advantage,—resulting from the crank and connecting-rod principle embodied,—of requiring little or no power at the commencement and finish of its movement from ahead to astern, or *vice versd*, thus allowing the engine to "get way on it" in readiness for the load, and to come easily to rest again. The small engine is sometimes made with two cylinders and cranks, and is also sometimes made reversible; but these are both very doubtful improvements.

In all steam reversing gears, arrangements should be made to prevent the accumulation of water in the cylinders of the small engine, in order that it may be in condition to start instantly when the hand-lever is moved.

Reversing gear weigh-shaft.—If the stresses on weigh-shafts were purely torsional, the diameters might vary as the cube-roots of the powers required to move the valves; but they have to withstand severe bending stresses, and must also remain fairly rigid under all conditions, and to do this, experience has shown that the diameters must vary more nearly as the square-roots of these powers, and there-fore directly as the diameters of the valve-rods.

For ordinary triple engines, with H.P. and L.P. valves "outside" (or forward and aft of their respective cylinders in the ordinary arrange-ment), and piston-valves to H.P. cylinder only, the diameter of weigh-shaft is given by the following rule:—

$$\text{Diameter of weigh-shaft} = 2 \times \left(\sqrt{\frac{L \times B \times p}{12,500}} + \cdot 25 \right)$$

But the quantity within the brackets is equivalent to the smallest diameter of valve-rod (*See* Rule 170), so that the rule for diameter of weigh-shaft may be written,—

Rule 179.

Diameter of weigh-shaft = 2 × smallest diameter of valve-rod.

In some cases it may be possible to save a little weight by tapering parts of the weigh-shaft, but, as a rule, it is better to use a practically parallel shaft, and, if weight is of great importance, make it hollow, as is commonly done in the case of large Naval engines; the necessary stiffness is then maintained without difficulty, and there is no fear of the bending stresses giving trouble.

When the H.P. and L.P. valves (or either of them) are "inside" of their respective cylinders, and the weight-shaft is thereby shortened the necessary stiffness will be given by a smaller shaft, and the diameter may be reduced in the proportion indicated by Rules 80 and 80a.

Size of reversing engine cylinders. — For ordinary triple engines as described above, this may be obtained from the general relation,--

Rule 180.
$$\frac{P \times A \times T}{L \times B \times p} = K \qquad . \qquad . \qquad . \qquad (1)$$

where L and B = length and breadth of one slide valve in inches ; the one that gives the highest value of $L \times B \times p$.

p = greatest difference of absolute pressures between which valve works (*See* page 160) in lbs. per sq. in.

P = Boiler pressure—lbs. per sq. in.

A = Area of piston or pistons of reversing engine—square inches.

T = Travel of piston or pistons, in feet, required to move gear over from "ahead" to "astern."

$K = \begin{cases} \cdot 33 \text{ for direct gear.} \\ 1 \cdot 6 \text{ for "all round" gear.} \end{cases}$

or, substituting smallest area (a) of valve-rod multiplied by 12,000, for $L \times B \times p$, the formula may be written with quite sufficient accuracy,—

Rule 180a.
$$\frac{P \times A \times T}{a \times 12,000} = K_1$$

where K_1 is $\begin{cases} \cdot 42 \text{ for direct gear.} \\ 2 \cdot 00 \text{ for "all round" gear.} \end{cases}$

In their fear of giving too little power in steam reversing gears, there is little doubt that engineers often run to the other extreme, and give much more than necessary, even when two-thirds boiler pressure is assumed. The values assumed for the constants in the above formula (1) give ample size, and the constant $1\cdot6$ may even be reduced to 1, if space and weight are of great importance.

With "all-round" gears the small engine should not make less than 15 revolutions in moving the gear over, and where possible 20 should be allowed, and up to 25 in large engines.

The pitch of the worm-wheel teeth should never be less than $1\frac{1}{2}$ inches, and for large engines $2\frac{1}{2}$ inches is usual; they should be machine-cut, and should have no more clearance than is absolutely necessary.

The worm is usually forged solid with its spindle, and may have a diameter of pitch cylinder given by,—

Rule 181.

Diameter of pitch cylinder of worm may be . 3 × pitch of teeth.

Rule 181a.

Width of face of worm-wheel   ,,   . 2 ×   ,

Rule 181b.

Length of worm (actual thread)   ,,   . 3 ×   ,,

Suitable proportions for weigh-shaft levers are given in Figs. 54 and 55.

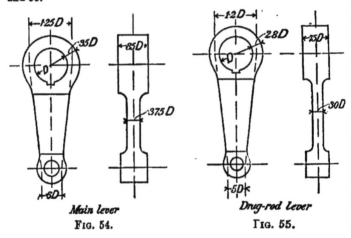

Main lever

FIG. 54.

Drag-rod lever

FIG. 55.

STEAM TURNING GEARS.

The best type of turning gear is the now all but universal double worm gear,—in which the second, or auxiliary, worm shaft is coupled direct to the crank-shaft of the small turning engine, which makes from 1200 to 2000 revolutions for one turn of the main engines, according to size. The size of cylinder, or cylinders, of turning engine should be such that, with say 50 lbs. of steam, the main engine can be turned one revolution in from 5 to 8 minutes, according to size, which gives,—with the above-mentioned ratios of gearing,—a speed of 250 revolutions per minute for the turning engine.

In the Navy the turning engines have usually two cylinders and two cranks, and are made reversible by means of links or a reversing valve, but one cylinder is quite sufficient, and a loose eccentric is all that is required for reversing.

In small vessels, where a special turning engine is not considered necessary, the auxiliary worm shaft may be driven from a donkey pump, or other auxiliary engine, by means of a pitch chain and wheels.

For ordinary three-crank triple engines, the sizes of cylinder for turning engine should be such as will satisfy the following relation,— the boiler pressure being assumed as 50 lbs. in all cases:—

Rule 182. $$\frac{A \times T}{D^3} = K,$$

where A = area of piston or pistons of turning engine, in sq. inches;
    T = travel of piston or pistons, in feet, while turning main
           engines through one revolution;
    D = diameter of crank-shaft of main engines;
    K = 14.

In Naval work the main wheels are commonly of cast-steel, but this is totally unnecessary,—as cast-iron is eminently suitable for the purpose; cast-iron has also the advantage of working well with wrought-steel worms, which cast-steel will not do.

However carefully a worm-wheel may be moulded there is sure to be more or less distortion in cooling, and it is therefore the cheapest in the long run to machine-cut the teeth.

The teeth should not be made too fine a fit; there is no objection to a good clearance.

The usual pitch for the teeth of the main wheel is as follows:—

For 6-inch to 8-inch crankshaft	.	. 2-inch pitch.
8½ " 11 " "	.	. 2¼ "
11½ " 14 " "	.	. 2½ "
14½ " 18 " "	.	. 3 "
above 18 " "	.	. 3½ "

When calculating the strength of the teeth, the number in gear at one time should not be taken at more than two.

Rule 183.—The length of tooth should not exceed ·6 × pitch, viz., ·26 × pitch outside the pitch circle, and ·34 × pitch within.

It is scarcely necessary to say that the pitch should be an even quarter or half inch, and that the diameter of the wheel must then be made such as will give the required number of teeth; also when the wheel is made in halves the number of teeth should be even.

The pitch of the auxiliary worm-wheel is usually 1½-inch to to 2¼-inch.

Rule 184.—The width of face of the wheels may be 1·75 × pitch of teeth when a solid worm is used, and 2·1 × pitch when the worm is loose on the spindle.

The diameter of the pitch cylinder of the worm may be,—

Rule 185.

When worm is forged with spindle . . 2·4 × pitch of teeth.

Rule 185a.

When worm is loose on spindle $\begin{cases} \text{forged steel} & . \quad 3 \times \quad " \\ \text{bronze} & . \; 3·3 \times \quad " \end{cases}$

Length of worm (actual thread) may be 3 × pitch of teeth.

Wheels up to 3 feet diameter are best made with a continuous plate, or disc, in lieu of arms; larger wheels may have six radial ribs on each side of the disc (the flanges counting as two, when the wheel is in halves), and the disc may then be lightened by cutting six circular, or nearly circular, holes in it.

The thickness of rim, of disc plate, and of ribs, may be same as thickness of tooth at pitch line.

Where possible the main wheel should be placed on a coupling. Where cast-steel brackets or bearings are used for the worm-shafts they should be bushed with gun-metal,—as "mild" steel shafts and cast-steel bearings have a great tendency to seize.

SCREW PROPELLERS.

Material of propeller.—For small merchant steamers propellers are almost invariably of cast-iron,—boss and blades being cast together in one piece. Experience has shown this material to be the cheapest in the end for this class of vessel, since,—though the life of the propeller seldom exceeds six years,—the first cost is low, and a spare one can be shipped in a few hours.

For larger merchant steamers, mail-boats, &c., the propellers are constructed of various materials, but the blades are usually cast separately, and attached to the boss by studs or screws ; the breakage of a blade can thus be made good without disturbing the boss,—a very important point, considering the practical impossibility of fitting a new boss properly without having the tail-shaft in the factory. Cast-steel blades have been used to some extent, but they cannot be recommended ; the first cost is considerably above that of cast-iron, and the average life is not more than four years.

Various attempts have been made to prevent corrosion,—and so prolong the life,—of iron and steel blades, by casting on coatings of white-metal (Johnstone's process), and by pinning on brass plates, but there is considerable difficulty in keeping out water from between the two metals, and the expenses,—direct and indirect,—are heavy.

There is therefore little doubt that, for first-class vessels, the gun-metal or bronze blade is the cheapest in the end ; and, owing to its smoother surface, &c., it is certainly the most *efficient* propeller for all vessels.

For Naval purposes propellers are almost invariably of gun-metal or bronze, and,—except sometimes in the case of torpedo boats,—the blades are always detachable.

Pitch of propeller.—For all ordinary cases,—take the intended speed of the ship in feet per minute (1 knot is equal to 101·33 feet per minute) ; increase it by one-tenth, and divide by the intended number of revolutions per minute ; the quotient will be the pitch in feet. For the large propellers of very bluff cargo boats,—increase the intended speed by one-twentieth only.

The general expression for the pitch, for any percentage (x) of apparent slip is then,—

Rule 186.

$$\text{Pitch of propeller} = \frac{S \times 101 \cdot 33}{R} \times \frac{100}{100 - x} = \frac{S}{R} \times \frac{10,133}{100 - x},$$

where S is the speed of vessel in knots, and R the revolutions per minute.

Apparent slip of propeller.—This is usually reckoned as a percentage of the nominal axial advance of the propeller (P × R), and amount in any given case is,—

Rule 187.

Apparent slip (per cent.) $= \dfrac{(P \times R) - (S \times 101\cdot33)}{P \times R} \times 100,$

where S is speed of ship in knots, P pitch of propeller in feet, and R revolutions per minute.

The apparent slip should generally be about 8 to 10 per cent., at full speed ; should it in any case be less than 5 per cent., it may be taken for granted that the propeller is not suited to the ship. In bluff cargo boats, however, the slip rarely exceeds 5 per cent., so that the preceding remark does not apply to them. A larger slip (say 12 to 18 per cent.) does not necessarily imply waste of power, and may be due to small diameter of propeller, but a larger percentage again would probably mean that the blade area or "surface" was too small. If this latter is the case, it at once becomes apparent when the indicated thrusts are plotted down and a curve drawn, for the curve will show a want of augmentation of thrust at the higher speeds. (Indicated thrust $= \dfrac{\text{I.H.P.} \times 33{,}000}{P \times R}$; for method of constructing curve, *see* page 21.)

Diameter of propeller.—For all ordinary cases the diameter of propeller should be that given by the following formula :—

Rule 188. Diameter of Propeller in ft. $= K \sqrt{\dfrac{\text{I.H.P.}}{\left(\dfrac{P \times R}{100}\right)^3}}$,

where P is the pitch in feet, R the revolutions per minute, and K a co-efficient, which has the values shown in the following Table:—

Table LII.—Propeller Co-efficients.

Description of Vessel.	Approximate speed in knots.	Number of screws.	Number of blades per screw.	Values of K.	Values of C.	Usual material of blades.
Bluff Cargo boats	8–10	Single	4	17 –17·5	19 –17·5	Cast-iron
Cargo boats —moderate lines	10–13	do.	4	18 –19	17 –15·5	Cast-iron
Pass. and Mail boats —fine lines	13–17	do.	4	19·5–20·5	15 –13	Cast-iron or steel.
Pass. and Mail boats —fine lines	do.	Twin	4	20·5–21·5	14·5–12·5	Cast-iron or steel.
Pass. and Mail boats —very fine	17–22	Single	4	21 –22	12·5–11	Gun-metal or bronze
Pass. and Mail boats —very fine	do.	Twin	3	22 –23	10·5– 9	Gun-metal or bronze
Naval vessels —very fine	16–22	do.	4	21 –22·5	11·5–10·5	Gun-metal or bronze
Naval vessels —very fine	do.	do.	3	22 –23·5	8·5– 7	Gun-metal or bronze
Torpedo boats —very fine	20–26	Single	3	24 –27	7– 5·5	Bronze or forged steel

Or, if $\left(\dfrac{P \times R}{100}\right)^{2\cdot3}$ be used instead of $\left(\dfrac{P \times R}{100}\right)^{2}$, the co-efficient (K) becomes practically constant, and the formula for four-bladed propellers becomes,—

$$\text{Diameter} = 7\cdot7 \sqrt{\dfrac{\text{I.H.P.}}{\left(\dfrac{P \times R}{100}\right)^{2\cdot3}}}.$$

For three-bladed screws,—multiply the diameter thus found by 1·05.

Blade area or surface of propeller.—For all ordinary cases satisfactory results will be obtained if the blade area be that given by the following rule :—

Rule 189.
Total developed area of blades in sq. ft. $= C \times \sqrt{\dfrac{\text{I.H.P.}}{\text{R.}}}$

where R is number of revolutions per minute, and C is a co-efficient having the values given in column six of Table LII.

In Naval vessels the centre of the propeller is usually more deeply immersed, in proportion to the diameter, and the revolutions are higher for a given power, than is the case in merchant vessels.

Number of blades.—Two bladed propellers are now seldom used on account of the vibration set up by the ever-varying resultant moment of the thrust reactions. Three bladed propellers work very steadily when the immersion is sufficient, and revolutions fairly high, but when the diameter of propeller becomes nearly as great as the draught of water, and revolutions are only moderate, four blades should be used, as being better balanced, and giving a more equable resultant moment of thrust reactions.

Thickness of propeller blades.—In determining the thickness of blade at the root, the two most important stresses to be considered are,—the bending stress due to the thrust reaction, and the tensile stress resulting from the centrifugal tendency of the mass of the blade. The first may be calculated from the mean normal thrust (see page 112), the blade being considered as a beam, fixed at one end and loaded at the other, the length of which is the distance from the root to the centre of pressure, and the load on which is the force, normal to the blade, due to one-third or one-fourth of the thrust reaction, according as the propeller is three or four bladed.

The stress resulting from the centrifugal tendency of the mass of the blade is also easily determined by the ordinary formula, viz.:—

$$C = \frac{wv^2}{gr}$$

where w is weight of blade in lbs.; v, velocity of centre of gravity in feet per second ; r, radius of centre of gravity in feet; and $g = 32$.

The material, whether of blade or of studs, is, of course, subject to the sum of these two stresses.

The thickness of blade can, however, be determined much more easily, and with a close approach to accuracy, by means of the following formula :—

Rule 190.
$$T = \sqrt{\frac{d^3}{n \times b} \times K}$$

where T is the thickness of blade at the centre of the shaft, or, in other words, the thickness the blade would have if continued through the boss to the centre of the shaft; d, the diameter of the tail shaft in inches; n, the number of blades; b, the breadth of blade in inches where it joins the boss, measured parallel to the shaft-axis, or, commonly, the length of boss; and K a co-efficient having values as follows :—

For cast-iron,	.	.	.	K = 4
,, cast-steel,	.	.	.	K = 1·5
,, gun-metal,	.	.	.	K = 2
,, high-class bronze,	.	.	K = 1·5

The thicknesses given by the above rule for gun-metal and bronze blades are such as are suited to propellers revolving at speeds usual in Naval work.

The great strength of cast-steel can only be partially taken advantage of, because propeller blades must not only be *strong* enough, but also *stiff* enough to prevent vibration; and, moreover, it corrodes rapidly.

See Table LIII., page 182, for results given by Rule 190.

The thickness of blades at the tip should be as follows :—

Rule 191.

Thickness of blade at tip—	Cast-iron,	.	.	·04D + ·4 inch.
,,	,,	Cast-steel,	.	.	·03D + ·4 ,,
,,	,,	Gun-metal,	.	.	·03D + ·2 ,,
,,	,,	High-class bronze,	.	·02D + ·3 ,,

where D is diameter of propeller in feet.

The following "rule of thumb," which is sometimes used, gives a very close approximation to the necessary thickness of blade :—

Thickness of blade (at shaft centre)—Cast-iron, ½ in. × No. of feet in diameter.

,, ,, ,, Cast-steel, ⅟₁₆ in. × No. of feet in diameter.

,, ,, ,, Gun-metal, ½ in. + ⅟₃₂ × No. of feet in diameter.

,, ,, ,, High-class bronze, ⅟₁₆ in. × No. of feet in diameter.

See Table LIIIa., page 183, for results given by Rule 191.

Table LIII.—Thickness of Propeller Blades.

Dia. of shaft in inches.	Solid cast-iron, 4 blades.		Loose bladed propellers.					
	Breadth of blade =length of boss=2·7×dia. of shaft.	Thickness of blade (at centre of shaft.)	Breadth of blade =·8×dia. of flange =1·8×dia. of shaft.	Thickness of blade at centre of shaft.				
				4 blades cast-iron.	4 blades steel or bronze.	3 blades steel or bronze.	4 blades gun-metal.	3 blades gun-metal.
inches.	inches.	inches.	inches.	inches.	inches.	inches.	inches.	inches.
6	16	3¾	10¾	4½	2¾	3¼	3¼	3⅝
7	19	4¼	12½	5¼	3¼	3⅝	3⅝	4¼
8	21½	5	14½	6	3⅝	4¼	4¼	4⅞
9	24	5½	16¼	6¾	4⅛	4¾	4¾	5½
10	27	6⅛	18	7½	4⅝	5¼	5¼	6⅛
11	30	6¾	19¾	8¼	5	5⅞	5⅞	6¾
12	32½	7⅜	21½	9	5½	6⅜	6⅜	7⅜
13	35	8	23½	9¾	6	6⅞	6⅞	7⅞
14	38	8½	25¼	10½	6⅜	7⅜	7⅜	8½
15	40½	9⅛	27	11¼	6⅞	8	8	9⅛
16	43	9¾	28¾	12	7⅜	8½	8½	9¾
17	46	10⅜	30½	12¾	7¾	9	9	10⅜
18	...	...	32½	13½	8¼	9½	9½	11
19	...	...	34¼	...	8⅝	10	10	11½
20	...	...	36	...	9⅛	10½	10½	12⅛

The thicknesses given in the above Table for cast-iron propellers assume that the metal has been at least twice melted; if metal that has been only once melted is used, the square of the thickness should be increased 10 per cent. Cast-steel and gun-metal are supposed to be to Admiralty specification.

If breadth of blade be made less than length of boss, thickness must, of course, be so increased that value of $b \times t^2$ shall always agree with that found from above Table.

The best method of setting out the longitudinal section of the blade is as follows:—

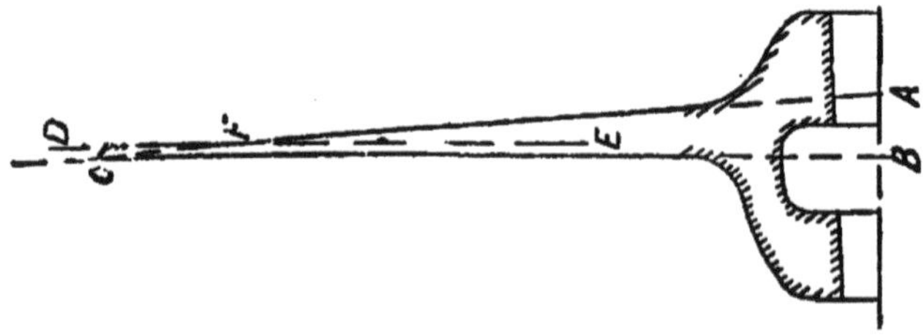

Fig. 56.

Determine AB (Fig. 56) from Table LIII., or by Rule 190; draw AC so as to give no thickness at tip; set off thickness at tip (from Table LIIIa.), and draw DE parallel to CB (at right angles to shaft axis), and then fill in the back of the blade near F so as to make DF die gradually into AC.

Table LIIIa.—Thickness of Propeller Blades at Tip.

Diameter of Propeller in feet.	Thickness at tip, in inches.			
	Cast-iron ·04D+·4	Cast-steel ·03D+·4	Gun-metal ·03D+·2	High-class Bronze ·02D+·3
7	11/16	5/8	7/16	7/16
8	3/4	5/8	7/16	7/16
9	3/4	11/16	7/16	7/16
10	13/16	11/16	1/2	1/2
11	13/16	3/4	1/2	1/2
12	7/8	3/4	9/16	9/16
13	15/16	13/16	9/16	9/16
14	15/16	13/16	5/8	9/16
15	1	7/8	5/8	5/8
16	1 1/16	7/8	11/16	5/8
17	1 1/16	15/16	11/16	5/8
18	1 1/8	15/16	3/4	11/16
19	1 3/16	1	3/4	11/16
20	1 3/16	1	13/16	11/16
21	1 1/4	1 1/16	13/16	3/4
22	1 1/4	1 1/16	7/8	3/4
23	1 5/16	1 1/8	7/8	3/4
24	1 3/8	1 1/8	15/16	13/16

Studs or screws for attaching blades to boss.—These are usually of naval brass (Muntz metal with addition of 1 per cent. tin), or of one of the stronger bronzes, for gun-metal or bronze propellers, and of mild steel for cast-iron or cast-steel propellers. They should be of size given by following formula:—

Rule 192.
$$a \times N \times r = \frac{T \times L}{K}$$

where a = area of one stud or screw at bottom of thread (sq. inches);
$\quad$ N = number of studs or screws for one blade (usually 7 to 11);
$\quad$ r = radius of studs or screws (in inches);
$\quad$ T = indicated thrust per blade = $\dfrac{\text{I.H.P.} \times 33,000}{\text{Pitch} \times \text{revs.} \times \text{No. of blades}}$;
$\quad$ L = ·6 × total length of blade (flange joint to tip);
$\quad$ K = $\begin{cases} 1700 \text{ for steel studs ;} \\ 1400 \text{ for Naval brass or bronze studs.} \end{cases}$

Blade flanges.—These should be of the following proportions :—

Rule 193.—Diameter of flange = 2·25 × diameter of tail shaft.

Rule 193a.

Thickness of flange $= \left\{ \begin{array}{l} ·85 \times \text{diam. of stud for G.M., bronze, or steel.} \\ 1·4 \times \qquad ,, \qquad\qquad ,, \quad \text{cast-iron.} \end{array} \right.$

Proportions of propeller boss.—The bosses of solid cast-iron propellers should be of the following proportions :—

Rule 194.—Length of boss = 2·7 × diameter of tail shaft.

Rule 194a.—Diameter of boss = 2·7 ×   ,,   ,,

Rule 194b.—The fore and aft section of the boss should be oval,—the principal radius being ·8 × diameter of boss.

The length of the tapered bore may be divided into three approximately equal parts, of which the two end ones will bear on the shaft, while the central one will be cored back.

The thickness of "shell" of the central, or cored out, portion of the boss should be,—

Rule 194c.

Thickness of boss = ·65 × thickness of blade at shaft axis.

Where the blade flanges are all within the sphere of the boss,—as is usual in Naval propellers,—the extreme diameter of boss is about 3·3 × diameter of tail shaft ; the length of such a boss should be the same as that of a cast-iron boss (Rule 194).

As previously mentioned under "Tail shafts," the taper to which the boss is bored out should be about 1 inch on the diameter for each foot of length, but never less than ¾-inch.

RADIAL PADDLE WHEELS.

These are now rarely employed except in small tugs, where first cost is a paramount consideration.

The effective diameter of a radial wheel is somewhat difficult to determine,—as it depends very much on the form of float, the amount of immersion, the waves set up by the wheel, &c.,—but it is usually measured between the centres of the opposite floats.

The slip of a radial wheel is from 15 to 30 per cent., according to the size of float.

For method of determining size of wheel *see* under "Feathering paddle wheels."

The area of one float, in square feet, may be found by the following rule :—

Rule 195.   Area of one float $= \dfrac{\text{I.H.P.}}{} \times \text{C}$

where **D** is the effective diameter in feet ; I.H.P. the total indicated horse-power ; and C, a multiplier varying from ·25 in tugs to ·175 in fast running light steamers.

Rule 195a.

The breadth of float is usually about $\dfrac{\text{Length of float}}{4}$.

Rule 195b. Thickness of float may be $\dfrac{\text{Breadth of float}}{8}$.

Rule 195c.—The number of floats varies directly with the diameter, and there should be one for every foot of diameter.

The floats should be of elm, or other tough strong wood that will withstand the action of the water.

The corners should be rounded, and the dipping edges beveled at the back, so as to enter the water as easily as possible.

FEATHERING PADDLE WHEELS.

To design a feathering wheel so that the floats shall enter edgeways when going at full speed, take P (Fig. 57, page 186), a point on the face of the float just entering the water, draw PA parallel to the water line, and cut off PA to represent the speed of the ship through the water, to some convenient scale ; draw PB tangent to the circle through P, whose centre is the centre of the wheel, and cut off PB to represent the speed of the wheel on that circle ; complete the parallelogram, and the resultant PR is the direction in which the float enters the water.

Produce RP, and draw parallel to it, at a distance from it equal to distance from gudgeon centre to face of float, a line cutting the circle of gudgeon centres in G.

Draw GH at right angles to PF, set off the breadth of the float, and mark off GL equal to the length of lever required.

Now draw another float whose face is vertical and immediately under the centre of the wheel, and the end of whose lever is M ; and with centres M and L, and radius GC, draw two arcs of circles intersecting in E. Then E is the centre of the eccentric pin, and LE the length of the radius rods ; and the ends of all the other float levers will lie in a circle struck from E as centre, with radius LE.

The diameter of wheel (in feet) is determined as follows :—

> Let K = speed of vessel in knots ;
> S = percentage of slip ;*
> R = intended number of revolutions per minute.

Then,—

Rule 196.—Diameter of wheel at axis of floats $= \dfrac{K\,(100+S)}{3\cdot14 \times R}$.

* Usually 15 to 25 per cent.

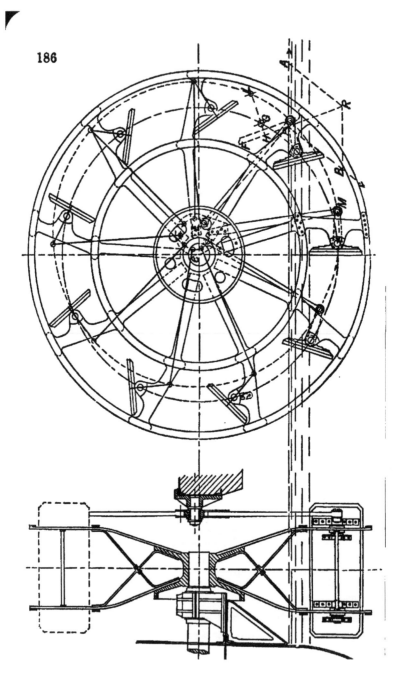

The diameter must also, of course, be such as will suit the structure of the ship, and will give the necessary immersion of the float.

Immersion of floats.—When a vessel works always in smooth water, the immersion of the top edge of float should not exceed $\dfrac{\text{Breadth of float}}{8}$; and for general service at sea an immersion of $\dfrac{\text{Breadth of float}}{2}$ is sufficient. If the vessel is intended to carry cargo, the immersion, when light, need not exceed 2 or 3 inches, and should not be more than the breadth of the float when at the deepest draught, —as the efficiency of the wheel falls off rapidly as the immersion increases.

Number of floats.—The number of floats should be the nearest whole number to the value given by the following formula :—

Rule 197.   Number of floats $= \dfrac{60}{\sqrt{R}}$

where R is number of revolutions per minute.

Area of float.—The area of float should be that given by the following formula :—

Rule 198.

Area of one float (in square feet) $= \dfrac{\text{I.H.P.} \times 33,000 \times K}{N \times (D \times R)^3}$

where D = diameter of wheel to axes of floats, in feet ;
     N = number of floats in one wheel ;
     R = number of revolutions per minute ;
whilst K has the following values:—
     For vessels plying always in smooth water .   .   . K = 1200
     For sea-going steamers   .   .   .   .   . K = 1400
     For tugs, and such craft as require to stop and start $\Big\}$ K = 1600
     frequently in a tide-way   .   .   .   .   .

Note.—It will be quite accurate enough, and more convenient, if the last four figures of the cube $(D \times R)^3$ be taken as ciphers.

Proportions of float.—The breadth and thickness of float may be fixed by the following rules :—

Rule 199.   Breadth of float $= \cdot 35 \times$ length.

Rule 200.

Thickness of wood float $= \dfrac{\text{Breadth}}{10 \text{ to } 12}$, according to nature of service.

Rule 200a.

Thickness of steel float $= \cdot 16 \times$ breadth in feet $+ \cdot 15$-inch.

When an iron or steel float is curved,—

Radius of curvature of float $=$ effective diameter of wheel.

Paddle-wheel frames.—These must be strong enough, in every case, to transmit the power of the engines to the water, and must also be stiff enough to work without undue springing or vibration. As the power of the engines is approximately proportional to the cube of the diameter of the shaft, the strength of the wheel frames should be in the same proportion, and the section of the arm would be given by the equation,—

$$t \times b^2 = \frac{K}{n} \times D^3,$$

where t and b are respectively thickness and breadth of arm, n total number of arms in one wheel, D diameter of inner, or engine, journal of paddle-shaft, and K a co-efficient.

But when the diameter of shaft is small, this gives a section of arm which would not be stiff enough for practical purposes without more cross-bracing than is usual in small wheels; in practice, small wheels are commonly made without any diagonal bracing at all,—the arms being made proportionately stronger,—as experience has shown that it is not advisable to use sections below a certain degree of lightness.

The formula, therefore, requires the addition of a constant quantity, and, if written as follows, gives very satisfactory results:—

Rule 201.　　　　$t \times b^2 = \frac{K}{n} \times (D^3 + 600).$

The value of K, for section of arm near boss, is 62,—and the other symbols are as defined above.

The formula may be used exactly as it stands for ordinary oscillating engines with sectional crank-shafts (Class 3, Table XXXIXa.), but the shafts of other types are made larger, on account of the greater bending stresses to which they are subject, and what may be called the "equivalent cube of diameter" must be obtained before Rule 201 can be applied. The equivalent D^3 is found by taking the cube of the actual diameter of the journal, and, if for example the engine be of Class 2 (Table XXXIXa.), taking $\frac{50}{58}$ths of it,—or multiplying it by 50 and dividing by 58.

Theoretically, the arms may be made lighter as they recede from the boss, but in practice it is usual to make them parallel,—partly for convenience of manufacture, and partly because, when diagonal stays are fitted, the arm is weakened near the inner ring by the necessary bolt holes.

When there is no outer rim, or ring, the portion of each arm beyond the inner rim must be strong enough to carry its own load without any help from the other arms, and must be considered as a beam fixed at one end and loaded at the other. As there is a tendency to contrary flexure of the arm within the ring, the arm is generally made widest at the ring,—tapering gradually towards the boss, and quickly towards the gudgeon centre or end. The type of wheel is heavy and expensive, and the floats are deprived of such protection as is given by the outer rims, and are therefore so much the more liable to injury.

Rule 202.—The ratio of breadth of arm to thickness is usually about 5 to 1. The arms must, of course, be locally widened out where the connection to the rim, or rims, is made. When there are two rims, from three to five bolts may be used at each joint, but with one rim only there should be six to ten bolts, according to size of wheel.

Nothing less than a ⅝-inch bolt should be used on ordinary wheels, and both heads and nuts should be square; also, all holes should be rymered out, and all bolts be driving fit.

Rule 203.

The section of inner ring may be ·7 × section of arm at boss.

Rule 203a.

The section of outer   ,,   ·6 ×   ,,   ,,

Rule 203b.

The section of only   ,,   1 ×   ,,   ,,

The proportions of breadth to thickness may be,—

Rule 204.   Inner ring . . $\dfrac{b}{t} = 5.$

Rule 204a.   Outer ring . . $\dfrac{b}{t} = 4.$

Rule 204b.   Only ring . . $\dfrac{b}{t} = 6\cdot5.$

In large wheels, the inner rims (or arms near inner rims) should be stayed back to the opposite sides of the boss; the section of stays may be,—

Rule 205.

Section of diagonal stay (if rectangular) ·7 × section of inner rim.

Rule 205a.

Section of diagonal stay (if circular)   ·5×   ,,   ,,

Rule 206.—Horizontal round ties, or distance rods, should also be fitted between each pair of gudgeon bearings, and as close as possible to them; they may have a diameter of 1·3 × thickness of arm.

It is a common practice to fix the centres of the gudgeons, on which the floats are hinged, slightly off the centre lines of the floats,—nearer the outer edges,—as the strains on the feathering gear are thereby reduced ; the amount is to a great extent optional with the designer, but varies in practice from one-tenth to one-twentieth of the breadth of the float.

The whole of the gudgeons and pins should be cased with brass, and should work in lignum-vitæ bushes ; unless the vessel is to work in sandy or muddy water,—when iron pins working in white-metal bushes, or pins cased with white-metal working in gun-metal bushes, will give better results.

The outer bearing for the paddle-shaft may (as stated in Rule 94) have a length of 1·5 to 2 × diameter of journal,—according to nature of service and consequent weight of wheel ; it should be strongly made and firmly fixed, as, in addition to the weight of wheel, the whole of the thrust is taken by it. The magnitude and direction of the resultant force on the bearing can be easily calculated, but, as its direction is always below the centre, the caps do not need to be very substantial ; the bearing should admit of adjustment in the line of the resultant.

The lubricant is mainly water, but oiling pipes and tallow boxes should be fitted.

A stuffing box should be fitted round the shaft where it passes through the ship's skin.

SEA VALVES.

All sea connections should be simple in construction, and of very ample strength in all parts, and the necks, especially, should be well bracketed to the flanges attaching the valves to the ship's side.

Sea suction, or inlet, valves.—In the Merchant Service these are usually ordinary cast-iron stop-valves, opening inwards ; in the Navy the Kingston valve is not now so much used, and ordinary stop-valves are also generally employed, but no material other than best gun-metal must be used in connection with them.

Gun metal strainer gratings should be fitted to all inlet valves, the mesh not exceeding half an inch in breadth, and the total area through the grating should be at least 20 per cent. greater than the net area of the pipe.

Some makers place the gratings of the larger valves on the inboard side of the valve, and provide means for their withdrawal and examination whilst the ship is still afloat ; if this is not done, there should be either a steam jet or a donkey delivery pipe fitted for clearing away any weed, &c. that may become fixed in the grating.

The number of valves on the ship's side should be as small as possible, and, where the nature of the service permits, one sea-valve may supply two or more systems of piping, shut-off valves being fitted on the inboard side of the sea-valve.

Where possible, it is better to take the water-service supply from the delivery side of the circulating pump than from an independent valve on the ship's side, as a better supply is obtained in this way.

The method of attachment to the ship's skin, recommended by Lloyd's, is shown in Fig. 58, and the Admiralty method in Fig. 59.

Blow-off valves.—These should always be gun-metal stop-valves opening inwards, and should have spigots passing through the ship's plating. If any guard cock is fitted, it should be placed between the valve on the boiler and the sea-valve, and independent of both.

The method of attachment recommended by Lloyd's is shown n Fig. 60 (page 192).

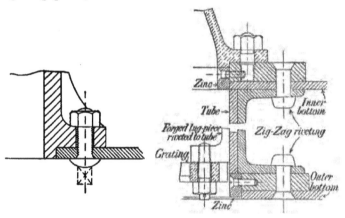

FIG. 58.   FIG. 59.

Where the valve has to open a communication through a double bottom, as is usually the case in Naval work, one of the Kingston type is generally employed, attached to the outer bottom as shown in Fig. 61 (page 192), and made tight by means of a stuffing box, where it passes through the inner bottom.

The use of cast-iron for boxes or casings of blow-off valves is forbidden by paragraph 121 of the Board of Trade rules.

Discharge valves.—Here also, with the object of reducing the number of openings in the ship's skin, the smaller valves should, where possible, discharge into the casings of the larger ones, the outlets from the latter being increased in area as necessary.

In the Merchant Service, lifting non-return valves with cast-iron casings are generally used ; but in the Navy, though the same type of valve is used, the only material permitted is gun-metal.

Provision should be made for hanging up the valve, but, when the circulating pumps are of the reciprocating type, it should not be possible to fix it on its seat in any way.

All discharge valves should have spindles passing through the covers, and cross handles for turning them round on their seats.

In some cases, where room is pinched, a swinging flap-valve may

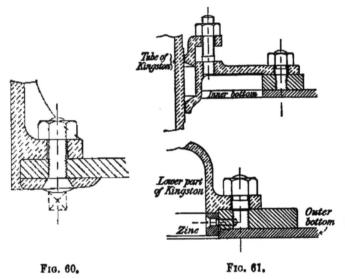

FIG. 60.   FIG. 61.

be employed with advantage; and where the pumps are of the centrifugal type, either valves of this class or sluice valves should be used, in order to reduce the resistances as much as possible.

There should always be a spigot passing through the skin plating of the vessel, and, where the valve is large, it is very desirable to have a doubling plate fitted inside of, and riveted to the skin plate.

The Board of Trade regulations referring to sea connections are as follows :—

137. All inlets or outlets in the bottom or side of a vessel near to, at, or below the deep-load water line other than the outlets of water-closet, soil, scupper, lavatory and urinal pipes, should have cocks or valves fitted between the pipes and the ship's side or bottom ; such cocks or valves should be attached to the skin of the ship, and be so arranged that they can be easily and expeditiously opened or closed at any time ; and the cocks, valves, and the whole length of the pipes should be accessible at all times.

Cocks or valves standing exceptional distances from the ship's plating, that is where the necks are longer than is necessary for making the joint, should not be passed without the sanction of the Board of

Trade, and one condition of their being passed is that they should be made of gun-metal, and well bracketed.

Although in the case of vessels which have already been granted passenger certificates it is very desirable that the cocks and valves above referred to should be so arranged that they can be opened and closed expeditiously by hand, and that they and the whole length of the pipes referred to should be accessible at all times, still in vessels in which there is a difficulty in attaining this object, a strict compliance with the above may be dispensed with, if after a consultation between the Principal Officer and the Surveyor they are satisfied that the difficulty is serious, and that the arrangements existing are on the whole safe.

The Principal Officer should in such cases always send a special report to the Board of Trade, with the Surveyor's report and statement of the arrangements on board, and of the reasons for authorising a departure from the ordinary conditions.

138. With a view to the prevention of accidents to boilers through the blow-off cocks being left open after the boiler is run up, and to prevent water getting accidentally or intentionally into the ship by cocks being left open, all blow-off cocks and sea connections below the plates or out of sight, should be fitted with a guard over the plug, with a featherway in the same, and a key on the spanner, so that the spanner cannot be taken out unless the plug or cock is closed. When cocks are in sight guards need not be fitted provided the spanners are secured to the plugs by pins. The spanners should not be shrunk on the heads of the plugs. One cock should be fitted to the boiler, and another cock on the skin of the ship or on the side of the Kingston valve.

139. In all cases where pipes are so led or placed that water can run from the boiler or the sea into the bilge, either by accidentally or intentionally leaving a cock or valve open, they should be fitted with a non-return valve and a screw not attached, but which will set the valve down in its seat when necessary ; the only exception to this is the fireman's ash cock, which must have a cock or valve on the ship's side, and be above the stoke-hole plates.

140. The exhaust pipe for the donkey engine should not be led through the ship's side, but should be led on deck or into the main waste steam pipe ; * and in all cases should have a drain-cock on it. This will prevent water getting into the pipe when the ship is deeply laden or rolling.

141. In the case of the outlets of watercloset, soil, scupper, lavatory and urinal pipes which are below the weather deck, there should be an elbow of good substantial metal other than cast-iron or lead ; and the pipe connected with this elbow should, if of lead, have a sufficient bend to provide for expansion in the pipe or any movement from the

* AUTHORS' NOTE.—This plan renders overhaul of main safety-valves difficult when vessel is in port ; a separate waste pipe should always be fitted.

13

working of the ship.   Pipes, no matter of what material they may be constructed, should never be fitted in a direct line between the apertures in the ship's side and its connection with the deck, or closet, or other fittings.   Unless the watercloset and scupper pipes and their outlets are fitted in the way hereby required, or in a manner that will in the opinion of the Surveyor and of the Principal Officer be more or equally efficient and safe, the Surveyor should refuse to grant a Declaration.   Where closets are fitted below the water line, as in the case of pumping closets, plans should be specially submitted for approval. For closets with their outlets above the weather deck, no special regulations are necessary.   The pipes and valves should be protected from the cargo by a substantial casing of wood or iron, which need not be watertight.

This clause applies only to the case of sea-going steamers coming under survey for the first time.

Lloyd's rules relating to sea connections are as follows : —

33. All discharge pipes to be, if possible, carried above the deep load line, and to have discharge valves fitted on the plating of the vessel in an accessible position.

34. No pipes to be carried through the bunkers without being properly protected.

40. All sea-cocks to be fitted on the plating of the vessel above the level of the stokehole and engine-room platforms, or attached to Kingston valves of a height sufficient to lift them up to the level of these platforms.

41. The bolts securing all cocks or sea connections to the plating of the vessel are to be tapped into the plating of the vessel or fitted with countersunk heads.

42. The blow-off cocks on the plating of the vessel are to be fitted with spigots passing through the plating, and a brass or gun-metal ring on the outside.   The cocks are to be so constructed that the key or spanner can only be taken off when the cock is shut.

See also paragraphs 2 and 4 of section 27 of Lloyd's Rules. Page 274.

PIPES AND PIPE ARRANGEMENTS.

The thicknesses of copper and brass tubes or pipes are usually specified or given in terms of one of the "wire gauges."   Table LIV. gives three of the most widely used of these "wire gauges," and their equivalents in decimals of an inch, but it is desirable in all new work to use only the Legal Standard Gauge (L.S.G.) as now by law established, as a multiplicity of standards tends only to cause confusion and annoyance.

TABLE LIV.—WIRE GAUGES AND THEIR EQUIVALENTS. 195

Table LIV.—Wire Gauges and their equivalents.

L. S. G. Legal Standard Wire Gauge.		B. W. G. Birmingham Wire Gauge.		W. W. G. Whitworth Wire Gauge.	
No. of Gauge.	Equivalent in ins.	No of Gauge.	Equivalent in ins.	No. of Gauge.	Equivalent in ins.
7/0	·500	5/0	·500	300	·300
6/0	·464	4/0	·460	280	·280
5/0	·432	3/0	·420	260	·260
4/0	·400	2/0	·380	250	·250
3/0	·372	1/0	·340	240	·240
2/0	·348	1	·300	220	·220
0	·324	2	·280	200	·200
1	·300	3	·260	180	·180
2	·276	4	·240	165	·165
3	·252	5	·220	150	·150
4	·232	6	·200	135	·135
5	·212	7	·180	125	·125
6	·192	8	·164	120	·120
7	·176	9	·148	110	·110
8	·160	10	·132	100	·100
9	·144	11	·120	95	·095
10	·128	12	·108	90	·090
11	·116	13	·096	85	·085
12	·104	14	·084	80	·080
13	·092	15	·072	75	·075
14	·080	16	·064	70	·070
15	·072	17	·056	65	·065
16	·064	18	·048	60	·060
17	·056	19	·040	55	·055
18	·048	20	·036	50	·050
19	·040	21	·032	45	·045
20	·036	22	·028	40	·040
21	·032	23	·024	38	·038
22	·028	24	·022	36	·036
23	·024	25	·020	34	·034
24	·022	26	·018	32	·032
25	·020	27	·016	30	·030
26	·018	28	·014	28	·028
27	·0164	29	·013	26	·026
28	·0148	30	·012	24	·024
29	·0136	31	·011	22	·022
30	·0124	32	·010	20	·020
31	·0116	33	·009	19	·019
32	·0108	34	·008	18	·018
33	·0100	35	·007	17	·017
34	·0092	36	·006	16	·016
35	·0084	37	·005	15	·015
36	·0076	38	·004	14	·014
37	·0068	39	·003	13	·013
38	·0060	40	·002	12	·012

The following Table shows the Legal Standard Wire Gauge and its equivalents in millimetres :—

Table LV.—Legal Standard Wire Gauge and Metric Equivalents.

Number of Gauge.	Equivalents.		Number of Gauge.	Equivalents.	
	Inch.	Millimetre.		Inch.	Millimetre.
7/0	·500	12·700	23	·024	·610
6/0	·464	11·785	24	·022	·559
5/0	·432	10·973	25	·020	·508
4/0	·400	10·160	26	·018	·457
3/0	·372	9·449	27	·0164	·4166
2/0	·348	8·839	28	·0148	·3759
0	·324	8·229	29	·0136	·3454
1	·300	7·620	30	·0124	·3150
2	·276	7·010	31	·0116	·2946
3	·252	6·401	32	·0108	·2743
4	·232	5·893	33	·0100	·2540
5	·212	5·385	34	·0092	·2337
6	·192	4·877	35	·0084	·2134
7	·176	4·470	36	·0076	·1930
8	·160	4·064	37	·0068	·1727
9	·144	3·658	38	·0060	·1524
10	·128	3·251	39	·0052	·1321
11	·116	2·946	40	·0048	·1219
12	·104	2·642	41	·0044	·1118
13	·092	2·337	42	·0040	·1016
14	·080	2·032	43	·0036	·0914
15	·072	1·829	44	·0032	·0813
16	·064	1·626	45	·0028	·0711
17	·056	1·422	46	·0024	·0610
18	·048	1·219	47	·0020	·0508
19	·040	1·016	48	·0016	·0406
20	·036	·914	49	·0012	·0305
21	·032	·813	50	·0010	·0254
22	·028	·711	...	...	...

Table LVI. has been drawn up as a general guide indicating the best thicknesses of pipes for various pressures and purposes, and may be taken as giving thoroughly reliable strengths for all ordinary marine work :—

TABLE LVI.—THICKNESSES OF COPPER PIPES. 197

Table LVI.—Thicknesses of Copper Pipes (L.S.G.).

Diameter of pipe in inches.	Steam pipes. Boiler pressures in lbs.						Auxiliary exhaust pipes.	Waste steam pipes.	Main water pipes.	Bilge suction and disch. feed suction and fire service.	Diameter of pipe in inches.	Main eduction and air-pump suction.
	200	180	155	125	85	50						
22	...	...	...	...	0	7	...	...	4	...	35	3
21	...	...	...	...	1	7	...	...	4	...	33	3
20	...	...	...	...	1	7	...	...	5	...	32	4
19	...	...	...	...	2	8	...	...	5	...	29	4
18	...	...	...	4/0	2	8	...	13	5	...	28	5
17	...	...	6/0	3/0	3	8	...	13	6	...	25	5
16	...	7/0	5/0	2/0	3	8	...	14	6	...	24	6
15	...	6/0	4/0	0	4	8	...	14	7	...	21	6
14	7/0	5/0	3/0	0	4	9	11	14	7	...	20	7
13	6/0	4/0	2/0	1	5	9	11	14	7	...	17	7
12	5/0	3/0	0	2	5	9	11	14	8	...	16	8
11	4/0	2/0	1	3	6	9	11	15	8	...	15	8
10	3/0	0	1	3	6	9	11	15	9	...	14	9
9	2/0	1	2	4	7	10	11	15	9	...	13	9
8	1	2	3	5	8	10	12	15	9	...	12	10
7	2	3	4	6	8	11	12	15	10	9	11	10
6	3	4	5	7	9	11	12	15	10	10	10	11
5	5	6	6	8	9	11	12	15	11	10	9	11
4	6	7	7	9	10	12	12	15	11	11	8	12
3	8	9	9	10	11	12	12	15	12	11	7	12
2	10	11	11	11	12	12	12	15	12	12	6	13
1	13	13	13	13	13	13	13	15	13	13	5	13

Blow-off and scum pipes.									
Diameter of pipe in inches, . .	1	1¼	1½	1¾	2	2¼	2½	2¾	3
Thickness, L.S.G.	10	10	10	9	9	8	8	7	7

Feed discharge pipes to be as steam pipes for 30 per cent. higher pressure ; but in no case to be taken lower than 125 lbs.

Receiver pipes to be as steam pipes for half the test pressure of the cylinder to which they lead steam ; but in no case to be taken lower than 50 lbs.

The above ganges refer to straight pipes only ; bends to be suitably strengthened.

The thicknesses of copper pipes requisite to comply with the Board of Trade rule (see page 204) are as follows :—

Table LVII.—Thickness of Copper Pipes by Board of Trade rule.

Diameter of pipe in inches.	150 lbs.		160 lbs.		170 lbs.		180 lbs.		190 lbs.		200 lbs.	
	Brazed.	Solid.	Brazed.	Solid.	Brazed.	Solid.	Brazed.	Solid.	Brazed.	Solid.	Brazed.	Solid.
1	13	16	13	16	13	16	12	16	12	16	12	15
1¼	12	16	12	15	12	15	12	15	12	15	12	14
1½	12	15	12	15	11	14	11	14	11	14	11	13
1¾	11	14	11	14	11	13	11	13	10	13	10	13
2	11	13	11	13	10	13	10	13	10	12	9	12
2¼	10	13	10	13	10	12	9	12	9	12	9	11
2½	10	12	9	12	9	12	9	11	9	11	8	11
2¾	9	12	9	12	9	11	8	11	8	10	8	10
3	9	11	9	11	8	11	8	10	8	10	7	9
3¼	9	11	8	10	8	10	8	10	7	9	7	9
3½	8	10	8	10	7	9	7	9	7	9	6	8
3¾	8	10	7	9	7	9	7	9	6	8	6	8
4	7	9	7	9	7	9	6	8	6	8	5	7
4½	7	9	6	8	6	8	5	7	5	7	5	6
5	6	8	5	7	5	7	5	6	4	6	4	5
5½	5	7	5	6	4	6	4	5	3	5	3	4
6	5	6	4	6	4	5	3	5	3	4	2	4
6½	4	5	3	5	3	4	2	4	2	3	1	3
7	3	5	3	4	2	4	2	3	1	3	1	2
7½	3	4	2	4	2	3	1	2	1	2	0	1
8	2	4	2	3	1	2	0	2	0	1	2/0	1
8½	2	...	1	...	0	...	0	...	2/0	...	2/0	...
9	1	...	0	...	0	...	2/0	...	2/0	...	3/0	...
9½	1	...	0	...	2/0	...	2/0	...	3/0	...	4/0	...
10	0	...	2/0	...	2/0	...	3/0	...	4/0	...	4/0	...
10½	0	...	2/0	...	3/0	...	4/0	...	4/0	...	...	...
11	2/0	...	3/0	...	4/0	...	4/0	...	5/0	...	...	...
11½	3/0	...	3/0	...	4/0	...	5/0	...	...	...	...	...
12	3/0	...	4/0	...	5/0	...	5/0	...	...	...	...	...
12½	4/0	...	4/0	...	5/0	...	...	...	...	...	...	...
13	4/0	...	5/0	...	5/0	...	...	...	...	...	...	...
13½	4/0	...	5/0	...	...	...	...	...	...	...	...	...
14	5/0	...	6/0	...	...	...	...	...	...	...	...	...
14½	5/0	...	...	...	...	...	...	...	...	...	...	...
15	6/0	...	...	...	...	...	...	...	...	...	...	...

"Note.—The Board of Trade rules (paragraph 119) state that "Feed pipes should "be made sufficient for a pressure 20 per cent. in excess of the boiler pressure."

TABLE LVIII.—THICKNESS OF WROUGHT-IRON PIPES. 199

The thicknesses of wrought-iron lap-welded pipes requisite to comply with the Board of Trade rule are shown in the following Table :—

Table LVIII.—Thickness of Wrought-Iron Pipes by Board of Trade rule.

Diameter of pipe in inches.	Thickness in inches (Decimal and nearest thirty-second above).											
	150 lbs.		160 lbs.		170 lbs.		180 lbs.		190 lbs.		200 lbs.	
5	·250	¼	·250	¼	·250	¼	·250	¼	·250	¼	·250	¼
5½	"	"	"	"	"	"	"	"	"	"	"	"
6	"	"	"	"	"	"	"	"	"	"	"	"
6½	"	"	"	"	"	"	"	"	"	"	"	"
7	"	"	"	"	"	"	"	"	"	"	"	"
7½	"	"	"	"	"	"	"	"	·250	¼	·250	¼
8	"	"	"	"	"	"	·250	¼	·253	9/32	·267	9/32
8½	"	"	"	"	·250	¼	·255	9/32	·269	9/32	·283	5/16
9	"	"	·250	¼	·255	9/32	·270	9/32	·284	5/16	·300	5/16
9½	"	"	·253	9/32	·269	9/32	·285	5/16	·300	5/16	·317	11/32
10	·250	¼	·267	9/32	·283	5/16	·300	5/16	·316	11/32	·333	11/32
10½	·2625	9/32	·280	9/32	·297	5/16	·315	11/32	·332	11/32	·350	⅜
11	·275	9/32	·293	5/16	·311	5/16	·330	11/32	·348	⅜	·367	⅜
11½	·2875	5/16	·307	5/16	·326	11/32	·345	⅜	·364	⅜	·383	13/32
12	·300	5/16	·320	11/32	·340	11/32	·360	⅜	·379	13/32	·400	13/32
12½	·3125	5/16	·333	11/32	·354	⅜	·375	⅜	·395	13/32	·417	7/16
13	·325	11/32	·347	⅜	·368	⅜	·390	13/32	·411	7/16	·433	7/16
13½	·3375	11/32	·360	⅜	·382	13/32	·405	13/32	·427	7/16	...	...
14	·350	⅜	·373	⅜	·396	13/32	·420	7/16	·443	15/32	...	...
14½	·3625	⅜	·387	13/32	·410	7/16	·435	7/16	...	...	...	...
15	·375	⅜	·400	13/32	·425	7/16	·450	15/32	...	...	...	...
15½	·3875	13/32	·413	7/16	·439	15/32	...	...	...	...	...	...
16	·400	13/32	·427	7/16	·453	15/32	...	...	...	...	...	...
16½	·4125	7/16	·440	15/32	...	...	...	...	...	...	...	...
17	·425	7/16	·453	15/32	...	...	...	...	...	...	...	...
17½	·4375	7/16	...	...	...	...	...	...	...	...	...	...
18	·450	15/32	...	...	...	...	...	...	...	...	...	...

Good proportions for the flanges of copper pipes for modern triple engines are given in the following Table :—

Table LIX.—Flanges for Copper Pipes.

Diameter of pipe in inches.	Diameter of flange.	Thickness of flange.		Diameter of bolts.	Radius of bolt circle.	Number of bolts.			Pitch of bolts. (nearest sixteenth.)		
		Steam, feed, blow-off, &c.	Auxiliary exhaust, water pipes, waste and eduction, &c.			Steam, feed, blow-off, &c.	Auxiliary exhaust and water pipes.	Waste and eduction pipes.	Steam, feed, blow-off, &c.	Auxiliary exhaust and water pipes.	Waste and eduction pipes.
ins.	ins.	in.	in.	in.	ins.				ins.	ins.	ins.
1	4	½	⅜	½	1⅜	4	4	3	1 15/16	1 15/16	2⅜
1¼	4¼	"	"	"	1½	4	4	3	2⅛	2⅛	2⅝
1½	4½	9/16	7/16	"	1⅝	4	4	4	2 5/16	2 5/16	2 5/16
1¾	4¾	"	"	"	1¾	5	4	4	2 1/16	2½	2½
2	5	"	"	"	1⅞	5	5	4	2 3/16	2 3/16	2⅝
2¼	5⅞	⅝	½	⅝	2¼	5	5	4	2⅝	2⅝	3 3/16
2½	6⅛	"	"	"	2⅜	6	5	4	2⅜	2 13/16	3⅜
2¾	6⅜	"	"	"	2½	6	5	4	2½	2 15/16	3½
3	6⅝	11/16	9/16	"	2⅝	6	5	5	2⅝	3 1/16	3 1/16
3¼	6⅞	"	"	"	2¾	7	6	5	2⅜	2¾	3¼
3½	7⅛	"	"	"	2⅞	7	6	5	2½	2⅞	3⅜
3¾	7⅜	"	"	"	3	7	6	5	2⅝	3	3½
4	7⅝	¾	⅝	"	3⅛	8	6	6	2⅜	3⅛	3⅛
4½	8⅛	"	"	"	3⅜	8	7	6	2 9/16	2 15/16	3⅜
5	8⅝	"	"	"	3⅝	9	7	7	2½	3⅛	3⅛
5½	9⅛	"	"	"	3⅞	9	8	7	2⅝	2 15/16	3⅜
6	10⅜	13/16	11/16	¾	4⅜	9	7	6	3	3 13/16	4⅜
6½	10⅞	"	"	"	4⅝	9	7	6	3 3/16	4	4⅝
7	11⅜	"	"	"	4⅞	10	7	6	3	4 3/16	4⅞
7½	11⅞	"	"	"	5⅛	10	8	7	3 3/16	3 15/16	4 7/16
8	12⅜	⅞	¾	"	5⅜	11	8	7	3 1/16	4⅛	4⅝
8½	12⅞	"	"	"	5⅝	11	9	7	3 3/16	3⅞	4⅞
9	13⅜	"	"	"	5⅞	12	9	8	3 1/16	4	4½
9½	13⅞	"	"	"	6⅛	12	9	8	3 3/16	4 3/16	4 11/16
10	14⅜	"	"	"	6⅜	13	10	9	3 1/16	3 15/16	4⅜
11	16¼	15/16	13/16	⅞	7 3/16	13	10	9	3 7/16	4½	5
12	17¼	"	"	"	7 11/16	14	11	10	3 7/16	4⅜	4⅞
13	18¼	"	"	"	8 3/16	16	12	10	3 5/16	4 5/16	5⅛
14	19¼	"	"	"	8 11/16	16	13	11	3 7/16	4 3/16	5
15	20¼	"	"	"	9 3/16	17	14	12	3⅜	4⅛	4⅞
16	21½	1	⅞	"	9 13/16	18	14	12	3 7/16	4⅜	5⅛
17	22½	"	"	"	10 5/16	19	14	12	3 7/16	4⅝	5⅜
18	23½	"	"	"	10 13/16	20	15	13	3⅜	4½	5 3/16

The flanges, of which particulars are given in the above Table, are designed either for coupling copper pipe to copper pipe, or copper pipe to gun-metal casting of corresponding strength, as indicated in Fig. 62; for connections with cast-iron pipes or valves the flange must be larger, unless studs are used.

Table LX. (page 202) gives suitable thicknesses for gun-metal pipes, elbows, T pieces, &c., and will be of assistance in designing gun-metal stop-valves, expansion stuffing boxes, &c.

The basis and method of construction of the Table are fully explained in the following memoranda :—

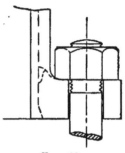

FIG. 62.

Memoranda.—H.P. includes feed and bilge delivery, blow-off, scum, steam, fire service, and all sea valves.

L.P. includes all suctions, water pipes not under pressure, exhaust, and waste steam.

The necks of all boiler fittings should be $\frac{1}{16}$-inch thicker than H.P. list.

For plain pipes or cylinders the list is calculated for working pressures of 180 lbs. and 90 lbs. per square inch, for the H.P. and L.P. columns respectively; the stress per square inch of material is taken at 2240 lbs., and a constant addition of ·05 inch is made to cover errors in casting; the formulæ are therefore $t = ·08$ rads. $+ ·05$ inch for H.P., and $t = ·04$ rads. $+ ·05$ inch for L.P.

For H.P. T pieces the square of the thickness (before addition of ·05 inch), has been increased by 33 per cent.,—as a T pipe with all three branches of equal diameter is considered to be that amount weaker than the corresponding plain cylinder without branch in side.

No increase is made in the thickness of L.P. T pieces (over plain pipes), as the pressure assumed, viz., 90 lbs., is already high.

Although brazed copper pipes have occasionally given way, owing to the metal in the neighbourhood of the seam having been " burned," or over-heated during the brazing, they are, on the whole, preferable to solid-drawn pipes, since the latter are not only frequently wanting in uniformity of thickness, but sometimes develop serious longitudinal splits without any warning, even after apparently satisfactory water-pressure tests.

When a brazed pipe is tested to destruction, the strength is rarely found to be more than about one-half of the theoretical strength, owing to the deterioration of the metal in the neighbourhood of the joint. Various suggestions have been made, and various devices tried with a view to overcoming this difficulty, but up to the present no really satisfactory solution has been found.

Table LX.—Thickness of Gun-metal Pipes, T Pieces, &c.

Diameter in inches.	Plain pipes. Calculated thickness. H.P.	Plain pipes. Calculated thickness. L.P.	Plain pipes. Thickness to be. H.P.	Plain pipes. Thickness to be. L.P.	T pieces. Calculated thickness. H.P.	T pieces. Calculated thickness. L.P.	T pieces. Thickness to be. H.P.	T pieces. Thickness to be. L.P.
inches.	inch.	inch.	inch.	inch.	inch.	inch.	inch.	inch.
1	·09	·07	¼	¼	·10	·07	¼	¼
1¼	·10	...	¼	¼	·11	...	¼	¼
1½	·11	·08	¼	¼	·12	·08	¼	¼
1¾	·12	...	¼	¼	·13	...	¼	¼
2	·13	·09	¼	¼	·14	·09	¼	¼
2¼	·14	...	¼	¼	·16	...	¼	¼
2½	·15	·10	¼	¼	·17	·10	¼	¼
2¾	·16	...	¼	¼	·18	...	¼	¼
3	·17	·11	¼	¼	·19	·11	¼	¼
3¼	·18	...	¼	¼	·20	...	¼	¼
3½	·19	·12	5/16	¼	·21	·12	5/16	¼
3¾	·20	...	5/16	¼	·22	...	5/16	¼
4	·21	·13	5/16	¼	·23	·13	5/16	¼
4½	·23	·14	5/16	¼	·26	·14	5/16	¼
5	·25	·15	5/16	¼	·28	·15	5/16	¼
5½	·27	·16	⅜	5/16	·30	·16	⅜	5/16
6	·29	·17	⅜	5/16	·33	·17	⅜	5/16
6½	·31	·18	⅜	5/16	·35	·18	⅜	5/16
7	·33	·19	⅜	5/16	·37	·19	⅜	5/16
7½	·35	·20	⅜	5/16	·40	·20	7/16	5/16
8	·37	·21	⅜	5/16	·42	·21	7/16	5/16
8½	·39	·22	7/16	5/16	·44	·22	7/16	5/16
9	·41	·23	7/16	5/16	·46	·23	½	5/16
9½	·43	·24	7/16	5/16	·49	·24	½	5/16
10	·45	·25	½	5/16	·51	·25	9/16	5/16
10½	·47	·26	½	⅜	·53	·26	9/16	⅜
11	·49	·27	½	⅜	·56	·27	9/16	⅜
11½	·51	·28	9/16	⅜	·58	·28	⅝	⅜
12	·53	·29	9/16	⅜	·60	·29	⅝	⅜
12½	·55	·30	9/16	⅜	·63	·30	⅝	⅜
13	·57	·31	⅝	⅜	·65	·31	11/16	⅜
13½	·59	·32	⅝	⅜	·67	·32	11/16	⅜
14	·61	·33	⅝	⅜	·70	·33	¾	⅜
14½	·63	·34	11/16	⅜	·72	·34	¾	⅜
15	·65	·35	11/16	⅜	·74	·35	¾	⅜
15½	·67	·36	11/16	⅜	·76	·36	13/16	⅜
16	·69	·37	¾	⅜	·79	·37	13/16	⅜
16½	·71	·38	¾	7/16	·81	·38	13/16	7/16
17	·73	·39	¾	7/16	·83	·39	⅞	7/16
17½	·75	·40	¾	7/16	·86	·40	⅞	7/16
18	·77	·41	13/16	7/16	·88	·41	⅞	7/16
18½	·79	·42	13/16	7/16	·90	·42	15/16	7/16

The Admiralty have tried winding all steam pipes over 8 inches in diameter with $\frac{3}{16}$-inch copper wire (making all bends, &c., in gun-metal), and have thereby about doubled the bursting pressure ; but this can only be looked upon as a temporary expedient. The results of some tests showed that a wired pipe stood just about the pressure it *ought* to have stood when unwired, had the copper not been injured in the brazing.

One result of the above-mentioned difficulties has been to give an impetus to the use of welded wrought-iron pipes, and for modern high pressures they are probably the best type that can be used. Welded steel pipes, with cover strips riveted over the joints, and solid-drawn steel pipes are also both being freely used now. The latter can at present be had up to about 8 inches diameter by 9 or 10 feet long.

Since the elastic limit of good sheet copper is reached at about 5500 lbs. per square inch, and the same limit for sheet copper more or less injured in brazing is undoubtedly considerably lower, great caution should be used in testing brazed pipes by water pressure, and the test pressure should be no higher than is absolutely necessary. Opinions will, of course, differ as to what relation the test pressure should bear to the working pressure, but there can be no doubt that the test of double the working pressure (demanded by the Board of Trade and Lloyd's), is much too high for modern working pressures, and is likely to do more harm, by permanently injuring the pipes, than good by revealing defects.

With a view to keeping down the thickness of copper necessary, and thus ensuring more reliable brazed joints, with less risk of burned copper, some makers have employed two and three separate smaller pipes in place of one larger one, and when copper pipes are insisted upon, the plan has its advantages.

The Board of Trade rules referring to pipes are as follows :—

116. Surveyors should pay particular attention to the examination and testing of steam pipes.

All new copper steam pipes should be tested by hydraulic pressure to not less than twice and not more than two and one half times the working pressure, unless the case has been specially submitted to the Board for consideration.

Wrought-iron lap-welded steam pipes should be tested by hydraulic pressure when new to at least twice the working pressure, but not to more than 3 times, unless the case has been specially submitted to the Board for consideration.

As regards old pipes the Surveyor may at any time he thinks it necessary, before he gives a declaration, have them tested by hydraulic pressure to satisfy himself as to any doubtful part, but they should be tested, with the lagging removed for examination, at least once in about every four years to not less than double the working pressure. A record of the test should be kept in the office boiler book.

There should be efficient means provided for draining all the steam pipes.

117. The working pressure of well made copper pipes when the joints are brazed is found by the following formula :—

$$\frac{6000 \times (T - \frac{1}{16})}{D} = \text{working pressure ;}$$

T = thickness in inches ;
D = inside diameter in inches.

When the pipes are solid drawn and not over 8 inches diameter substitute in the foregoing formula $\frac{1}{32}$ for $\frac{1}{16}$.

118. The internal pressure on wrought-iron pipes made of good material, which are lap-welded and are a sound job, may be determined by the following formula, provided that the thickness is not less than $\frac{1}{4}$ inch :—

T = thickness in inches ;
D = diameter inside in inches ;

$$\frac{6000 \times T}{D} = \text{working pressure.}$$

119. Feed pipes should be made sufficient for a pressure 20 per cent. in excess of the boiler pressure.

120. In all cases in which a socket expansion joint is fitted to a bent steam pipe, the Surveyor should require a fixed gland and bolts to be fitted, in order to prevent the end of the pipe being forced out of the socket. This regulation should be complied with in all cases of bent pipes fitted with socket expansion joints. It is also desirable that fixed glands and bolts should be fitted to the expansion joints of straight steam pipes, as cases have occurred, particularly with small straight pipes, in which the ends of the pipes have been forced out of the sockets.

Tables LVII. and LVIII. are calculated in accordance with the rules given in paragraphs 117 and 118 of the above rules.

Lloyd's rule referring to pipes is as follows :—

32. In all steam pipes provision is to be made for expansion and contraction to take place without unduly straining the pipes, and all main steam pipes are to be tested by hydraulic pressure to twice the working pressure, in the presence of an Engineer-Surveyor.

Pipe arrangements.—The greatest care should be taken in scheming arrangements of steam and exhaust pipes to keep them as far as possible in the same horizontal plane; a "pocket" in which water can collect must never be permitted under any circumstances. Neither should a steam pipe which has been led horizontally for some distance be suddenly bent up into the vertical; *all* sharp bends are sources of danger, and to be avoided, but sharp rising bends are specially so.

The provision of proper arrangements for draining all steam pipes, regulator valve boxes, shut-off valve boxes, &c., is also a matter requiring the closest attention,—as water that has once left the boilers and entered the steam pipes will never, with any ordinary arrangement of pipes, drain back to the boiler in the face of the issuing steam. Where the pipes are long it is very necessary to provide collectors of some sort,—as traps do not act with the necessary rapidity ; the inlet pipes to the collectors should be large, and so placed as to arrest and lead away the rapidly moving water.

The question of the arrangements necessary to allow for the expansion of such pipes as vary much in temperature is a difficult one, and can only be very generally treated here.

Steam pipes for modern pressures are so thick and so rigid that bends are of little or no use for giving elasticity, and stuffing-boxes should be provided wherever expansion and contraction will take place in ordinary work.

If expansion stuffing-boxes and glands are not entirely of gun-metal, as is usual in Naval work, the glands should be of that metal, and the bodies so bushed or lined with it as to prevent the possibility of rusting up and consequent jamming.

These stuffing-boxes should be cast with the necessary flanges or brackets for securing them to the bulkheads or deck-beams, and the fixed end of each length of piping (the end furthest away from the stuffing-box), should be anchored in a similar way ; if the pipe does not pass through a bulkhead to which it can be fixed by means of a "bulkhead flange," this anchorage may be obtained by casting a bracket on some T pipe or stop-valve, or by inserting a special short length of gun-metal pipe with the necessary bracket cast on it.

Care must be taken that bulkheads and other parts of the structure of the ship receiving thrusts from large steam pipes, are sufficiently strengthened or stiffened by means of brackets or angle-bars, or by other suitable methods.

In the best class of work the end of the pipe that enters the stuffing-box is always made as a separate piece in cast gun-metal, and this method has the advantage that the pair of flanges at the junction between the gun-metal pipe and the copper or iron one, form a convenient "guard flange," or flange to take the "guard" bolts which prevent the pipe from being blown out of the stuffing-box. These guard bolts should never be omitted except the pipe be very short and rigid, and the end attachments also very rigid,—as fatal accidents have occurred in the past owing to their absence.

It is also very important that those in charge of the fitting-up of the pipes on board ship should have full and clear information (a special small scale tracing or diagram is best) as to the amount of expansion anticipated, and to be provided for, at each stuffing-box.

Cases sometimes occur in which a certain amount of relief is given by a long elbow, and the length in the direction of the general lead of the pipe is hardly so great as to necessitate a stuffing-box.

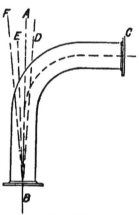

If AB (Fig. 63) is long (say not less than 10 diameters), and AC is too short to make a stuffing-box at C necessary, the strains at B may be reduced one-half by drawing the joint C together when cold, so as to bring AB into the position DB,—DA being equal to half the anticipated expansion of AC; then, when the pipe is hot, AB will assume the position EB instead of FB,—where AE is equal to one-half AF.

FIG. 63.

STOP VALVES.

These should always be made with the spindle a separate piece, distinct from the valve, and so attached as to allow a little "play,"—thus enabling the valve to accommodate itself to the seat when the latter is slightly "out of truth."

Steam stop valves, above about $3\frac{1}{2}$ inches diameter, should have external and accessible spindle nuts, carried in crossheads or bridges; for water valves this is not so necessary, but it is usually best to make only one type of valve for both purposes,—only varying the strengths of the parts.

Screw glands should not be used for valves over 1 inch in diameter, and, when used, should always be fitted with some means of locking.

Ordinary gland studs should be fitted with "check" or "lock" nuts and split pins.

Hand-wheels for stop valves should never be turned and polished; they are better left rough and served with twine.

For large valves (over 10 inches in diameter say) a forged cross-handle (four arms), with the ends turned over at right angles, either towards or away from the valve, is better than a wheel; many engineers prefer this type of handle to the hand-wheel for all valves above about 4 inches diameter, and there is much to be said in its favour.

For boiler-room valves a similar forged handle, but with only two arms, one bent outwards and the other straight, is the best

For very large stop valve chests the best material is, undoubtedly, cast-steel. Trouble is occasionally caused by unsound castings, but this cannot be regarded as unavoidable.

In Naval practice, all stop valves for water, oil, air, &c., are entirely of gun-metal, and steam valves of small and moderate sizes are constructed in the same way, whilst the larger sizes have commonly cast-steel chests with gun-metal seats, &c.

The number of joints may often be considerably reduced by casting the expansion stuffing-boxes along with the stop valve chests.

The handling of large shut-off valves may be much facilitated by the addition of a small by-pass valve.

The whole of the valves in any vessel should open or close by turning the hand-wheel in the same direction. The Admiralty rule is to close with a right-hand motion. To remove all doubt it is a good plan to cast on the upper surface of all hand-wheels the words " open " and " shut," thus :—OPEN←—X—→SHUT.

FUELS, COMBUSTION, &c.

Coal.—The chief varieties of coal and their leading characteristics are as follows :—

(1) *Anthracite*, consisting almost entirely of free carbon, generally jet black in appearance, but sometimes greyish like black lead, has a specific gravity generally of about 1·5, but sometimes as high as 1·9 ; it burns without emitting flame or smoke, but requires a strong draught to burn at all. It is capable of evaporating (theoretically) nearly 16 times its weight of water, but to obtain good results from it careful stoking is necessary, as when suddenly exposed to heat it does not cake but becomes very friable, breaks up into small pieces, and falls through the bar-spaces if disturbed much. The fires should be worked light when using it, and the coal carefully spread. The heat is very intense and local, so that furnaces intended to burn it should be high in the crowns.

(2) *Dry bituminous coal* contains from 70 to 80 per cent. of carbon, and about 15 per cent. of volatilizable matter ; its specific gravity is from 1·3 to 1·45. It burns easily and swells considerably while being converted into coke. The harder kinds do not burn so readily, nor do the pieces stick together so easily when burning, and are generally well adapted for marine boilers.

(3) *Bituminous caking coal*, containing from 50 to 60 per cent. of carbon, is generally of about the same specific gravity as the dry bituminous ; it contains, however, as much as 30 per cent. of volatilizable matter, and consequently develops hydro-carbon gases ; it burns with a long flame, and sticks together in caking, so as to lose all trace of the original forms of the pieces. It requires special means to prevent smoke.

(4) *Cannel coal*, or *long flaming coal.*—This is seldom used for steam purposes, as it gives off large quantities of smoke, and is very scarce. It is the best coal for the manufacture of gas.

(5) *Lignite*, or *brown coal*, is of later formation than the other

coals, and in some instances approaches to a peaty nature. It contains, however, when good, from 56 to 76 per cent. of carbon, and has a specific gravity from 1·20 to 1·35. It also contains large quantities of oxygen, and a small quantity of hydrogen. The commoner kinds of lignite are poor, and contain as little as 27 per cent. of carbon, and therefore are not suitable for steaming purposes.

Wood.—Dry wood contains on an average about 50 per cent. of carbon, 41 of oxygen, and 6 of hydrogen.

Patent fuels.—These usually consist of coal-dust, mixed with a little coal-tar and pressed into hard bricks ; their value depends very much on the class of coal from which they are made.

The value of a fuel is determined by its chemical composition. All fuels contain more or less carbon, most have also hydrogen and oxygen in various proportions, and some contain small quantities of nitrogen, sulphur, &c., in addition.

The ordinary symbols for these substances and their combining weights are as follows :--

Carbon	— symbol C	— combining weight	12.
Hydrogen	„ H	„	1.
Oxygen	„ O	„	16.
Nitrogen	„ N	„	14.
Sulphur	„ S	„	32.

When they combine chemically one with another, as in combustion, they invariably do so in the ratio of their combining weights, or of some multiple of the same. Thus,—

Carbonic oxide	— symbol CO	$= C_{12} + O_{16}$.
Carbonic acid	„ CO_2	$= C_{12} + O_{32}$.
Water .	„ H_2O	$= H_2 + O_{16}$.
Olefiant gas .	„ CH_2	$= C_{12} + H_2$.
Marsh gas .	„ CH_4	$= C_{12} + H_4$.

Atmospheric air is a mechanical mixture of nitrogen and oxygen in the proportion of 77 parts by weight of the former to 23 of the latter.

The heat devoloped by any substance during combustion is called its *total heat of combustion,* and is measured in *units of heat* or *thermal units.*

The British thermal unit is the amount of heat required to raise the temperature of one pound of pure water one degree Fahrenheit when at or near its greatest density (39·1° F.).

The Mechanical equivalent of heat is the number (772) of foot-pounds of energy required to raise the temperature of one pound of water one degree Fahrenheit ; or in other words,—

1 British thermal unit = 772 foot pounds,
or 1 Horse-power = 33,000 foot-pounds = 42·74 thermal units.

To fully burn 1 lb. of carbon (or convert it into carbon dioxide or " carbonic acid "), 2·7 lbs. of oxygen, or 12 lbs. of air are required, and a total heat of combustion of 14,500 units is developed,—*i.e.,* sufficient to evaporate 15 lbs. of water from and at 212°. If, however, the air supply is restricted or deficient, there is a tendency to the production of carbon monoxide or "carbonic oxide,"—a product requiring only half the amount of oxygen for its formation,—and the total heat developed is then only 4400 units.

Each pound of hydrogen requires 8 lbs. of oxygen or 36 lbs. of air for its complete combustion, and developes in combining 62,032 units of heat,—sufficient to evaporate 64 lbs. of water from and at 212°.

Sulphur exists only in small quantities in good coal, and its total heat of combustion is only about 4000 units per pound.

When two substances combine chemically, as in combustion, the weight of the products of combustion is the sum of the weights before combination,—*e.g.*, 1 lb. of carbon uniting with 12 lbs. of air gives 3·7 lbs. of carbonic acid diluted by 9·3 lbs. of nitrogen ; and the temperature of the products is found by dividing the total heat of combustion by their weight multiplied by their specific heat.

The specific heat of any substance is the quantity of heat required to raise the temperature of a pound of it one degree Fahrenheit, and is measured by the ratio that this quantity bears to the quantity required to raise the temperature of a pound of water one degree,— *i.e.*, to the British thermal unit ; thus, the specific heat of hydrogen, at a pressure of 30 inches of mercury, is 3·4, of carbonic acid ·216.

When oxygen and hydrogen are both present in a fuel they combine to form water (or rather steam) without developing any heat, and it is therefore only when there is an *excess* of hydrogen that any heating effect is produced. The only effect of nitrogen is to reduce the intensity of combustion and lower temperature of products.

The following rules are deduced from the above considerations :—

Rule 207.—Total heat of combustion of one pound of fuel,—in thermal units
$$\left. \right\} = 14,500 \left\{ C + 4\cdot28\left(H - \frac{O}{8} \right) \right\}.$$

Rule 208.—Theoretical evaporative power of one pound of fuel,— expressed in pounds of water
$$\left. \right\} = 15 \left\{ C + 4\cdot28\left(H - \frac{O}{8} \right) \right\}.$$

Rule 209.—Number of pounds of air required to burn one pound of fuel
$$\left. \right\} = 12C + 36\left(H - \frac{O}{8} \right).$$

Ordinary coal or coke requires for its complete combustion 12 lbs. of air per pound of the fuel, but in practice, with chimney draught, twice this quantity must be supplied, as much air passes through the furnace in an uncombined state.

At the temperature of 62° F. and at the sea-level, the volume of 1 lb. of air is 13·14 cubic feet ; therefore 315 cubic feet of air are necessary for the proper combustion of 1 lb. of coal or coke in an ordinary furnace. If artificial or forced draught is employed, this quantity may be reduced to about 250 cubic feet, more or less, according to the force of the draught.

The following Table gives additional information as to the composition, total heat of combustion, and evaporative value of various fuels:—

Table LXI.—Composition and Value of Fuels.

Description	Carbon	Hydrogen	Oxygen	Sulphur	Ash, &c. including Nitrogen	Total Heat of Combustion. Units.	Evaporative Power from and at 212°. Pounds.	One ton occupies in Cubic Feet.
Welsh—Ebbw Vale,	87·78	5·15	0·89	1·02	3·66	16221	16·79	...
" Powell's Duffryn,	88·26	4·66	0·60	1·77	4·71	15788	16·34	...
" Llangennech,	94·97	4·26	3·50	0·42	6·85	14682	15·20	...
" Graigola,	84·87	3·84	7·19	0·45	1·91	14130	14·63	42·7
" Average,	83·87	4·79	4·15	1·43	5·89	14858	15·52	45·3
Newcastle Average,	82·12	5·31	5·69	1·24	5·12	14820	15·32	47·4
Derbyshire Average,	79·68	4·94	10·28	1·01	4·06	13860	14·34	47·4
South Yorkshire,	81·88	4·83	7·47	0·54	2·95	14296	14·71	46·0
Lancashire Average,	77·90	5·32	9·53	1·44	6·18	13918	14·56	45·2
Scotch "	78·53	5·61	9·69	1·11	5·03	14164	14·65	42·0
Irish Anthracite,	80·03	2·80	...	6·76	11·03	13802	14·5	35·7
American Anthracite,	88·54	...	...	0·04	8·60	...	...	42·35
" Bituminous,	73·21	...	...	0·42	11·27	...	...	42·44
French Anthracite,	86·17	2·67	2·85	...	8·56	14088	14·53	40·00
" Hard Bituminous,	88·56	4·88	4·38	...	2·19	15525	16·10	42·75
" Caking "	87·73	5·08	5·65	...	1·54	15422	16·00	42·75
Chilian Coal, "	63·56	5·43	14·84	2·50	14·13	11080	11·68	...
Indian " Average,	70·20	...	...	...	22·9	...	...	...
Patent Fuel—Warlichs,	90·02	5·56	...	1·62	6·01	16495	17·07	...
" " Average,	88·40	4·97	2·79	1·26	6·01	15000	15·66	34·4
Lignite—Russian,	73·72	6·09	20·19	ash neglected		14263	14·7	...
" Poor kinds,	66·51	4·72	28·77			11444	12·0	...
Coke—Best Durham,	85 to 92	...	...	0·25 to 2·0	4 to 12	12882	13·80	...
Woods—Beech,	49·36	6·01	42·69	...	1·91	7800 average when dry	8·1	94·4
" Oak,	49·64	5·92	41·16	...	3·26			94·4
" Birch,	50·20	6·20	41·62	...	1·96			106·1
" Fir,	51·79	6·28	41·98	...	...			...
" Willow,	49·96	5·96	39·56	...	4·33	9961	10·30	...
Peat—Fairly dry,	59·80	5·80	29·6	0·3	4·7	20240	20·88	124·7
Petroleum,	84·7	13·1	2·2					

Rate of combustion.—In the mercantile marine, when working economically with chimney draught only, the coal burned per square foot of grate per hour varies from about 15 lbs., with long bars (6 feet to 6 feet 6 inches), to about 20 lbs. when the length of the bars does not exceed 1·33 × diameter of furnace, and all calculations for grate areas of merchant steamers should be based on these figures ; for, although larger quantities may be *consumed* by forcing the fires, or may, perhaps, be completely burned when the wind is strong and draught good, there will be, on the other hand, many days when the quantity burned will be less than the average given above.

The higher efficiency of the shorter bars is largely due to the better air supply, since, with a given diameter of furnace, the area at the mouth of the ash-pit is the same whether the bars are long or short ; and it is no doubt for this reason that Mr Macfarlane Gray's rule,—that *the consumption of coal per foot of grate is very nearly proportional to the diameter of the furnace,*—is found to be so nearly correct in everyday practice. Another cause of the higher efficiency of short bars is the fact that bars over about 5 feet in length cannot be properly stoked by hand,—the result being that the fire burns into holes at the back end, and allows cold air to rush in,—thus depriving the other parts of the fire of their due proportion of air and reducing the efficiency of the furnace. In practice the efficiency of a grate over 4 feet in length is nearly inversely as its length.

The rate of combustion also naturally depends a good deal on the quality or class of coal burned, but by far the most important factor is the strength of the draught.

Chimney or funnel.—The draught obtained with any chimney depends mainly on its sectional area, its height, and the difference of absolute temperature between the gases in it and the external air.

Professor Rankine gave the following formulæ for chimney draught :—

Let w be the weight of fuel burned in a given furnace per second in pounds.

V_0 the volume at 32° of the air supplied per pound of fuel.

τ_0 the *absolute* temperature at 32° Fahr., which is 461° + 32°,

τ_1 the absolute temperature of the gas discharged by the chimney, whose sectional area is A ; then,—

Velocity of the current in the chimney in feet per second is

$$= \frac{w \times V_0 \times \tau_1}{A \times \tau_0}.$$

The density of that current in pounds to the cubic foot is very nearly

$$= \frac{\tau_0}{\tau_1}\left(0\cdot0807 + \frac{1}{V_0}\right);$$

that is to say, from 0·084 to 0·087 × $(\tau_0 \div \tau_1)$.

Let l denote the whole length of the chimney, and of the flue leading to it, in feet ;

m its "hydraulic mean depth;" that is, its area divided by its perimeter; which, for a square or round flue and chimney, is one quarter of the diameter;

f, a coefficient of friction, whose value for currents of gas moving over sooty surfaces is estimated by Peclet at 0·012;

g, a factor of resistance for a passage of air through the grate, and the layer of fuel above it; whose value, according to the experiments of Peclet on furnaces burning from 20 to 24 pounds of coal per square foot of grate per hour, is 12.

Then according to Peclet's formula,—

The *head* required to produce the draught in question is

$$-\frac{\mu^2}{2g}\left(1 + G + \frac{f.l}{m}\right),$$

which, with the values assigned by Peclet to the constants, becomes

$$-\frac{\mu^2}{2g}\left(13 + \frac{0·012 + l}{m}\right).$$

When the *head* is given the value of μ may be calculated, and then,

Weight of fuel which the furnace is capable of burning *completely* per hour

$$-\frac{\mu \times A \times \tau_0}{V_0 \times \tau_1}.$$

It is usual to reckon the *head* by taking one inch of water as the unit; then,—

$$\text{Head in inches of water} = 0·192 \times h \times \frac{\tau_0}{\tau_1}\left(0·0807 + \frac{1}{V_0}\right).$$

Mr Thornycroft has found, by careful experiment with steam launches and torpedo boats working with a *plenum* (that is, with a closed stokehole into which air is forced), "that of the initial pressure, the resistance of the tubes accounts for about seven-tenths of the whole, the resistance of the fire and fire-bars being only about one-tenth;" and that "the pressure in the funnel, as measured, was sensibly equal to atmospheric pressure."

Professor Rankine also stated that if H be the height of the funnel, τ_2 the absolute temperature of the external air, then,—

Head produced by chimney draught

$$-H\left(0·96\frac{\tau_1}{\tau_2} - 1\right).$$

or, taking *h* as the head,

Height of chimney required to produce a given draught

$$-h \div \left(0·96\frac{\tau_1}{\tau_2} - 1\right).$$

The velocity of the gas in the chimney is proportional to $\sqrt{h}$, and therefore to $\sqrt{0·96\,\tau_1 - \tau_2}$.

The density of that gas is proportional to $\dfrac{1}{\tau_1}$.

The weight discharged per second is proportional to velocity $\times$ density, and therefore, to $\dfrac{\sqrt{0\cdot96\tau_1-\tau_2}}{\tau_1}$; which expression becomes a maximum, when $\tau_1=\dfrac{25}{12}\tau_2$. Therefore the best chimney draught takes place when the absolute temperature of the gas in the chimney is to that of the external air as 25 to 12.

When this condition is fulfilled $h=H$.

That is, the height of the chimney for the best draught is equal to the *head* expressed in hot gas, and the density of the hot gas is half that of the air.

From the above it appears that the best temperature of gases at the base of the funnel is about 600°, and that therefore nearly one-fourth of the total heat of combustion is absorbed in creating the draught. With modern high pressures, and correspondingly high temperatures, it is not generally possible to keep the gases below about 750°—that is,—with the proportions of heating surface to grate surface, size of tubes, &c., commonly employed. During a Naval forced draught trial, the temperature at base of funnel is not usually much below 1200°.

A convenient and easily applied rule for size of chimney is as follows:—

Rule 210.—Height of chimney in feet $=\ \cdot007\left(\dfrac{C}{A}\right)^2$.

Rule 210a.—Area of chimney in square feet $=\dfrac{C\times\cdot084}{\sqrt{H}}$.

Where $H=$ Height in feet.
$A=$ Area of section in square feet.
$C=$ Consumption in lbs. per hour on grates.

Where the draught depends on the chimney alone, 40 feet should be regarded as the minimum height; chimneys measuring 90 feet from the dead plates have recently been fitted in some Naval ships, and in one recent case chimneys of 100 feet have been fitted in a large mail-boat.

Ordinarily, in the merchant service, the area of chimney is about $\frac{1}{8}$th to $\frac{1}{7}$th of the grate area, whilst in the Navy they are seldom larger than $\frac{1}{5}$th, and vary from $\frac{1}{9}$th to $\frac{1}{5}$th, according to the air-pressure intended to be used.

The thickness of funnel plates may be approximately as follows :—

Upper plates ... $\cdot1$-inch $+(\cdot025$ inch for each foot of diameter).
Middle ,, ... $\cdot125$,, $+(\cdot025$,, ,,).
Lower ,, ... $\cdot15$,, $+(\cdot025$,, ,,).

If the funnel is stiffened by angle, tee, or channel rings or hoops, somewhat thinner plates may be used.

The funnels of Naval ships are made considerably lighter, the plates rarely exceeding $\frac{3}{16}$-inch in thickness.

When a number of boilers are served by the same funnel, and some of the furnaces are much further away from the base of the funnel than others (as is sometimes unavoidably the case in Naval Vessels) dampers for regulating the force of the draught should be fitted in each "leg" of the uptakes.

Similarly,—proper arrangements (flues, screens, &c.) should be fitted to distribute the air in the stokeholes, when the "closed stoke-hole" system, of forced draught is used.

Smoke-box doors should be made "three thick,"—and arrangements should be made to obtain a good current of air between the inner and central plates, whilst the space between the central and outer plates should be filled with some good non-conducting substance.

Forced Draught.—In the Navy the "closed stoke-hole" system of forced draught is now generally used ; as the name implies, the whole stoke-hole is made air-tight, and the fans force air down into it,—the men working under pressure and entering and leaving through air locks.

Of the "closed ash-pit" systems, Mr Howden's is the only one that is used to any considerable extent. In this system the air is led by pipes from the fans to each furnace separately, and can thus be shut off from any one before opening the door, so that the rush of cold air, which is such an objectionable feature in the closed stoke-hole system, is entirely avoided. Mr Howden also heats the air supply to 180° to 200° by passing it amongst a number of tubes placed in the uptake, and thus no doubt materially increases the efficiency of the furnace, since this increase of temperature so facilitates combustion as to give a gain much in excess of the actual amount of heat stored in the air.

The great advantage of artificial or forced draught is that it is at all times completely under control, and quite independent of wind or weather, thus ensuring a uniform efficiency of furnace with all ordinary qualities of coal, and with almost any rate of combustion that may be required. In Naval ships it is a necessity,—since the chimney may be entirely destroyed at any moment.

Pressures of air in stoke-holes, &c., are usually reckoned in inches of water (often called for shortness "inches" simply), and are measured by a "water-gauge,"—an instrument consisting of a bent glass tube of U form, partially filled with water, and having the open end of one leg exposed to the pressure to be measured, and the other in communication with the outer air. The pressure is indicated by the difference in level of the water in the two legs, and is read off from a scale attached to the instrument.

The following Table shows the rates of combustion and of evaporation usual on sea trials of various classes of vessel, under the conditions named :—

TABLE LXII.—RATES OF COMBUSTION AND EVAPORATION. 215

Table LXII.—Rates of combustion and evaporation.

Description of vessel	Type of boiler, grate, &c.	Nature and force of draught	Pounds of coal per ft. of grate per hour.	Pounds of water per pound of coal.	Pounds of water per ft. of grate per hour.
Ordinary Merchant Steamer	Ordinary cylindrical, long bars	Chimney only	15	9·15	187
"	Ordinary cylindrical, short bars	" "	20	9·25	185
Passenger or Mail Steamer	Ordinary cylindrical medium bars	Chimney, and ·5–·75 in. water	25–30	9–8·75	225–262
Cruiser or Battle-ship	Ordinary cylindrical, very long bars	Chimney, and ·5 in. water	30	8	240
" "	" "	Chimney, and 2 in. water	42	7·75	325
Torpedo Gunboat	Locomotive type, square grate	Chimney, and 2·5 in. water	55	7·2	396
Torpedo Boat	" " "	Chimney, and 3 in. water	67	6·8	456
"	" " "	Chimney, and 6 in. water	96	6·6	633
"	Water-tube boiler	Chimney, and 2 in. water	46	8·7	400
Paddle Steamer	Rectangular box, 7 foot bars	Chimney only	22	7·7	170
"	Rectangular box, 6 foot 6 inch bars, brass tubes	Chimney, and ·5 in. water	27	7·2	195
"	Ordinary cylindrical, 6 foot 6 inch bars, brass tubes	Chimney, and ·5 in. water	25	8·8	220

Table LXIII.—Coal per I.H.P, and I.H.P. per foot of Grate.

Description of vessel	Type of boiler and grate, nature and force of draught, and working pressure.	Type of engines and ratio of cylinders.	Pounds of coal per I.H.P. per hour.	I.H.P. per foot of grate.	Sq. ft. of heating surface per I.H.P.	Sq. ft. of heating surface per ft. of grate.
Ordinary Merchant Steamer	Ordinary cylindrical; bars 6 ft. 6 to 6 ft. 6 in.; chimney only; 160 lbs.	Triple, cyls. 7 to 1	1·63	9·2	2·5	23
,,	Ordinary cylindrical; bars 4 ft. 6 in. to 5 ft.; chimney only; 160 lbs.	"	1·61	12·4	2·5	31
Passenger or Mail St'r.	Ordinary cylindrical; 5 ft. 6 in. bars.; chimney + ·5 to ·75 in. water; 160 lbs.	Triple, cyls. 6·75 to 1	1·66–1·74	15–17·2	2·2	33–37
Cruiser or Battle-ship	Ordinary cylindrical; bars 7 ft.; chimney + ·5 in. water; 150 lbs.	Triple, cyls. 5 to 1	2	15	2	30
,,	Ordinary cylindrical; bars 7 ft.; chimney + 2 in. water; 150 lbs.	"	2·32	18	1·67	30
Torpedo Gunboat	Locomotive type; square grate; chimney + 2·5 in. water; 150 lbs.	"	2·61	21	1·63	34
Torpedo Boat	Locomotive type; square grate; 3 in. water; 150 lbs.	Triple, cyls. 4·8 to 1	2·8	24	1·66	40
,,	Locomotive type; square grate; 6 in. water; 150 lbs.	"	3·2	30	1·83	40
,,	Water tube boiler; 2 in. water; 200 lbs.	"	2·1	22	2·73	60
Paddle Steamer	Rectangular box boiler; 7 ft. bars; chimney only; 30 to 35 lbs.	Jet condensing simple	3·4	6·5	3·7	24
,,	Rectangular box; 6 ft. 6 in. bars; brass tubes, chimney + ·5 in. water; 80 to 35 lbs.	"	3·6	7·5	3·2	24
,,	Ordinary cylindrical; 6 ft. 6 in. bars; chimney + ·5 in. water; brass tubes; 120 lbs.	Compound, cyls. 3·25 to 1	2·5	10	2·6	26

With inferior coal, such as steamers are often obliged to use on foreign stations, the figures given in columns 5 and 6 may be reduced as much as 20 per cent., whilst on the other hand, with picked coal and highly skilled stokers, they may be increased nearly 10 per cent.

The feed temperatures usual in the respective classes of vessels are assumed, and allowance should be made for any special feed-heating apparatus that may be fitted.

The evaporative efficiency of the boilers of cruisers and battle-ships appears lower than it really is, as a large amount of steam is consumed by the numerous auxiliary engines, but is left out of consideration here on account of the great difficulty of forming any reliable estimate of its quantity.

By way of a comparison with the figures given in the above Table, it may be noted in passing that a goods locomotive consumes about 40 lbs. of coal per foot of fire-grate per hour, and an express passenger engine from 65 to 80 lbs.; and, as the evaporative efficiency of the locomotive boiler is high, the water evaporated per square foot of grate per hour may be taken as about 420 lbs. for the goods engine, and from 650 to 750 lbs. for the passenger engine.

Table LXIII. (page 216) shows the average amount of coal consumed per I.H.P. on sea trials of various types of steamship, the I.H.P. ordinarily obtained from each square foot of grate in the same circumstances, and also the usual proportions of heating surface.

For the following very valuable Table (page 218),—giving most complete details as to the performances of the machinery of a number of first-class Torpedo-boats,—the authors are indebted to the courtesy of Messrs J. I. Thornycroft & Co., the eminent Torpedo-boat builders of Chiswick, London, W.

The steam used by the various auxiliary engines is assumed to be all consumed by the main engines.

The thanks of the Authors are also due to Messrs Yarrow & Co., the eminent Torpedo boat builders of Poplar, London, E., for the very interesting summary of their present practice, as regards 1st and 2nd Class Torpedo boats, given in Table LXIIIb. (page 219).

Table LXIIIa.—Performances of Torpedo Boat Machinery. (Messrs Thornycroft & Co.)

A. Type of Engine.	B. Ratio of Cylinders.	C. Coal per foot of grate per hour.	D. Water per lb. of coal from and at 212°.	E. Ditto, steam from 100°.	F. Water per foot of grate per hour from and at 212°.	G. Ditto, steam from 100°.	H. Steam pressure.	I. Pounds of coal per I.H.P. per hour.	J. I.H.P. per foot of grate.	K. Square feet of heating surface per I.H.P.	L. Square feet of heating surface per foot grate.	M. Air pressure in inches of water.	N. Pounds of steam per I.H.P. per hour.	O. Type of Boiler.	P. Distinguishing number of engine.	Q. Description of vessel.
	Equivalent values	$\frac{IJ}{F\,D}$	$\frac{F}{C}$	$\frac{N}{I}$	CD	$\frac{NJ}{CE}$		$\frac{C}{J}$	$\frac{LK}{CI}$	$\frac{L}{J}$	KJ		$\frac{IE}{GJ}$			
Double Compound.	1 : 2·68	69·3	7·7	6·64	533	460	120	3·92	17·7	1·78	31·6	3	26	Locomotive type.	1	First-class Torpedo Boats.
		87·5	7·65	6·58	670	576	120	8·95	22·13	1·43	31·6	5	26		2	
	1 : 2·8	95·2	7·61	6·54	724	622	120	3·98	23·95	1·32	31·6	5	26		2	
		97·2	7·54	6·48	732	630	120	3·98	24·42	1·29	31·6	5·5	25·75		3	
		82	7·6	6·55	623	536	130	8·94	20·8	1·82	37·9	3·2	25·75			
Triple Compound.	1 : 2·04 : 5·05	45·4	9·77	8·35	443	379	200	2·12	21·46	2·82	60·53	2·1	17·65	Thornycroft Water tube.	4	
		40·7	9·95	8·5	406	346	200	1·98	20·54	2·95	60·53	1·7	16·8		5	
		43·75	9·81	8·39	429	367	200	2·02	21·64	2·8	60·53	1·3	16·9		6	
	1 : 1·96 : 4·84	66·2	8·92	7·87	585	488	250	2·44	27·11	2·12	57·46	2·6	18		7	
		74·1	8·55	7·11	633	527	250	2·77	26·78	2·15	57·46	3·5	19·7		8	

Table LXIIIb.—Messrs Yarrow's Standard Torpedo boat practice.

		1st Class Torpedo Boat.	2nd Class Torpedo Boat.
Standard Sizes {	Length	130 feet	60 feet
	Breadth	13 feet 6 inches	9 feet 3 inches
Displacement		85 to 90 tons	About 16 tons
Load carried		25 tons	4 tons
Average { Triple engines		23 to 23½ knots	About 17¼ knots
speed { Quadruple ,,		Up to 24½ knots	
Type of engine		Inverted triple, surface condensing, three-crank ; or quadruple four-crank	Inverted triple surface condensing three crank
Cylinders and { Triple		14½″—21″—32″ 16″ Stroke	8″—12″—17½″ 9″ Stroke
stroke { Quadruple		14″—20″—27″—36″ 16″ Stroke	
Ratios of { Triple		1 : 2·09 : 4·87	1 : 2·25 : 4·78
cylinders { Quadruple		1 : 2·04 : 3·72 : 6·61	
Type of boiler .		By preference, Loco. type, for durability and easy access ; but Water-tube boilers of "Yarrow" type* when high trial trip speed is first consideration.	
Boiler { Triple		160 to 180 lbs. .	160 to 180 lbs.
pressure { Quadruple		200 to 220 lbs.	
Air pressure		Average under 3 inches, but occasionally as high as 5 inches.	
Coal burned per square foot of grate per hour		60 to 100 lbs.	
Heating surface per square foot of grate		With Locomotive boilers 45 to 50 square feet.	
Revolutions { Triple		About 410	About 560
per minute			
at full power { Quadruple		About 435	
I.H.P. per square foot of grate		Triple engines and Locomotive boilers, 35 to 38 I.H.P. Quadruple engines and Loco. boilers, 37 and 40 I.H.P.	

*When water-tube boilers of the "Yarrow" type are used, the grate-bar area and heating surface obtained are about 50 per cent. greater than with Locomotive type boilers, and the quantity of coal burned per square foot of grate is much reduced.

With a 60 foot boat, fitted with triple engines and tubulous boiler, Messrs Yarrow have recently obtained a speed of 19¾ knots ; and with a 130 foot boat, fitted with quadruple engines and Locomotive type boiler,—a speed of 24½ knots on a two hours continuous run ; and they believe these speeds to be the highest recorded for these classes of boat.

From one of their tubulous boilers, weighing 3 tons, they have obtained 300 I.H.P.,—a performance which they also believe to be the highest on record.

EVAPORATION, HEATING SURFACE, &c.

The efficiency of the heating surface of a boiler depends on,—its position, the material composing it and its thickness, the condition of the surfaces, and the circulation of the water in contact with it.

(1) **Position.**—The most efficient heating surface is that of the furnace crowns and combustion chambers ; its high efficiency is due partly to the great difference of temperature between the two sides of the plates, partly to the freedom of the surfaces from deposit of soot or ash, and also in some measure to the fact that mineral or earthy matter deposited from the water does not readily adhere to vertical surfaces or to surfaces over which there is a good circulation of water.

Next in order of efficiency come the upper surfaces of the tubes, more especially of the *upper* tubes at the ends next the combustion chambers, for the reason that the flame and heated gases always seek the highest possible course and rapidly lose heat in passing through the tubes, entering them at say 2400° and leaving them under 800°, and also because the bubbles of steam disengage themselves more easily when they can rise directly from a surface.

When a "nest" of tubes is made very deep (*i.e.* when there are many rows one above another) it very often happens, for the reason just given, that the lower rows do little or no work, and would be much better absent from the boiler.

The surfaces below the level of the fire-bars, and the front tube-plates are generally left out of consideration as of no value.

(2) **Material.**—Amongst ordinary metals suitable for boiler construction, copper is by far the most efficient conductor of heat, its efficiency compared with iron or steel being about as 3 to 1; for various reasons, however, its use is now almost confined to the fire-boxes of torpedo boats, even brass tubes being now very seldom used except for the boilers of fast paddle steamers.

Its high conductivity so reduces the possible difference in temperature between the fire side and the water side of a plate as to very materially diminish the racking strains that are ordinarily produced during the operation of getting up steam, and for this reason it is peculiarly suitable for the tube-plates of torpedo boat or other boilers in which the changes of temperature are sudden and great.

(3) **Condition of surfaces.**—A perfectly smooth and clean metallic surface does *not* give satisfactory results as a heating surface, but is apt to cause sudden and violent, though intermittent ebullition, and

for this reason, few boilers work satisfactorily and steadily until they have acquired a thin film of scale or deposit.

Since the cooling of the products of combustion takes place mainly in the tubes, it is there that the soot is principally deposited, and tubes must be swept more or less frequently, according to the class of coal used, if the efficiency of the surface is to be maintained.

The external surfaces of the tubes, especially on the upper sides, are peculiarly liable to the accumulation of deposit where water containing salt or other mineral matter is used, as the upward circulation is naturally sluggish and much impeded in large "nests" of tubes.

(4) **Circulation.**—The proper circulation of the water over the heating surfaces is of the very greatest importance, and no boiler in which the design and arrangements are not such as to promote such a circulation will ever steam satisfactorily or evaporate anything like the quantity of water that it would do with proper arrangements. The principal evils that result from defective circulation are, deposit of scale on the heating surfaces, over-heating, and consequent buckling of plates and leaking of tubes, and irregular and much diminished evaporative efficiency, oftentimes accompanied by intermittent fits of "priming."

Special difficulty has been found in getting a proper circulation of water over the back tube-plates (or in the case of locomotive type boilers, *front* tube-plates) of Naval boilers, mainly on account of the close pitch of the tubes and consequent contraction of the water ways; and similar difficulty has also been experienced with the large flat (almost square) tops of the fire-boxes in boilers of the locomotive type.

In this latter case, the cure is found in one or two water tubes, similar in function to "Galloway" tubes, a cure which is just as effectual now as it was found to be many years ago in the case of large double-ended cylindrical boilers. A better arrangement still would probably be to divide the fire-box completely, making two distinct furnaces.

The ideal condition of heating surface, as regards circulation, is undoubtedly such as is found in the Herreshoff and Thornycroft boilers.

Diameter of tubes.—The diameter of tube has a considerable influence on the efficiency of the tube surface, for the contents increase as the square of the diameter, whilst the surface increases as the diameter only. Thus if a 4-inch tube be substituted for two 2-inch ones, the absorbing surface will remain the same, but twice the quantity of gas will be passed, if the velocity of flow remain the same. If the velocity of flow be reduced in the 4-inch tube (to one-half) so as to pass only the same quantity of gas as the two 2-inch ones, the 4-inch tube is still at a disadvantage,—inasmuch as the mean distance of the molecules of gas from its surface is greater than in the 2-inch ones.

When artificial or forced draught is used the diameter of tube should be smaller than for natural draught,—if length of tube and velocity of gas are to remain the same,—since the gases have a higher temperature and must be divided into smaller threads or streams if they are not to escape into the chimney at a wastefully high temperature : the draught, being of a positive nature, may be relied on to overcome the extra friction and prevent the closing up of the tubes by soot with undue rapidity.

The diameters of tube in ordinary use are as follows :—

(a) Merchant steamer.   Chimney draught.   3¼ inch or 3½ inch.
(b)   ,,   Artificial   ,,   3 inch.
(c) Cruiser or battleship.   ,,   ,,   2¼ inch.
(d) Torpedo gunboat.   Loco type boilers.   1¾ inch.
(e) Torpedo boat.   ,,   ,,   ,,   1½ inch to 2 inch.

Quantity of water evaporated per pound of coal.—This of course depends not only on the quality of the coal, but on the type of boiler, and strength of draught, and on the skill of the stoker. The average evaporation of water that may be expected with good coal, under the various conditions indicated, is shown in columns 5 and 6 of Table LXII.

Equivalent evaporation from and at 212° F.—In order to compare evaporative results obtained with different temperatures of feed and pressures of steam, it is necessary to eliminate the effects of the varying conditions by reducing all the results to one common standard.

The standard generally employed is "the equivalent evaporation from and at 212° F.,"—i.e., the number of pounds of water that would be evaporated in each case, per pound of coal per hour, if the feed water were supplied at 212° F., and completely evaporated under the pressure (one atmosphere) due to that temperature.

On reference to the Table "Properties of Saturated Steam," page 333, it is seen that 966 (practically) thermal units are required to evaporate 1 lb. of water from 212°, and under a pressure of one atmosphere ; so that, if the number of thermal units imparted to the water by each pound of coal during the test be determined, and divided by 966, the quotient will be the equivalent evaporation from and at 212°. The quantity of heat imparted by each pound of coal is determined by referring to column 7 of the same table, opposite to the proper pressure, and there reading off the total heat, from 32°, contained in each pound of steam, deducting from this the number of units, above 32°, contained in each pound of feed-water as supplied, and then multiplying the remainder by the number of pounds of water evaporated, per pound of coal per hour, during the test. This operation is concisely expressed by the formula,—

$$N_{212} = N_T \left\{ \frac{H - (t - 32)}{966} \right\} ;$$

where N_{212} = Number of lbs. water evaporated per lb. coal per hour, from and at 212°.

N_T = Number of lbs. water evaporated per lb. coal under test conditions.

H = total heat, from 32°, per lb. of steam at test temperature and pressure.

t = temperature at which feed was supplied during test.

PROPORTIONS OF BOILERS.

Grate surface.—From the information given in Tables LXII. and LXIII., it will be an easy matter to determine the I.H.P. per foot of grate that may be anticipated (or the grate area that must be provided for any given I.H.P.) in the case of proposed new boilers of any of the types mentioned.

Size of furnace.—The furnaces of cylindrical boilers should not be less than 30 inches in diameter, nor more than 48 inches, unless in exceptional cases : wherever possible a diameter of not less than 40 inches should be given, as, with smaller furnaces,—owing to thickness of fire being practically constant for all diameters,—the space above the fuel is much contracted, and the combustion less perfect in consequence. As previously stated, long bars cannot be properly worked by hand, and are not so efficient as short ones : whenever possible they should be limited to a length of 5 feet.

Number of furnaces.—

Boilers up to 9 feet diameter may have 1 furnace.
    ,,   ,,   13 ft. 6 ins.   ,,   2 furnaces.
    ,,   ,,   15 feet   ,,   3   ,,
    ,, beyond 15 feet   ,,   4   ,,

If the boilers are double-ended the number of furnaces will, of course, be double the figure given above.

Heating surface.—For the proper absorption of the heat developed in the furnaces and combustion chambers, the amount of heating surface necessary in the various types of boiler, and under the various conditions mentioned, is approximately that given in columns 6 and 7 of Table LXIII. ; the amounts there given are ample for the purpose, and any addition will not only be ineffective (unless in very special circumstances), but will do positive harm by crowding up the water spaces, and spoiling the circulation.

Steam room.—The steam room allowed in the various types of boiler, working at or about the pressures named, may be as follows :—

Table LXIV.—Steam Room in Cubic Feet per I.H.P.

Description of Boiler, &c.	Type of Engines.	Working pressure.	Steam room c. ft. per I.H.P.
{ Ordinary cylindrical ; mercantile, . .	Triple ; screw,	160 lbs.	·38 to ·46
{ Ordinary cylindrical ; Naval { N.D., . { F.D., .	,, ,,	,,	·35 ·30
Ordinary cylindrical,	Compound; paddle,	120 lbs.	·45
Rectangular box, .	Simple ; paddle,	30 to 35 lbs.	·65
{ Locomotive type (torpedo boat, &c.), .	Fast running triple or quadruple,	175 to 220 lbs.	·13 to ·28

In cylindrical boilers the top row of tubes should be not less than ·3 × diameter of boiler from the top ; if higher the contraction of water surface is apt to cause priming.

Table LXV.—Dimensions and Weights

Rules on which Constructed.	Shell.					Working Pressure.	Furnaces.		
	Material.	Diameter.		Length.		Thickness.		Number.	Mean Diameter.	Thickness.
		Ft.	in.	Ft.	in.	Ins.	Lbs.		Ins.	Ins.
Admiralty, . . .	Steel	16	6	10	3	1¼	155	4	44	9/16
" . . .	"	16	0	18	0	1 3/16	155	8	42	9/16
" . . .	"	14	6	17	6	1 3/32	130	6	45	½
" . . .	"	14	6	16	6	1 1/16	130	6	41	½
" . . .	"	13	4	9	7	1	155	3	41	9/16
" . . .	"	13	1	18	6	1	155	6	42	9/16
" . . .	"	13	1	9	6	1	155	3	42	9/16
" . . .	"	12	6	9	6	1	155	3	36	½
" . . .	"	11	6	17	3	⅞	155	4	42	9/16
" . . .	"	10	11	17	6	29/32	155	4	36	19/32
Board of Trade and Lloyd's, .	"	14	6	10	6	1 5/16	160	3	42	½
" " " .	"	14	3	16	0	1 5/16	160	6	38½	19/32
" " " .	"	14	3	9	3	1 5/16	160	3	38½	19/32
" " " .	"	14	0	17	0	1¼	160	6	40	21/32
" " " .	"	14	0	10	6	1 3/32	155	3	42	9/16
" " " .	"	13	9	19	0	13/16	80	4	48	½
" " " .	"	13	6	16	5	1 7/32	165	6	40	9/16
" " " .	"	13	3	16	3	1 5/32	150	6	34½	17/32
" " " .	"	13	0	16	0	1 3/16	165	4	45	19/32
" " " .	"	13	0	11	0	1 5/32	150	2	48	17/32
" " " .	"	13	0	10	6	1⅛	150	3	36	½
" " " .	"	13	0	10	0	1 3/32	150	3	38	21/32
" " " .	"	12	9	11	0	1⅛	160	3	40	17/32
" " " .	"	12	3	16	6	1⅛	160	4	38	½
" " " .	"	12	0	16	3	1⅛	150	4	39	9/16
" " " .	"	12	0	9	1½	1⅛	150	2	39	9/16
" " " .	"	11	9	15	6	1 1/32	160	4	39	19/32
" " " .	"	11	9	9	6	1 3/32	150	2	39	9/16
Board of Trade, . .	"	16	0	11	0	⅞	80	3	48	21/32
" " . . .	"	14	9	16	6	1 11/32	160	6	42	9/16
" " . . .	"	14	9	16	6	¾	80	6	42	21/32
" " . . .	"	13	9	15	0	¾	90	4	45	21/32
" " . . .	"	13	9	10	6	1 3/32	142	3	38	7/16
" " . . .	"	12	9	15	0	1	110	4	42	½
" " . . .	"	12	3	10	6	1 3/32	160	2	42	17/32
" " . . .	"	12	3	9	6	1⅜	200	2	42	19/32
" " . . .	"	11	9	15	0	1	150	4	40	½
" " . . .	"	11	9	11	0	1	150	2	46	15/32
" " . . .	"	10	6	9	3	⅞	150	2	33	½
" " . . .	"	8	6	8	6	⅝	150	2	30	½
" " . . .	"	8	6	6	0	5/16	80	2	33	7/16
" " . . .	"	8	0	8	6	9/16	90	1	42	17/32
Lloyd's, ...	"	15	6	11	0	¾	85	4	36	17/32
"	"	14	3	10	3	¾	80	3	42	½
"	"	12	0	9	6	21/32	80	2	41	½
"	"	11	6	9	6	13/16	140	2	40	½
"	"	11	2	18	3	1 1/32	150	4	38	½
"	"	10	6	9	3	⅞	150	2	37	½
"	"	9	3	9	1½	11/16	155	2	31	15/32
"	"	9	0	8	6	21/32	150	2	32	7/16

of Boilers actually made.

No. of Combustion Chambers.	Tubes.					Total Heating Surface.	Weight of			Weight p. 100 ft. of Total Heating Surface.		
	Number.	Diameter.	Length.		Surface.		Boiler.	Water.	Total.	Boiler.	Water.	Total.
		Ins.	Ft.	In.	Sq. ft.	Sq. ft.	Tons.	Tons.	Tons.	Tons.	Tons.	Tons.
2	596	2¼	7	0	2676	3176	46·8	22·5	69·3	1·47	·71	2·18
4	1200	2¼	6	7	5105	5905	70·1	43·6	113·7	1·19	·74	1·92
3	700	2½	6	9	3335	3960	49·77	30·45	80·22	1·26	·77	2·03
3	804	2¼	6	4½	3288	3868	47·6	29·62	77·22	1·23	·76	1·99
3	346	2⅝	6	10	1602	1913	27·4	15·1	42·5	1·43	·79	2·22
6	708	2⅝	6	9	3218	3835	46·8	26·7	73·5	1·22	·70	1·92
3	354	2⅝	6	9	1609	1917	26·4	13·9	40·3	1·38	·72	2·10
3	336	2½	6	6	1400	1685	24·8	13·8	38·6	1·47	·82	2·29
4	576	2½	6	4	2355	2770	33·0	19·8	52·8	1 19	·71	1·91
1	524	2½	6	10½	2305	2530	29·6	16·6	46·2	1·17	·66	1·83
3	232	3¾	8	0	1820	2250	40·0	23·0	63·0	1·77	1·02	2·80
3	590	3¼	6	3	3074	3620	57·75	30·6	88·35	1·60	·84	2·44
3	295	3¼	6	3	1537	1910	35·0	17·5	52·5	1·83	·91	2·74
3	466	3¼	6	9	2840	3375	57·1	30·0	87·1	1·69	·88	2·57
3	226	3¼	7	3½	1484	1870	37·2	19·5	56·7	1·99	1·04	3·03
4	388	3¼	6	10	2388	2885	40·5	32·5	73·0	1·40	1·13	2·53
3	436	3¼	6	4	2350	3000	51·0	26·0	77·0	1·70	·86	2·56
3	468	3¼	6	4½	2490	3000	45·0	26·0	71·0	1·50	·86	2·36
2	464	3¼	6	3	2417	2810	46·7	28·6	75·3	1·66	1·02	2·68
2	176	3¾	7	6	1270	1590	31·3	17·2	48·5	1·96	1·09	3·05
3	152	3¾	7	0	1024	1360	30·4	16·5	46·9	2·24	1·21	3·45
3	166	3½	7	0	1085	1345	27·5	15·8	43·3	2·04	1·17	3·21
3	170	3½	7	10½	1220	1625	30·0	16·0	46·0	1·85	·98	2·83
1	344	3¼	6	7	2033	2380	39·7	22·55	62·25	1·67	·94	2·61
2	364	3¼	5	5½	1775	2260	38	28·2	66·2	1·68	1·24	2·92
1	182	3¼	6	1½	1000	1240	23·1	13·04	36·14	1·86	1·05	2·91
2	344	3¼	6	1½	1755	2040	32·0	21·0	53·0	1·56	1·03	2·6
2	142	3¼	6	6	880	1085	23·25	12·2	35·45	2·14	1·12	3·26
3	250	3¼	6	6	1360	1800	32·75	22·75	55·5	1·83	1·25	3·08
1	556	3¼	6	6	3260	3750	58·2	32·0	90·2	1·55	·85	2·4
3	392	3¼	6	2	2021	2500	40·2	28·25	68·45	1·61	1·13	2·74
2	420	3¼	5	7½	1968	2310	34·7	26·0	60·7	1·50	1·12	2·62
3	192	3¾	7	3	1340	1730	31·2	20·2	51·4	1·8	1·16	2·96
2	380	3¼	5	10	1860	2270	35·7	23·3	59·0	1·57	1·03	2·60
1	180	3½	7	4½	1195	1480	24·5	15 0	39·5	1·65	1·01	2·66
1	192	3¼	6	6	1045	1300	27·0	13 0	40·0	2·07	1·00	3·07
2	348	3¼	6	0	1740	2100	31·4	20·3	51·7	1·50	1·00	2·50
1	128	3¾	8	0	990	1250	22·6	14·0	36·6	1·80	1·12	2·92
2	118	3¼	6	3	614	810	16·2	10·0	26·2	2·0	1·23	3·23
2	104	3	5	11	473	600	11·5	6·68	18·18	1·93	1·11	3·04
1	64	2½	4	0	163	250	6·25	4·7	10·95	2·50	1·88	4·38
1	80	2½	5	9	324	420	8·00	5·8	13·8	1·90	1·36	3·26
2	274	3½	7	6	1858	2250	33·65	23·6	58·25	1·50	1·05	2·55
3	198	3¼	7	0	1251	1570	25·2	18·8	44·0	1·60	1·20	2·80
2	172	3¼	6	4	980	1205	20·6	12·6	33·2	1·66	1·00	2·66
2	168	3¼	6	6	913	1145	20·0	10·7	30·7	1·74	·93	2·68
1	208	3¼	7	0	1310	1685	34·3	21·2	55·5	2·03	1·30	3·33
1	110	3½	6	0	642	800	16·2	9·0	25·2	2·02	1·12	3·15
1	104	3¼	6	3	541	692	13·2	8·0	21·2	1·90	1·15	3·05
2	100	3¼	6	0	500	655	12·6	6·5	18·6	1·92	1·00	2·95

Paddle engines should theoretically have considerably larger steam spaces than the faster running screw engine, but in practice weight of machinery is of such great importance that everything is cut down to the lowest limit, and many compound paddle engines, working at pressures of 100 to 120 lbs., have not more than ·4 cubic foot per I.H.P.

Water spaces, &c.—The spaces between the furnaces themselves, between the furnaces and the shell, and between the combustion chambers, although sometimes made as narrow as 5 inches, are better 5½ inches, and even 6 inches when possible ; the space between back of combustion chamber and end of boiler should be 6 inches at the bottom, increasing to 9, 10, or even 12 inches at the top.

A suitable pitch for tubes is 1·35 to 1·4 × external diameter of tube ; where weight is of great importance, as in Naval vessels, the tubes are sometimes pitched a little closer than this, but with water spaces so contracted there is always risk of priming. A fair average pitch is given by the rule,—

$$\text{Pitch of tubes} = D + \frac{\sqrt{D}}{2} + \text{·35 inch} ;$$

where D is outside diameter of tube in inches.

Whenever possible the tubes should be placed in horizontal and vertical rows, and not arranged in any diagonal or zigzag fashion.

The clear space between the nests of tubes should never be less than 10 inches, and when possible should be 10½ or 11 inches,—in order that a man may be able to get down : manholes between the furnaces and tubes can then be dispensed with,—a very important point,—for the life of a boiler depends to a very considerable extent on the number of openings in its shell,—the leakage from such openings being most destructive. For the same reason, longitudinal seams below the water level should be avoided as much as possible.

Table LXV. gives particulars (dimensions and weights) of a number of boilers that have been actually constructed in accordance with different rules as indicated. The weights given are those of "bare boiler," without either furnace fittings or boiler mountings.

Tables LXVI. and LXVIa. show the various dimensions and weights necessary, with the different types of boiler indicated, to obtain a constant heating surface (of 2000 square feet).

TABLE LXVI.—COMPARISON OF BOILERS, ETC. 227

Table LXVI.—Comparison of Boilers for 80 lb. pressure by Board of Trade Rules

General Description.	No. of Boilers.	Shell Diam. (Ft. in.)	Shell Lgth. (Ft. in.)	Furnaces No.	Furnaces Diam. (Ins.)	Furnaces Lgth. (Ft. in.)	Tubes No.	Tubes Diam. (Ins.)	Tubes Lgth. (Ft. in.)	Tubes Surf. (Sq. ft.)	Total Heating Surface (Sq. ft.)	Total Capacity (C. ft.)	Weight Boiler (Tons)	Weight Water (Tons)	Weight Total (Tons)
Double-ended boiler, having two-combustion chambers, each common to *opposite* furnaces,	1	12 0	14 0	4	39	5 6	376	3¼	5 6	1725	2000	1582	{26·9 iron, 23·75 steel}	}29{	{46·9, 43·75}
Double-ended boiler, having two-combustion chambers, each common to furnaces at *same* end,	1	12 0	14 0	4	39	5 3	376	3¼	5 3	1643	2000	1582	{27·15 iron, 24·0 steel}	}20{	{47·15, 44·0}
Double-ended boiler, having four combustion chambers, one to each furnace,	1	12 0	14 0	4	39	5 3	376	3¼	5 3	1643	2000	1582	{27·4 iron, 24·2 steel}	}20{	{48·0, 44·8}
Single-ended boiler, having three combustion chambers, one to each furnace,	1	14 6	9 7	3	42	6 6	274	3¼	6 9	1665	2000	1582	{29·75 iron, 26·15 steel}	}20{	{49·75, 46·15}
Two single-ended boilers, each having two combustion chambers, one to each furnace,	2	11 0	8 4	4	38	5 10	316	3¼	5 10	1650	2000	1582	{38·7 iron, 30·6 steel}	}20{	{58·7, 50·6}

Table LXVIa.—Comparison of Steel Boilers for 160 lbs. pressure by Board of Trade Rules.

General Description.	No. of Boilers.	Shell. Diam. Ft. in.	Shell. Lgth. Ft. in.	Furnaces. No.	Furnaces. Diam. Ins.	Furnaces. Lgth. Ft. in.	Tubes. No.	Tubes. Diam. Ins.	Tubes. Lgth. Ft. in.	Tubes. Surface. Sq. ft.	Total Heating Surface. Sq. ft.	Total Capacity. C. ft.	Weight. Boiler. Tons.	Weight. Water. Tons.	Weight. Total. Tons.
Double-ended boiler, having two combustion chambers, each common to *opposite* furnaces,	1	12 0	14 0	4	41	5	376	3¼	5 6	1715	2000	1582	35·8	21	56·8
Double-ended boiler, having two combustion chambers, each common to furnaces at *same* end,	1	12 0	14 0	4	41	5	376	3¼	5 3	1680	2000	1582	36	21	57
Double-ended boiler, having four combustion chambers, one to each furnace,	1	12 0	14 0	4	41	5	376	3¼	5 2	1610	2000	1582	36·4	21·5	57·9
Single-ended boiler, having three combustion chambers, one to each furnace,	1	14 0	9 7	3	45	6	274	3¼	5 9	1640	2000	1582	37·25	22	59·25
Two single-ended boilers, each having two combustion chambers, one to each furnace,	2	11 0	8 4	4	39	5	340	3¼	5 10	1645	2000	1582	39·2	21·4	60·6

STEEL BOILERS.*

The following is a summary of the Admiralty, Board of Trade, and Lloyd's rules relating to the construction of steel boilers :—

Admiralty tests of material.—All steel to be Siemens-Martin. Every plate, &c., used is to be tested, and must comply with the requirements stated below :—

Table LXVII.—Admiralty Tensile Tests.

Description of Material.	Minimum ultimate tensile strength, tons per square inch.	Maximum ultimate tensile strength, tons per square inch.	Minimum elongation in 8 inches per cent.
Not exposed to flame & not flanged,	27	30	20%
,, ,, but flanged,	25	28	25%
Rivet bars, ...	25	27	25%
Fire box, &c., and steam-pipe plates,	24	26	26%
Corrugated or ribbed furnace, .	23	25	27%
Tube strips, ...	21	25	25%

For bending tests the specimens are to be heated to a low cherry red, and then cooled in water at 82° F.

Strips of plate $1\frac{1}{2}$ inches wide must bend double in press,—inner radius being $1\frac{1}{2}$ times thickness of plate. For pieces of rivet bar inner radius to be $1\frac{1}{2}$ dia. of bar ; and for strips from tubes, $\frac{1}{2}$ inch.

Plates exposed to flame are also to be tested by welding and forging, some of the welds being broken in the testing machine to ascertain degree of perfection.

Angle, tee, and bar steel is to stand such other forge tests as the overseer may direct.

Samples from each batch of rivets are also to stand the following tests :—To be bent double (cold), inner radius of bend being equal to diameter of rivet ; to be bent double (hot) and hammered till the two parts of shank meet ; head to be flattened (hot) without cracking until its diameter is $2\frac{1}{2}$ times diameter of shank ; and shank to be nicked on one side, and bent over to show quality of metal.

Tubes under $\frac{3}{16}$-inch thick must stand drifting out at each end, when hot, to 20% larger diameter, and expanding out, when cold, to 15% larger diameter,—in each case without fracture,—and tubes of $\frac{3}{16}$-inch thick and over must expand half these amounts. Sections of thin tubes 2 inches long to bear hammering down to $1\frac{1}{4}$-inch, and to bear flattening till sides touch.

Board of Trade tests of material.—Strips 2 inches wide should be cut from at least one of every four ordinary plates, from each end of each plate over 15 feet in length, and from each corner of each plate over 20 feet × 6 feet, or over $2\frac{1}{2}$ tons in weight ; where more than one test piece is taken from a plate the *mean* result is to be adopted.

* All reference to iron plates and iron rivets is excluded from this section.

Table LXVIIa.—Board of Trade Tensile Tests.

Description of material.	Minimum ultimate tensile strength, tons per square inch.	Maximum ultimate tensile strength, tons per square inch.	Elongation on 10 inches, per cent.
Plates not exposed to flame, . . .	27	32	About 25 %, not less than 18 %.
Plates that are exposed to flame, .	26	30	...
Rivet bars, . . .	26	30	Not less than 25 %.
Stay bars, . .	27	32	About 25 %, not less than 20 %.
Rivets, . . .	27	32	Contraction of area about 60 %.

When stay or rivet bars do not exceed 1 inch diameter, 1 in each 20 may be tested.

When stay or rivet bars do not exceed 1½ inch diameter, 1 in each 12 may be tested.

When stay or rivet bars are larger than 1½ inch diameter, 1 in each 8 may be tested.

For bending tests the specimens should be 2 inches wide and 10 inches long, and should be bent double,—the inner radius being 1½ times the thickness of the plate ; for plates not exposed to flame, the tests should be made on the plate in its ordinary condition, and for plates that are exposed to flame, the specimens should be heated to a cherry red, and cooled out in water at 80° F.

Lloyd's tests of material.—As far as practicable, one tensile and one temper test is to be made of the product of each furnace charge ; but when a great number of charges are in question, one tensile and one temper test to every ten plates will suffice,—if all prove satisfactory.

The ultimate tensile strength of all material—plates, bars, and rivets—is to be not less than 26 tons per square inch, nor more than 30 tons per square inch, with an elongation of not less than 20 per cent. in 8 inches.

For bending tests the specimens are to be heated to a low cherry red, and quenched in water at 82° F., and then bent to a curve, the inner radius of which is not greater than 1½ times the thickness of the plate or bar.

The temper test is to be applied to samples taken from *every* plate ntended to be used for furnaces or combustion chambers.

Board of Trade rules for cylindrical shells.—In all calculations for strength of shells the *minimum* strength of plate, as disclosed by the tests, must be used.

To ascertain the strength of shell, the relative sectional areas of plate and rivet must first be determined by the following formulæ :—

Rule 211. $\dfrac{(\text{Pitch} - \text{Dia. of Rivet}) \times 100}{\text{Pitch}} = \left\{ \begin{array}{l} \text{Percentage of strength} \\ \text{of plate at joint as com-} \\ \text{pared with solid plate,} \end{array} \right.$

For maximum permissible pitch of rivets, &c., see formulæ and sketches at end of this section. Pages 245 to 249.

Rule 212.

$\dfrac{(\text{Area of rivet} \times \text{No. of rows of rivets}) \times 100}{\text{Pitch} \times \text{thickness of plate}} = \left\{ \begin{array}{l} \text{Percentage of strength} \\ \text{of rivets as compared} \\ \text{with solid plate.} \end{array} \right.$

If the rivets are in double shear, multiply the percentage thus found by 1·75.

In consequence of the low shearing strength of steel rivets the Board require that in all types of joint the *nominal* rivet section shall be $\frac{23}{28}$ of the net plate section ; thus, in order that rivet section may be considered to have the same strength as plate section their relation must be :—

In lap joints,—

(Area of rivet × No. of rows of rivets) × 23 = (Pitch – diameter of rivet) × thickness of plate × 28.

And in butt joints,—

(Area of rivet × No. of rows of rivets × 1·75) × 23 = (Pitch-diameter of rivets) × thickness of Plate × 28.

The working pressure per square inch that may be allowed on the safety valves is then given by,—

Rule 213.—Working Pressure = $\dfrac{S \times \% \times 2T}{D \times F}$

where S = tensile strength of material in lbs. per square inch ;
  % = the smaller of the two percentages, found by Rules 211 and 212, divided by 100 ;
  T = thickness of plate in inches ;
  D = inside diameter of boiler in inches (inside diameter of outer strake, if any) ;
  F = factor of safety from following Table if % refers to plate section ;
  F = 4·5 if % refers to rivet section.

Various penal additions are provided in the rules for cases in which the workmanship is of inferior character.

If it is proposed to use steel for superheaters particulars should be submitted to the Board for consideration.

Table LXVIII.—Board of Trade factors of safety.

When all rivet holes are drilled in place after bending; *all* seams fitted with double butt-straps, each at least five-eighths the thickness of the plates they cover; all seams at least double riveted; and boilers open to inspection during construction.	F = 4·5
To be added when circumferential seams are lap and double riveted.	·1
To be added when longitudinal seams are lap and double riveted.	·2
To be added when longitudinal seams are lap and treble riveted.	·1
To be added, when boiler is of such length as to fire from both ends, unless middle circumferential seams are treble riveted.	·3

Compensating rings should be fitted around all manholes and openings, of at least the same effective sectional area as the plate cut out, and in no case should rings be of less thickness than plate to which they are attached. Manholes in shells of cylindrical boilers should have their shorter axes placed longitudinally.

It is very desirable that compensating rings round openings in flat surfaces be of ∟ or ⊤ iron.

The neutral parts of shells under steam domes must be efficiently stiffened and stayed.

Lloyd's rules for cylindrical shells.—

Rule 213a.—Working Pressure (lbs. per sq. in.) $= \dfrac{C \times (T-2) \times B}{D}$

where D = mean diameter of shell in inches.

T = thickness of plate in *sixteenths of an inch.*

$$C = \begin{cases} 20— \text{ when longitudinal seams have double butt-straps of equal width.} \\ 19\cdot25 \text{ when longitudinal seams have double butt-straps of unequal width.} \\ 18\cdot5 \text{ when longitudinal seams are lap joints.} \end{cases}$$

B = least percentage of strength of longitudinal joint, found as follows :—

For plate at joint, ... $B = \dfrac{p-d}{p} \times 100$

For rivets at joint, ... $B = \dfrac{n \times a}{p \times t} \times 85$

where p = pitch of rivets in inches ;

t = thickness of plate *in inches ;*

d = diameter of rivet holes in inches ;

n = number of rivets used per pitch in longitudinal joint ;

a = sectional area of rivet in square inches.

When rivets are in double shear 1·75a is to be used in place of a.

Inside butt-straps are to be at least three quarters the thickness of the shell plate.

For the shells of superheaters or steam-chests enclosed in the uptakes, or exposed to the direct action of flame, the values of C should be only two-thirds of those given above.

Proper deductions are to be made for openings in shells ; all man-holes in shells to have compensating rings.

Shell plates under domes are to be adequately stayed.

Board of Trade rules for flat surfaces.--The highest permissible pressures on flat surfaces are found by the following formula :—

Rule 214. Working pressure (lbs. per square inch) $= \dfrac{C \times (T+1)^2}{S-6}$

where T = thickness of plate in *sixteenths of an inch ;*
       S = surface supported in square inches ;
       C = constant having values given in following table :—

Table LXIX.—Board of Trade Constants for Flat Surfaces.

Description of attachment of stay.	Plates not exposed to heat or flame.	Plates exposed to heat or flame and in contact with steam.	Plates exposed to heat or flame and in contact with water.
Stays with nuts and doubling strips, the latter of width equal to ⅔rds pitch and thickness equal to plates to which they are riveted.	200	...	...
Stays with nuts and riveted washers,—latter having diameter equal to ⅔rds pitch and thickness equal to plates to which they are riveted.	187·5	...	...
Stays with nuts and washers,—latter having diameter three times that of stay and thickness ⅔rds that of plate they cover.	125	75	...
Stays fitted with nuts only.	112·5	67·5	...
Stays screwed into plate and fitted with nuts.	...	...	100
Stays screwed into plate and riveted over.	...	39·6	66

When plates are stiffened by L or T bars and greater pressure is required the case should be submitted to the Board for consideration.

When the circular flat end of a cylindrical shell is in question, S in the formula may be taken as the area of the square inscribed in the circle through centres of rivets securing end,—provided angle ring or flange is of sufficient thickness.

The following Tables are calculated by means of the Board of Trade rules for flat surfaces :—

Table LXX.—Flat Surfaces of Combustion Chambers (B. of T.).

Pressure in lbs. per square inch	Stays screwed into plates and fitted with nuts.					
	½-inch plates.		⁷⁄₁₆-inch plates.		⅝-inch plates.	
	Pitch.	Surface.	Pitch.	Surface.	Pitch.	Surface.
150	7·74	60·0	8·52	72·6	9·31	86·6
155	7·63	58·2	8·39	70·5	9·16	84·0
160	7·52	56·6	8·27	68·5	9·03	81·6
165	7·42	55·1	8·16	66·6	8·91	79·3
170	7·32	53·6	8·05	64·8	8·78	77·2
175	7·23	52·3	7·94	63·1	8·67	75·1
180	7·14	51·0	7·84	61·5	8·56	73·2
185	7·05	49·8	7·75	60·0	8·45	71·4
190	6·97	48·6	7·65	58·6	8·34	69·6
195	6·89	47·5	7·56	57·2	8·25	68·0
200	6·82	46·5	7·48	56·0	8·15	66·5

Table LXXI.—Flat Surfaces of Boiler ends in Steam Spaces (B. of T.).

Pressure in lbs. per square inch.	Stays fitted with nuts and riveted washers, latter having diameter equal to ⅔rds pitch, and thickness equal to plate they cover, plates shielded from flame.							
	¹¹⁄₁₆-inch plate.		¾-inch plate.		¹³⁄₁₆-inch plate.		1-inch plate.	
	Pitch.	Surface.	Pitch.	Surface.	Pitch.	Surface.	Pitch.	Surface.
150	15·84	251·0	16·94	287·2	18·05	326·0	19·16	367·2
155	15·59	243·0	16·67	278·1	17·76	315·6	18·85	355·5
160	15·35	235·6	16·42	269·6	17·49	306·0	18·56	344·6
165	15·12	228·7	16·17	261·7	17·23	296·9	18·28	334·4
170	14·90	222·2	15·94	254·2	16·98	288·2	18·02	324·7
175	14·69	216·0	15·72	247·1	16·74	280·3	17·76	315·6
180	14·49	210·2	15·50	240·4	16·51	272·6	17·52	307·0
185	14·30	204·6	15·30	234·0	16·29	265·4	17·29	298·9
190	14·12	199·4	15·10	228·0	16·08	258·6	17·06	291·1
195	13·94	194·4	14·91	222·3	15·88	252·1	16·84	283·8
200	13·77	189·7	14·72	216·9	15·68	246·0	16·64	276·9

Table LXXIa.—Flat Surfaces of Boiler ends in Steam spaces (B. of T.).

Pressure in lbs. per square inch.	Stays fitted with nuts and doubling strips, the latter of width equal to ⅔rds pitch, and thickness same as plate to which they are riveted; plates shielded from flame.							
	⅞-inch plate.		⅞-inch plate.		⅞-inch plate.		1-inch plate.	
	Pitch.	Surface	Pitch.	Surface	Pitch.	Surface	Pitch.	Surface
150	16·35	267·3	17·49	306·0	18·63	347·3	19·78	391·3
155	16·09	258·9	17·21	296·3	18·34	336·3	19·46	378·9
160	15·84	251·0	16·94	287·2	18·05	326·0	19·16	367·2
165	15·61	243·6	16·69	278·7	17·78	316·3	18·88	356·3
170	15·38	236·6	16·45	270·7	17·53	307·2	18·60	346·0
175	15·16	230·0	16·22	263·1	17·28	298·6	18·34	336·3
180	14·96	223·8	16·00	256·0	17·04	290·4	18·09	327·1
185	14·76	217·9	15·79	249·2	16·81	282·7	17·84	318·4
190	14·57	212·3	15·58	242·8	16·59	275·4	17·64	310·2
195	14·38	207·0	15·38	236·7	16·38	268·5	17·39	302·4
200	14·21	202·0	15·19	231·0	16·18	262·0	17·17	295·0

Table LXXII.—Lloyd's Constants for Flat Surfaces.

Description of attachment of stay.	Value of C.
Screw stays with riveted heads,—	
Plates ⁷⁄₁₆-inch thick and under, . . .	90
Plates over ⁷⁄₁₆-inch thick, . . .	100
Screw stays fitted with nuts,—	
Plates ⁷⁄₁₆-inch thick and under, . . .	110
Plates over ⁷⁄₁₆-inch and under ⁹⁄₁₆-inch, . .	120
Plates ⁹⁄₁₆-inch and over, ...	135
Stays fitted with double nuts, ...	175
Stays fitted with double nuts and with washers of diameter equal to ⅓rd the pitch, and of thickness equal to half the plate they cover, . . .	185
Stays fitted with double nuts and with riveted washers of diameter equal to ⁹⁄₁₆ths pitch, and of thickness equal to half the plate they cover, . . .	200
Stays fitted with double nuts and with riveted washers of diameter equal to ⅔rds pitch, and of thickness equal to plate they cover, ...	220
Stays with double nuts and doubling strips,—latter having breadth equal to ⅔rds distance between rows of stays, and thickness equal to plate to which they are riveted, ...	220
Stays with double nuts and doubling strips,—latter having breadth equal to ⅔rds pitch of stays, and thickness equal to plate to which they are riveted, ...	240

Lloyd's rules for flat surfaces. — The strength of flat plates supported by stays is given by the following formula :—

Rule 214a.

$$\text{Working pressure (lbs. per square inch)} = \frac{C \times T^2}{P^2}$$

where T — thickness of plate in *sixteenths of an inch ;*
     P — greatest pitch in inches ;
     C — constant having values given in preceding Table.

In the case of front plates of boilers in the steam space the values of C must be reduced 20 per cent. if the plates are not shielded from the direct action of heat.

For tube plates in the nest of tubes, C must be taken as 140, and P as the *mean* pitch of stay tubes.

For the wide spaces between nests of tubes, P is to be taken as the horizontal distance from centre to centre of the bounding rows of tubes and C, as follows :—

Pitch of Stay Tubes in bounding rows.	When tubes have no nuts outside plates.	When tubes are fitted with nuts outside plates.
Where there are two plain tubes between each stay tube, . . .	120	130
Where there is one plain tube between each stay tube, . . .	140	150
Where every tube in these rows is a stay tube, and each alternate one has a nut, ...	...	170

Table LXXIII. is calculated in accordance with Rule 214a and the constants given in Table LXXII.

Table LXXIII.—Flat Surfaces by Lloyd's Rules.

Pressure in lbs. per square inch.	Screw stays fitted with nuts. Pitches in inches.			Plates protected from fire. Pitches in inches.											
				Double nuts and riveted washers. Diameter = 2/5 pitch. Thickness = 1/2 plate. C=200.				Double nuts and riveted washers or strips. Diameter = 2/3 pitch. Width = 2/3 pitch of rows. Thickness = thick. of plate. C=220.				Double nuts and riveted strips. Width = 2/3 pitch. Thickness = thick. of plate. C=240.			
	1/2-inch plates. C=120	9/16-inch plates. C=135	5/8-inch plates. C=135	Thickness of plate.				Thickness of plate.				Thickness of plate.			
				13/16	7/8	15/16	1	13/16	7/8	15/16	1	13/16	7/8	15/16	1
150	7·15	8·54	9·48	15·01	16·17	17·32	18·47	15·74	16·95	18·16	19·38	16·44	17·71	18·97	20·24
155	7·04	8·40	9·33	14·76	15·90	17·04	18·17	15·49	16·68	17·87	19·06	16·18	17·42	18·67	19·91
160	6·93	8·26	9·18	14·53	15·65	16·77	17·89	15·24	16·42	17·59	18·76	15·92	17·15	18·37	19·59
165	6·82	8·14	9·04	14·31	15·41	16·51	17·61	15·01	16·17	17·32	18·47	15·68	16·88	18·09	19·30
170	6·72	8·02	8·91	14·10	15·18	16·27	17·35	14·79	15·93	17·06	18·20	15·45	16·63	17·82	19·01
175	6·62	7·90	8·78	13·90	14·97	16·04	17·10	14·58	15·70	16·82	17·94	15·22	16·40	17·57	18·74
180	6·53	7·79	8·66	13·70	14·76	15·81	16·86	14·37	15·48	16·58	17·69	15·01	16·17	17·32	18·48
185	6·44	7·69	8·54	13·52	14·56	15·60	16·64	14·18	15·27	16·36	17·45	14·81	15·95	17·08	18·22
190	6·36	7·59	8·43	13·34	14·36	15·39	16·42	13·99	15·07	16·14	17·22	14·61	15·73	16·86	17·98
195	6·28	7·49	8·32	13·17	14·18	15·19	16·20	13·81	14·87	15·93	16·99	14·42	15·53	16·64	17·75
200	6·20	7·39	8·22	13·00	14·00	15·00	16·00	13·63	14·68	15·73	16·78	14·24	15·34	16·43	17·53

Board of Trade rule regarding compressive stresses on
tube plates.—

Rule 215.

Working pressure (lbs. per square inch) $= \dfrac{(D-d)\, T \times 16,000}{W \times D}$

where D = least horizontal distance between centres of tubes, in
 inches ;
 d = inside diameter of ordinary tube, in inches ;
 T = thickness of tube plate, *in inches ;*
 W = outside width of combustion chamber, in inches,—from
 tube plate to back of box, or from tube plate to tube
 plate in double-ended boilers, with common combustion chambers.

Lloyd's rule regarding compressive stresses on tube-plates.—

Rule 215a. $T = \dfrac{P \times W \times D}{1600 \times (D-d)}$,

where P = working pressure in lbs. per square inch ;
 W = width of combustion chamber over plates,—in inches ;
 D = horizontal pitch of tubes in inches ;
 d = inside diameter of plain tube in inches ;
 T = thickness of tube-plate in *sixteenths of an inch.*

Board of Trade rule for girders supporting combustion chamber
tops, &c.—

Rule 216.

Working pressure (lbs. per square inch) $= \dfrac{C \times d^2 \times T}{(W-P)\, D \times L}$,

where W = width of combustion chamber,—in inches ;
 P = pitch of supporting bolts   ,,
 D = distance from centre to centre of girders,—in inches ;
 L = length of girder in feet ;
 d = depth of girder,—in inches ;
 T = thickness of girder   ,,
 $C = \begin{cases} 550 \text{ when girder has one supporting bolt,} \\ 825 \quad ,, \quad\;\; ,, \quad \text{two or three ,,} \\ 935 \quad ,, \quad\;\; ,, \quad \text{four} \qquad ,, \end{cases}$

Lloyd's rule for girders supporting combustion chamber tops,
&c.—

Rule 216a.

Working pressure (lbs. per square inch) $= \dfrac{C \times d^2 \times T}{(L-P)\, D \times L}$,

where **L** = width between tube-plates or tube-plate
       and back plate of chamber ;
    P = pitch of stays in girder ;
    D = distance from centre to centre of girder ;   } All in inches.
    d = depth of girder at centre ;
    T = thickness of girder at centre ;

$$C = \begin{cases} 6,600 & \text{when each girder has one stay,} \\ 9,900 & \text{,, ,, two or three stays,} \\ 11,000 & \text{,, ,, four or five ,,} \\ 11,550 & \text{,, ,, six or seven ,,} \\ 11,880 & \text{,, ,, eight or more ,,} \end{cases}$$

Admiralty requirements as regards stays.—At the test pressure steel stays under 1½ inch diameter may carry a load of 16,000 lbs. per square inch of net section, and those of 1½ inch diameter and upwards, 18,000 lbs. Rivets or bolts used for securing stays must be at least 25 per cent. stronger than the stays.

Board of Trade rules relating to stays.—Steel stays that have been welded, or worked in the fire, must not be used.

Rule 217.—Solid steel stays may be allowed a working stress of 9000 lbs. per square inch of net section.

Dished ends, unless of thickness required for a flat end, must be stayed ; but where they are theoretically equal to the pressure needed, when considered as portions of spheres, the stays may carry a working load of 14,000 lbs. per square inch of net section. If they are not theoretically equal to the pressure needed they must be stayed as flat surfaces.

The areas of diagonal stays are found as follows:—Find the area of direct stay necessary; multiply it by the length of the diagonal stay; and divide product by length of perpendicular from end of diagonal stay to surface supported; quotient will be area of diagonal stay.

When gusset stays are used their area should be in excess of that found in above way.

Lloyd's rules relating to stays.—No steel stays are to be welded.

Rule 217a.

    Stays not exceeding 1¼ inch smallest diameter may carry 8000 lbs. per square inch net section.

    Stays above 1¼ inch smallest diameter may carry 9000 lbs. per square inch net section.

    Stay tubes may carry 7500 lbs. per square inch of net section.

Table LXXIV.—Surface that may be supported by one Stay. (Board of Trade rule.)

Diameter of stay.	No. of threads per inch.	Area at bottom of thread, sq. in.	Maximum number of square inches of plate that can be supported by one stay when strained to 9000 lbs. per square inch.										
			150 lbs.	155 lbs.	160 lbs.	165 lbs.	170 lbs.	175 lbs.	180 lbs.	185 lbs.	190 lbs.	195 lbs.	200 lbs.
1⅜	10	1·221	73·3	70·9	68·7	66·6	64·6	62·8	61·1	59·4	57·8	56·3	54·9
1½	9	1·448	86·9	84·1	81·4	79·0	76·6	74·5	72·4	70·4	68·6	66·8	65·1
1⅝	9	1·727	103·6	100·3	97·1	94·2	91·4	88·8	86·3	84·0	81·8	79·7	77·7
1¾	8	1·985	119·1	115·2	111·6	108·3	105·1	102·1	99·2	96·6	94·0	91·6	89·3
1⅞	8	2·310	138·6	134·1	129·9	126·0	122·3	118·8	115·5	112·4	109·4	106·6	108·9
2	7	2·593	155·6	150·6	145·9	141·4	137·3	133·4	129·6	126·1	122·8	119·7	116·7
2¼	7	3·356	201·4	194·9	188·8	183·0	177·7	172·6	167·8	163·3	158·9	154·9	151·0
2½	6	4·107	246·4	238·5	231·0	224·0	217·4	211·2	205·3	199·8	194·5	189·5	184·8
2¾	6	5·054	303·2	293·5	284·3	275·7	267·6	259·9	252·7	245·9	239·4	233·3	227·4
3	5	5·914	354·8	343·4	332·7	322·6	313·1	304·1	295·7	287·7	280·1	272·9	266·1
3¼	5	7·040	422·4	408·8	396·0	384·0	372·7	362·1	352·0	342·5	333·3	324·9	316·8

Table LXXIVa.—Surface that may be supported by one Stay. (Lloyd's Rules.)

Diameter of stay.	No. of threads per inch.	Area at bottom of thread, sq. in.	Maximum number of square inches of plate that can be supported by one stay when strained to 9000 lbs. per square inch.										
			150 lbs.	155 lbs.	160 lbs.	165 lbs.	170 lbs.	175 lbs.	180 lbs.	185 lbs.	190 lbs.	195 lbs.	200 lbs.
1⅜	10	1·221	65·1	63·0	61·0	59·2	57·5	55·8	54·3	52·8	51·4	50·1	48·8
1½	9	1·448	77·2	74·7	72·4	70·2	68·1	66·2	64·8	62·6	60·9	59·4	57·9
1⅝	9	1·727	92·1	89·1	86·3	83·7	81·3	78·9	76·7	74·7	72·7	70·8	69·1

Table LXXIV. is calculated in accordance with Rule 217, and Table LXXIVa. in accordance with Rule 217a ; above 1⅝-inch Lloyd's rule is same as B. of T. rule, and Table LXXIV. serves for both.

Board of Trade rules relating to furnaces.—Cylindrical furnaces with longitudinal joints welded, or made with a butt-strap double riveted, or with double butt-straps single riveted,—when rivet holes are drilled in place after bending, and slightly countersunk from the outside,—may carry the working pressure given by,—

Rule 218.

$$\text{Working pressure (lbs. per sq. in.)} = \frac{99,000 \times (\text{thick. of plate in ins.})^2}{(\text{Length in ft.} + 1) \times \text{dia. in ins.}}$$

provided always that the pressure so found does not exceed that given by the following formula,—which limits the crushing stress per square inch of material to 4400 lbs. :—

Rule 218a.

$$\text{Working pressure (lbs. per sq. in.)} = \frac{8800 \times \text{thickness in inches}}{\text{diameter in inches}}.$$

When the furnace is made with rings the length is to be measured between the rings.

When the furnaces are corrugated and machine made,—as made by the Leeds Forge Co., or ribbed and grooved, as made by Messrs J. Brown & Co., Sheffield,—if they are practically circular in section, if the lengths of plain parts at ends do not exceed 6 inches, and if the plates are not less that $\frac{5}{16}$ inch thick,—the working pressure may be that given by,—

Rule 218b.

$$\text{Working pressure (lbs. per sq. in.)} = \frac{14,000 \times \text{thickness in inches}}{D},$$

where $D = \begin{cases} \text{for corrugated furnace,—outside diameter in inches at} \\ \quad \text{bottom of corrugations;} \\ \text{for ribbed and grooved furnace,—outside diameter in} \\ \quad \text{inches over plain portions.} \end{cases}$

In corrugated furnaces,—

Pitch of corrugations should not exceed 6 inches.
Depth   ,,   (from top outside to bottom inside) should not be less than 2 inches.
Plates at ends not to be unduly thinned in flanging.

16

In ribbed and grooved furnaces,—

Pitch of ribs should not exceed 9 inches.
Ribs should project at least 1⅝ inches above plain parts.
Grooves should not exceed ¾ inch in depth.
Ends should be rolled slightly thicker than plain parts and not reduced at any part, by flanging, below thickness of plain parts.

Lloyd's Rules relating to furnaces.—The pressures that may be carried by plain cylindrical furnaces are given by the following formula :—

Rule 218c.

$$\text{Working pressure (lbs. per square inch)} = \frac{89,600 \times T^2}{L \times D} \;;$$

where T = thickness of plate *in inches* ;
D = outside diameter of furnace in inches ;
L = length of furnace in feet.   If strengthening rings are fitted, the length between the rings is to be taken.

The pressure, however, must not exceed that given by,—

Rule 218d. $\dfrac{C \times T}{D}$ = lbs. per square inch ;

where $C = \begin{cases} 8000 \text{ for plates } \frac{9}{16} \text{ inch thick and under.} \\ 8800 \text{ for plates over } \frac{9}{16} \text{ inch thick.} \end{cases}$

The working pressure that may be carried by furnaces fitted with Adamson rings is given by,—

Rule 218e.

$$\text{Working pressure (lbs. per square inch)} = \frac{C \times T}{D} \;;$$

where $C = \begin{cases} 10,400 \text{ for one Adamson ring (at about mid-length).} \\ 11,400 \text{ for two Adamson rings.} \end{cases}$

The working pressures that may be carried by various special types of furnace are as follows :—

Rule 218f.

$$\text{Working pressure (lbs. per square inch)} = \frac{C \times (T-2)}{D} \;;$$

where T = thickness of plate in *sixteenths of an inch* ;
$D = \begin{cases} \text{outside diameter of corrugated furnace in inches ;} \\ \quad\quad\;\; \text{,,} \quad\quad\quad \text{ribbed} \quad\quad\text{,,} \quad\quad\text{,,} \\ \quad\quad\;\; \text{,,} \quad\quad\quad \text{plain parts of Holmes' patent furnace;} \end{cases}$

C = 1000 for steel corrugated furnaces when tensile strength of material is under 26 tons, and corrugations are 6 inches pitch and 1½ inches deep ;

C = 1259 for steel furnaces corrugated on Fox's or Morison's plans, when tensile strength of material is between 26 and 30 tons ;

C = 1160 for ribbed furnaces (ribs 9 inches apart) ;

C = 912 for spirally corrugated furnaces ;

C = 945 for Holmes' patent furnaces, when corrugations are not more than 16 inches pitch, nor less than 2 inches high.

Board of Trade rules relating to steel boilers generally.—

Hydraulic tests.—New boilers should be tested by hydraulic pressure, to twice the working pressure, in the presence and to the satisfaction of the Board's Surveyors. In the case of old boilers in a vessel, the hydraulic test should never exceed twice the calculated working pressure of the boiler, but must always exceed the working pressure.

Evaporators, generators, feed make-ups, &c., where the evaporation of water under pressure is an essential feature, should be regarded as steam boilers, whether the evaporation is effected by heat from coal gas, from steam, or from any other source, and the strength, quality of material, and method of construction of such apparatus, should be in accordance with the regulations for steam boilers, and they should be examined by the Surveyor on each occasion the vessel is surveyed for passenger certificate in the same manner as other boilers on board the vessel.

The mountings, &c., should as a general rule be similar to those required in the case of boilers on board passenger vessels, but the question of safety valves will in each case be considered by the Board, and full particulars to enable this to be done should be submitted by the Surveyor as early as possible.

The particulars of evaporators, their safety valves, &c., should be recorded on the declaration in the same manner as is now done in the case of other auxiliary boilers.

Local heating of plates should be avoided.

Annealing.—All plates that are punched, flanged, or locally heated must be carefully annealed afterwards.

Welding.—Steel plates that have been welded should not be passed to carry a tensile stress ; in other cases they should be efficiently annealed after welding.

Makers of steel.—When the steel is not to be made by any of the following makers the case will receive the special consideration of the Board, and this should be specially noted by the Surveyors.

Messrs W. Beardmore & Co.
  ,,   J. Brown & Co.
  ,,   C. Cammel & Co.
  ,,   D. Colville & Sons.
  ,,   The Consett Iron Co., Ltd.
  ,,   The Glasgow Iron and Steel Co., Ltd.
  ,,   The Landore Steel Co.
  ,,   The Leeds Forge Co.
  ,,   The Moor Steel and Iron Co., Ltd.
  ,,   Palmer's Shipbuilding and Iron Co., Ltd.
  ,,   The Steel Co. of Scotland.
  ,,   The Weardale Iron and Coal Co.
  ,,   The West Cumberland Iron and Steel Co.
— For plates and angles, stay and rivet bars.

  ,,   The Lanarkshire Steel Co., Ltd.
  ,,   Nettlefolds.
  ,,   John Spencer & Sons, Ltd.
  ,,   Wright, Butler, & Co.
— For stay and rivet bars.

Boiler tracings, &c.—Difficulty has been experienced with regard to the survey of steel boilers owing to the fact that some makers were not aware, at the time the boilers were commenced, that a Board of Trade certificate would be necessary, and the makers have therefore omitted to submit tracings until the boilers have been nearly completed. Tracings of boilers may therefore be received for examination upon payment of the usual fee of £2, and the Surveyors may proceed as far as witnessing the hydraulic test before any further instalment of the survey fee is paid. Engineers and boiler makers should be advised of this arrangement.

Donkey Boilers that are in any way attached to, or connected with, the main boilers, or with the machinery used for propelling the vessel, should be surveyed and be fitted the same way as the main boilers, and have a water and steam gauge, and all other fittings complete, and, as regards safety valves, should comply with the same regulations as the main boilers.

Launch Boilers.—The boilers of steam launches forming part of the statutory boat capacity of passenger steamers should as regards construction, strength, material, safety valves, and other fittings comply with the same regulations as the main boilers.

Lloyd's Rules relating to steel boilers generally.—

Hydraulic test.—Boilers to be tested by hydraulic pressure, in the presence of an Engineer-Surveyor, to twice the working pressure, and carefully gauged while under test.

Rivet holes, &c.—All the holes in steel boilers should be drilled, but if they are punched the plates must be annealed afterwards.

Annealing.—All plates that are dished or flanged, or in any way heated in the fire for working,—except those that are subjected to a compressive stress only,—are to be annealed after the operations are completed.

BOARD OF TRADE RIVETED JOINTS.

The following sketches of riveted joints, and formulæ for determining their various proportions, are given in an appendix to the Board of Trade Rules ; the formulæ are here given in a form differing slightly from that adopted in the Rules :—

F in the following formulæ stands for the factor of safety (for which see Table LXVIII., page 232), and r for percentage of plate left between rivet holes.

ORDINARY CHAIN AND ZIGZAG RIVETED JOINTS.

$$\left. \begin{array}{c} \text{Percentage of plate} \\ \text{left between holes} \end{array} \right\} = \frac{100 \,(\text{pitch} - \text{diameter of rivet})}{\text{pitch}} = r.$$

$$\left. \begin{array}{c} \text{Nominal per-} \\ \text{centage of} \\ \text{rivet section} \end{array} \right\} = \frac{\left. \begin{array}{c} 31 \cdot 944 \,\text{Butt} \\ 18 \cdot 254 \,\text{Lap} \end{array} \right\} \times \text{area of riv.} \times \text{No. rivets in pitch} \times F}{\text{pitch} \times \text{thickness of plate}}.$$

To find pitch so that nominal rivet section and net plate section may be of equal strength :—

$$\text{Pitch} = \frac{\text{area of rivet} \times \text{No. rivets in pitch} \times F}{\left. \begin{array}{c} 3 \cdot 130 \,\text{Butt} \\ 5 \cdot 478 \,\text{Lap} \end{array} \right\} \text{thickness of plate}} + \text{diameter of rivet.}$$

To find pitch and diameter of rivet :—

$$\text{Diameter of rivet} = \frac{\left. \begin{array}{c} 3 \cdot 986 \,\text{Butt} \\ 6 \cdot 975 \,\text{Lap} \end{array} \right\} \times r \times \text{thickness of plate}}{(100 - r) \times \text{No. rivets in pitch} \times F}.$$

$$\text{Pitch} = \frac{\left. \begin{array}{c} 398 \cdot 6 \,\text{Butt} \\ 697 \cdot 5 \,\text{Lap} \end{array} \right\} \times r \times \text{thickness of plate}}{(100 - r)^2 \times \text{No. rivets in pitch} \times F}.$$

Also, when diameter of rivet is found first :—

$$\text{Pitch} = \frac{100 \times \text{diameter of rivet}}{100 - r}.$$

When double butt-straps are used each strap must have a thickness of ⅝ths of the plate it covers.

A single butt-strap must have a thickness equal to ⅛ × thickness of plate it covers.

Distance from centre of rivet to edge of plate (E in Figs.) = $1\frac{1}{2}$ × diameter of rivet.

Distance between rows of rivets :—

(*a*) **Chain riveted joints (Figs. 2, 4, 9, 11),—**

V = not less than 2 × dia. of rivet, preferably $\dfrac{(4\ \text{dia. rivet}) + 1}{2}$.

(*b*) **Zigzag riveted joints (Figs. 3; 5, 10, 12),—**

$$V = \frac{\sqrt{(11\ \text{pitch} + 4\ \text{dia. rivet})\,(\text{pitch} + 4\ \text{dia. rivet})}}{10}.$$

Diagonal pitch (Figs. 3, 5, 10, 12),—

$$p_{\text{D}} = \frac{6\ \text{pitch} + 4\ \text{dia. rivet}}{10}.$$

RIVETED JOINTS WITH ALTERNATE RIVETS IN OUTER, OR OUTER AND INNER, ROWS LEFT OUT.

Percentage of plate left between holes = $\dfrac{100\ (\text{pitch} - \text{dia. of rivet})}{\text{pitch}}$

Nominal percentage of rivet section = $\dfrac{\left.\begin{array}{c}31\cdot944\ \text{butt} \\ 18\cdot254\ \text{lap}\end{array}\right\} \times \text{area of riv.} \times \text{No. rivets in pitch} \times F}{\text{pitch} \times \text{thickness of plate.}}$

Percentage of combined plate and rivet section =

$$\frac{100\ (\text{pitch} - 2\ \text{dia. rivet}}{\text{pitch}} + \frac{\text{Percentage of rivet section}}{\text{No. rivets in pitch}}.$$

Double butt-straps for this type of joint must each be of thickness given by,—

Thickness of butt-strap = $\dfrac{5 \times \text{thick. of plate} \times (\text{pitch} - \text{dia. of rivet})}{8 \times (\text{pitch} - 2\ \text{dia. of rivet})}$.

Distance from centre of rivet to edge of plate (E in Figs) = $1\frac{1}{2}$ × diameter of rivet.

Distance between rows of rivets :—

(*a*) **Chain riveted joints (Fig. 15),—**

$\left. \begin{array}{l} V = \dfrac{\sqrt{(11\ \text{pitch} + 4\ \text{dia. rivet})\,(\text{pitch} + 4\ \text{dia. rivet})}}{10} \\[2em] \text{or } V = \text{not less than } 2 \times \text{dia. of rivet, and preferably} \\[1em] \qquad \dfrac{(4 \times \text{dia. rivet}) + 1}{2} \end{array} \right\}$ The greater of these two values to be used.

(*b*) **For joint K** (Fig. 15), —

$V_1 = 2 \times$ dia. of rivet as a minimum, but the value $\dfrac{(4 \times \text{dia. of rivet}) + 1}{2}$ is preferable.

(*c*) **Zigzag riveted joints** (Fig. 19), —

$$V = \sqrt{(\tfrac{1}{11}\text{ pitch} + \text{dia. rivet})(\tfrac{1}{10}\text{ pitch} + \text{dia. rivet})}.$$

Diagonal pitches :—

(*a*) **Diagonal pitch** (Figs. 17, 18, 19, 20), —

$$p_{\scriptscriptstyle D} = \tfrac{6}{10}\text{ pitch} + \text{dia. of rivet.}$$

(*b*) **For joint J** (Fig. 19), —

$$P_{\scriptscriptstyle D} = \frac{3\text{ pitch} + 4\text{ dia. rivet}}{10}.$$

Distance between inner rows of rivets (Fig. 19), —

$$V_1 = \frac{\sqrt{(11\text{ pitch} + 8\text{ dia. rivet})(\text{pitch} + 8\text{ dia. rivet})}}{20}.$$

Maximum Pitches for Riveted Joints.

When the work is of first-class quality, the pitches may, so far as safety is concerned, be those given by the following formula, but it is not always advisable to go quite up to these limits. The pitch must never exceed 10 inches, even with the thickest plates.

Maximum pitch $= (C \times \text{thickness of plate}) + 1\tfrac{5}{8}$

where C is a constant having the values given in the following Table :—

Number of rivets in one pitch.	Values of C for lap joints.	Values of C for double butt-strap joints.
1	1·31	1·75
2	2·62	3·50
3	3·47	4·63
4	4·14	5·52
5	...	6·00

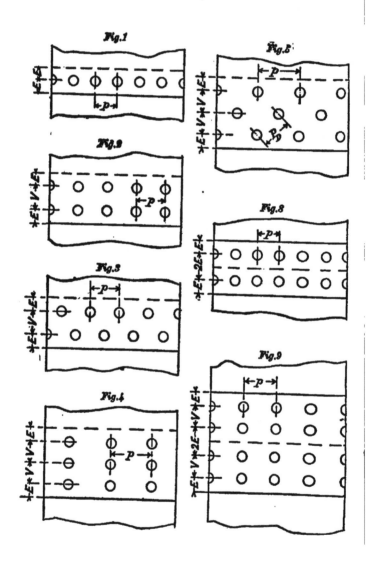

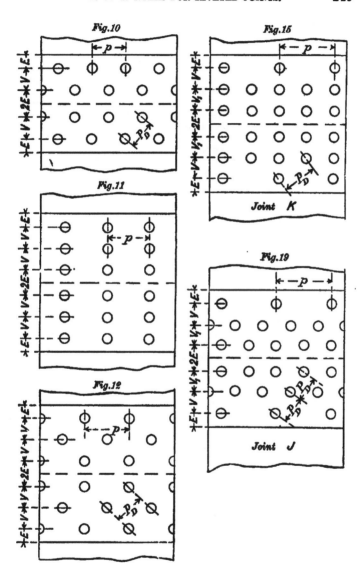

Fig.10

Fig.15

Joint K

Fig.11

Fig.19

Joint J

Fig.12

SUPERVISION OF BOILER WORK.

The following instructions to boiler-maker overseers are those usually given in Admiralty specifications :—

The boilers will be subject to the supervision of an overseer, who will be directed to attend on the premises of the contractors during the progress of the work on the boilers, to examine the material and workmanship used in their construction, to witness the prescribed tests, and to see that this specification, as regards the boilers and work in connection, is conformed to in all respects by the contractors. The extent of supervision is described in the following paragraphs extracted from Admiralty instructions to overseers, and the contractors are to afford him every facility for their proper execution.

The plates and other material used in the construction of the boilers to be subjected to such tests as may be directed in the specification. Every plate used is to be carefully examined by the overseer for laminations, blisters, veins, and other defects, and to ensure that it is of the proper thickness and brand. No plate, angle, &c., which from any cause is considered by the overseer to be unfit for the intended use is to be fitted.

During the construction of the various parts of the boilers, the overseer is to satisfy himself that the dimensions as shown on the approved drawings are being adhered to by the contractors.

Whenever plates are flanged or welded, or in any case where iron or steel is worked in such manner that it is particularly liable to suffer in strength unless carefully handled, the overseer is to be present if possible on all occasions during the time the work on each article is in progress, and he is to fully satisfy himself that it is sound before he allows any part to be put in the boilers.

Samples of the rivets being used for the boilers are to be taken by the overseer during the progress of the work and tested as specified hereafter, and any batches of rivets found defective are to be rejected. Before rivets are put in, the overseer is to see that the plates are brought properly together, and that the holes are fair with one another. He is not to allow drifting on any account, but he is to see that they are carefully rimed fair where necessary. He is also to make sure during the progress of the work that the rivets fill the holes completely, and that the heads are properly set up, well formed, and finished.

The overseer is to see that all internal parts of the boiler are riveted with rivets having heads and points of approved shape, and that any seams he considers necessary are riveted on the fire side. No snap heads are to be allowed in the internal parts. Any proposal for hydraulic riveting the internal parts is to be submitted to the Admiralty, with sketch of the proposed heads and points. In all parts where the rivets are not closed by hydraulic riveting machinery he is to see that the rivet holes are countersunk and that coned rivets are used. All holes in the plates, angles, &c., are to be drilled, and not punched, and are to be drilled in place after bending. The clearance between rivet hole and rivet before closing is not to be greater than approved by the overseer.

The overseer will see that the particulars of the form, dimensions, and pitch of the various stays shown on the drawings are adhered to, and samples of them are to be tested as directed in this specification ; and he will be guided by his experience as a workman in testing and judging of the soundness of the forging and construction of the various stays.

He is to see that palm stays if fitted are forged from the solid and not welded, that all short stays are nutted on all flat surfaces except where otherwise approved and screwed to a pitch of eight threads per inch for stays of 1 inch diameter and above, that the holes for the screwed stays in the water spaces are drilled and tapped together after the furnaces and combustion chambers have been riveted in place in the boiler, that the combustion chamber stays are drilled square to the bevel of the combustion chamber plates, and that no bevel washers are inside the chamber.   Any girder stays used for combustion chambers are to be well bedded on to the tube plates to the satisfaction of the overseer.

The overseer is to see that the arrangement of the zinc plates shown on the approved drawings is adhered to, that the metallic surfaces in contact are filed bright, and that means are adopted to secure a firm grip of the clips by which the plates are attached.

The overseer is to witness the testing, in all cases, of the boiler tubes, in accordance with this specification, before they are put in the boiler.

When the boilers are reported to the overseer by the contractors as being completed, ready for testing by water pressure, the overseer is to witness a preliminary test of them in accordance with the specification, carefully observing with the assistance of gauges and straight edges whether any bulging or deflection of the plates has taken place.

The official test will be conducted on all occasions in the presence of an inspecting officer.   A test pressure gauge is supplied to the overseer from the Admiralty, and the official test is to be made with this gauge.

After the boilers have been tested by water pressure the overseer is to see that they are properly cleaned inside and outside, and then well painted with red lead.   It is important that the whole surface of the boilers should be thoroughly cleansed of scale formed in manufacture before any paint is put on them.   The boilers are not to be exposed to the weather till they are so painted, and properly cleaned and closed up to his satisfaction.

The overseer is to make himself fully acquainted with the progress of the whole of the work in its various stages, to satisfy himself that every part is sound before it is allowed to be put in the boilers, and to see that the following instructions for the treatment of mild steel are strictly complied with.

Treatment of mild steel.—All plates or bars which can be bent *cold* are to be so treated ; and if the whole length cannot be bent cold, heating is to be had recourse to over as little length as possible.

All plates of the boilers are to be flanged by hydraulic pressure, and in as few heats as possible.

In cases where plates or bars have to be heated, the greatest care should be taken to prevent any work being done upon the material after it has fallen to the dangerous limit of temperature known as a "blue heat" say from 600° to 400° Fahrenheit. Should this limit be reached during working, the plates or bars should be reheated.

Plates or bars which have been worked while hot are to be subsequently annealed simultaneously over the whole of each plate or bar.

All plates for boilers and steam-pipes are to be treated as follows for removal of scale :—Previous to work being commenced on them, they are to stand for eight hours in a mixture of 19 of water to 1 of hydrochloric acid. They should be placed on edge and not laid flat. On removal from acid bath, they should be thoroughly brushed and washed in fresh water, and then placed on edge to dry. This treatment is to be carried out on the premises where the boilers or pipes are made.

In cases where any bar or plate shows signs of failure or fracture in working it is to be rejected. Any doubtful cases are to be referred to the Admiralty.

BOILER MOUNTINGS.

Stop-valves.—The diameter of the boiler stop-valves is generally fixed from the previously determined size of main steam pipe at engines. When the branch pipes from two or more separate but similar boilers join together into one main steam pipe, of diameter D, the size of each branch pipe may be that given by,—

Rule 219.—Diameter of Branch pipe to each boiler $= D \sqrt{\dfrac{4}{3n}}$

where n is the number of boilers or of branches.

For the boilers of merchant steamers where the rate of combustion is from 15 to 20 lbs. of coal per foot of grate per hour, the area of the stop-valve, or of the branch steam-pipe, may be,—

Rule 220.—Area of stop-valve (sq. ins) =

$$\left(\frac{\text{Grate area}}{4} + \frac{\text{Heating surface}}{100} \right) \times \sqrt{\frac{100}{\text{Working pressure.}}}$$

A more generally convenient form of rule,—since it takes into consideration the rate of combustion, or the extent to which the boiler is forced,—is,—

Rule 221.

Area of stop-valve (sq. ins.) $= \dfrac{\text{I.H.P.}}{19} \times \sqrt{\dfrac{100}{\text{Working pressure.}}}$

This rule gives about,—

1 sq. in. stop-valve area for each 10·4 I.H.P. at 30 lbs. working pressure.
1   ,,   ,,   ,,   14·7   ,,   60 lbs.   ,,
1   ,,   ,,   ,,   18·0   ,,   90 lbs.   ,,
1   ,,   ,,   ,,   19·0   ,,   100 lbs.   ,,
1   ,,   ,,   ,,   23·4   ,,   150 lbs.   ,,
1   ,,   ,,   ,,   25·7   ,,   180 lbs.   ,,

Where, as in Naval ships, the power taken out of any boiler may be temporarily increased, say 20 to 30 per cent. above natural draught full power, during an emergency, the designer must use his judgment as to what increase in area of stop-valve is advisable ; using the above formula (Rule 221), and supposing I.H.P. to represent the mean power to be developed on forced draught trial, present practice for 150 lbs. would be fairly represented by,—

Rule 221a.

$$\text{Area of stop-valve (sq. ins.)} = \frac{\text{I.H.P.}}{26} \times \sqrt{\frac{100}{\text{Working pressure.}}}$$

When the pipes are below 6 inches in diameter, the proportionally greater friction in the smaller pipes should be allowed for by progressively diminishing the denominator of the I.H.P. fraction in the formula as the diameters of the pipes decrease.

The more important points connected with the design and construction of stop-valves have already been dealt with (pages 206 and 207), and it is only necessary to add here that boiler stop-valves should have turned spigots fitting accurately into the holes in the boiler plates,—which should be carefully cut out by means of a bar and cutter, and *not* by hammer and chisel,—and that the flanges and necks should be extra strong and well ribbed, and the bolts or studs (bolts with heads inside the boiler are best) by which they are attached to the boiler, larger and more numerous than in pipe flanges of similar size : a ⅝-inch bolt should be the smallest size used for attaching boiler mountings,—even for the smallest valve or cock.

In Naval work the boiler stop-valves are generally made entirely of gun-metal, and are of the non-return type : non-return valves are also placed at the various bulkheads, in order to localize as far as possible the effects of injury to boilers or pipes by shot, &c.

Internal steam pipes.—Where the steam room in a boiler is small relatively to the I.H.P. taken from the boiler, internal pipes with closed ends, and provided with sufficient narrow transverse slits, (saw-cuts) or small holes, to give a clear area equal to twice that of the pipe section, should be fitted. They are best made of sheet brass,—*not* copper. The number and arrangement of the pipes must be determined in accordance with the conditions of the case, but the Admiralty usually specify two pipes running the full length of the boiler ; so that, with the stop-valve at one end, there are two pipes leading to it, each of about half the sectional area of the valve or of the branch steam pipe from the boiler, and with the stop-valve on the shell at mid-length there are four shorter pipes converging to it.

The steam is then gently collected from a large area, and strong currents, which might induce priming, are avoided.

Safety-valves.—The size of safety-valve should be such that it is capable of discharging all the steam that can be generated in the boiler, without allowing the pressure to rise more than 10 per cent. above that to which the valve is nominally loaded; it therefore depends mainly on area of grate, rate of combustion, and working pressure. A convenient and easily applied rule is,—

Rule 222.—Area in square inches of each of two valves =

$$\left(\frac{\text{Grate area}}{20} + \frac{\text{Heating surface}}{200}\right) \times \sqrt{\frac{100}{\text{Working pressure.}}}$$

Wherever possible, the valves should be placed with the spindles vertical, as the action of the valve (which is very sensitive to, and easily affected by any increase in the friction of its various parts) is then more certain. For the same reason all the parts should be a very easy fit, and the rubbing surfaces of the spindle should be draw-filed and polished with emery cloth in the same direction.

The whole of the parts must be cleaned from time to time, when in use, and the greatest care must be taken to avoid making any dints or burrs on the working surfaces.

The spring should be so proportioned that its initial compression is not less than half the diameter of the valve; this result can be obtained with various proportions of spring, and when height is limited a shorter spring of larger diameter and larger section of steel may be used.

In Naval practice, the valves, valve-boxes, &c. are usually made entirely of gun-metal, and the valve is generally made as a separate piece (not cast with the spindle), while the easing gear is fitted to lift the valve from below. This construction has the advantage of preventing any bending or springing of the spindle, that may arise from the ends of the spring not being quite true and square by its axis, from affecting the tightness of the valve on its seat; and also permits examination of the valves to be made without disconnecting and taking down easing gear shafts, &c.

The valve faces should always be flat, *not* angled to 45 or 60 degrees.

When the design of the valve permits, it is a great convenience to those who will afterwards have charge of the machinery, to have a thread cut on the upper end of the spindle, so that, before taking the valve to pieces for examination or re-grinding, a nut may be put on to prevent the release or expansion of the spring when the joint at the base of the spring tower is broken; spring, tower, spindle and valve can then be readily and quickly removed in one piece.

When loose valve seats are used (as with cast-iron chests) they should be securely fixed in place by a flange, or lugs, and studs or screws.

The type of valve recommended by the Board of Trade is that shown in Fig. 64.

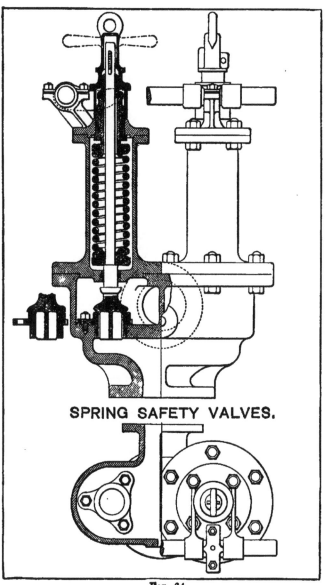

SPRING SAFETY VALVES.

FIG. 64.

The valves, seats, spindles, compressing screws and nuts, spring washers, spindle bushes and cotters, studs and nuts for valve seats, and bushes for bearings of easing gear shafts, should be of good quality brass or gun-metal.

The valve is shown both with and without lip, but valves without lip are recommended as being less violent in action.

The point of the compressing screw should be well rounded and should enter from $\frac{3}{8}$ to $\frac{3}{4}$ inch into the spring washer.

Spiral springs.—The size of steel required by the Board of Trade is given by,—

Rule 223.
$$d = \sqrt[3]{\frac{S \times D}{C}}$$

where d = diameter or side of square of steel *in inches*.

D = diameter of coil (centre to centre of wire) in inches.

S = load on spring in pounds.

$C = \begin{cases} 8000 \text{ for round steel.} \\ 11000 \;,, \text{ square } \;,, \end{cases}$

In Naval work the values adopted for C are commonly 11,000 and 15,000 for round and square steel respectively.

Mr Traill, in his work on "Steam Boilers," recommends that d should always equal $\dfrac{D}{5}$: when this proportion of spring is adopted the above rule becomes,—

$$\left. \begin{aligned} \sqrt{\frac{S}{1600}} &= d \\ \text{and } 1600 d^2 &= S \end{aligned} \right\} \text{for round steel.}$$

$$\left. \begin{aligned} \sqrt{\frac{S}{2200}} &= d \\ 2200 d^2 &= S \end{aligned} \right\} \text{for square steel.}$$

The relation between load on, and compression of, spiral springs is given by,—

Rule 224.   Compression in inches $= \dfrac{D^3 \times S \times N}{d^4 \times a}$

where N = number of coils;

d = diameter or side of square of wire in *sixteenths of an inch;*

$a = \begin{cases} 26 \text{ for round steel,} \\ 32 \text{ for square steel;} \end{cases}$

and the other symbols have the same meanings as above.

In order to obtain correct results with the above formula, great care must be taken to count *free* coils only ; coils at the ends of the spring, which touch one another, must not be counted.

TABLE LXXV.—SAFETY VALVE SPRINGS, ETC.   257

Table LXXV. is obtained by the aid of the above formulæ.

Table LXXV.—Safety Valve Springs by Board of Trade Rules.

Dimensions in inches.		Working loads in lbs.		Dimensions in inches.		Working loads in lbs.	
Diameter or side of square of steel.	Mean diameter of coil.	Round steel.	Square steel.	Diameter or side of square of steel.	Mean diameter of coil.	Round steel.	Square steel.
¼	1¼	100	137	25/32	3 29/32	976	1342
9/32	1 13/32	126	174	13/16	4 1/16	1056	1452
5/16	1 9/16	156	214	27/32	4 7/32	1139	1566
11/32	1 23/32	189	260	7/8	4 3/8	1225	1684
3/8	1 7/8	225	309	29/32	4 17/32	1314	1806
13/32	2 1/32	264	363	15/16	4 11/16	1406	1933
7/16	2 3/16	306	421	31/32	4 27/32	1501	2064
15/32	2 11/32	351	483	1	5	1600	2200
½	2½	400	550	1 1/32	5 5/32	1701	2339
17/32	2 21/32	451	620	1 1/16	5 5/16	1806	2483
9/16	2 13/16	506	696	1 3/32	5 15/32	1914	2631
19/32	2 31/32	564	775	1 1/8	5 5/8	2025	2784
5/8	3 1/8	625	859	1 5/32	5 25/32	2139	2941
21/32	3 9/32	689	947	1 3/16	5 15/16	2256	3102
11/16	3 7/16	756	1039	1 7/32	6 3/32	2376	3267
23/32	3 19/32	826	1136	1¼	6¼	2500	3437
¾	3¾	900	1237	...	...	...	...

Mr Traill, in his above-mentioned work, fixes the values of a at 22 and 30 from experiments he has made, whilst Rankine gave 20 to 23 for round and 28·6 to 32·9 for square steel, and Mr Hartnell's experiments (quoted by Unwin) place the values for round steel at 21 to 24·8,—the higher figure for ¼ inch wire, and the lower for ⅜ inch.

The superior transverse elasticity of the round sections, and especially of small round sections, compared with square ones, is probably due to the fact that the round wires are often drawn, whilst the square ones are only rolled; also the square section of steel is probably more affected by the coiling process than the round one.

For Board of Trade and Lloyd's rules relating to safety valves, see end of this section.

Feed check valves.—The sizes of feed pipes should be determined in accordance with the rules laid down in the section on "Feed pumps" (pp. 139-144), and the internal diameter of the valve seat of check valve should be at least ¼ inch more than the diameter of pipe so found. Feed check valves should be very strongly made, and entirely of the best gun-metal; the necks by which they are attached to the boilers should be specially strong, and well ribbed to

17

the flanges. The spigots should be turned, and the holes in the boiler plates carefully bored to suit. The working faces of the valves, should be flat, and proportioned in accordance with Rule 155 (page 141). The spindles should be very stout, and provided with square threaded screws ; if space allows the nuts to be placed in external crossheads or bridges, they are better so arranged.

The Admiralty require that the main feed check valve shall always be placed at the right hand, as one faces the boiler and the auxiliary valve at the left hand.

Internal feed pipes.—The feed water should be led by an internal pipe (of brass) placed two or three inches below the water level, to a part of the boiler where there is a descending current, and there delivered downwards through a number of fine transverse slits or small holes. Care must be taken that this internal pipe is always filled with water, and not partly with water and partly with steam,—as in the latter case a severe hammering action would be set up, the joints would be started, and the pipe very quickly destroyed.

Blow-off, and scum, valves.—These should also be stoutly made valves of gun-metal,—ribbed as above directed for feed check valves,—and are best fitted so that the pressure tends to hold the valve on its seat, in order to reduce the risk of leakage as much as possible. Now that fresh water is so much used, and loss made up by means of evaporators, and from reserve tanks, the blow-off valve is much less necessary than formerly, and in some recent Naval vessels it has been entirely omitted.

The scum valve should have an internal pipe leading to it from a circular " scum pan " or dished plate of sheet brass (about 15 inches diameter) fixed near to the centre of the water surface, and at about the lowest working level; or, if preferred, the internal pipe may reach a little beyond the centre of water surface, and may be closed at the end, and have simply a few longitudinal slits in its upper side to admit the scum.

The clear area through the blow-off valve may be that given by,—

Rule 225.

$$\text{Area of blow-off pipe (square inches)} = 1 + \frac{\text{Tons of water in boiler}}{5}.$$

The area of scum pipe may be one-third that of blow-off pipe.

Water gauge.—All boilers, excepting small donkey boilers, should have two sets of water gauge fittings, and large double-ended boilers are better with three sets,—one of which may be fitted direct to the shell.

The gauge cocks and glass tubes should be carried on gun-metal stand-pipes, from which copper pipes should be led to the steam and water spaces of the boilers.

The internal diameter of the stand-pipe and its connections should not be less than 1 inch, and for large boilers 1½ inch is desirable, and the connections should be made of extra stout piping,—say not thinner than 9 L.S.G.

No cocks or valves should be fitted where these connections join the boiler,—as serious accidents have occurred through their use,—and care should be taken not to place the inlets near to any opening through which there may be a current of steam or water, since the level of the water in the glass may be affected thereby,—a difference of pressure of one-tenth of a pound causing an alteration in level of 2·7 inches.

Also, in fixing the level at which the water is to stand in the glass, the lowering of level due to the cooling and consequent contraction of the water in the stand-pipe connection should not be forgotten,—though it is always an error on the side of safety.

Rankine gave the following rule for determining the volume of water at different temperatures:—

Let the volume of the water at its temperature of maximum density (39·2° F.) be represented by unity, and let its volume at T° be V, then,—

$$V = \frac{1}{2}\left(\frac{T+461}{500} + \frac{500}{T+461}\right).$$

He calls this an approximate rule, but states that the error is only one four-hundredth.

Circulating apparatus.—It is very desirable to have large rigidly constructed boilers fitted with some apparatus for circulating the water whilst getting up steam ; if hydrokineters are not fitted, connections should be made with one of the auxiliary steam pumps.

In almost all cylindrical boilers the rate of evaporation may be improved by fitting a proper arrangement of circulating plates ; the first cost is something, and they are rather in the way when the boiler is being cleaned, but the gain is often worth the cost and trouble.

An air-cock should be fitted at the highest point of each boiler.

Admiralty proportions of safety valves.—For 150 lbs. pressure the Admiralty are at present fitting valves in the larger classes of vessel, with a total gross area of about ·4 square inch per foot of grate.

The Board of Trade Rules referring to boiler mountings, &c., are as follows:—

110. No boiler or steam chamber should be so constructed, fitted, or arranged that the escape of steam from it through the safety valves required by the Act of Parliament can be wholly, or partially, intercepted by the action of another valve.

A stop valve should always be fitted between the boiler and the steam pipe, and, where two or more boilers are connected with a steam receiver or superheater, between each boiler and the super-heater or steam receiver. The object of this is obvious, viz., to avoid the failure of all the boilers through the failure of one. The necks of stop valves should be as short as practicable.

Declarations should not be given for new steamers, for steamers that have not had passenger certificates before February 1871, nor for steamers that have had new boilers since February 1871, unless stop valves are fitted between the boilers and the steam pipes.

111. Each boiler should be fitted with a glass water-gauge, at least three test cocks, and a steam gauge. Boilers that are fired from both ends, and those of unusual width, should have a glass water gauge and three test cocks at each end or side, as the case may be. When a steamer has more than one boiler, each boiler should be treated as a separate one, and have all the requisite fittings.

When the upper and lower glass water-gauge cocks are not attached directly to the shell of the boiler, but a stand pipe or column is fitted, cocks or valves, although the former are more desirable, should as a general rule be fitted between the boiler and the stand pipes, &c. Cocks or valves need not, however, be insisted on in cases where the columns, stand pipes, &c., are of moderate length and of extra strength, provided that the diameter of the bore at any part is not less than 3 inches.

If the column, stand pipes, &c., are of less diameter than 3 inches, and the pipes are bolted to the boiler without the intervention of cocks or valves, the arrangement need not be objected to, if otherwise satisfactory, providing there is no difficulty in keeping the passages at the ends clear, and ascertaining that they are so. To do this it will be necessary that the passage in the part of the column between the top and bottom gauge-glass cocks be cut off or closed, which may be done permanently, or by the interposition of a cock or valve at that part. The latter is a convenient and desirable arrangement even when cocks are fitted on the boiler.

In the case of high pressures, it is desirable that the cocks or valves which prevent the escape of water or steam from the boiler, be fitted with handles which can be expeditiously manipulated from a convenient position.

It is desirable in all cases that test cocks be fitted directly to the skin of the boiler, and when the passage in the part of the column to which the glass water-gauge cocks are attached (if so fitted) is permanently cut off, the test cocks *must* be fitted directly to the skin of the boiler.

The Surveyors should satisfy themselves by actual examination whether the glass-water gauges of the boilers of the vessels they survey are fitted with automatic valves or fittings, as the existence of such fittings cannot always be ascertained by external examination. In all cases full particulars of automatic gauges should be submitted for consideration before the gauges are passed.

The Board of Trade Rules referring to Safety-Valves are as follows:—

123. The Engineer Surveyor shall declare, amongst other things, the limits of the weight to be placed on the safety-valves; that the safety-valves are such, and in such condition as required by the Act, and

that the machinery is sufficient for the service for the time he fixes, and is in good condition for that time.

The locked-up valves, *i.e.*, those out of the control of the engineer when steam is up, should have an area not less, and a pressure not greater, than those which are not locked up, if any such valves are fitted.

Cases have come under the notice of the Board of Trade in which steamships have been surveyed and passed by the Surveyors, with pipes between the boilers and the safety-valve chests. Such arrangement is not in accordance with the Act, which distinctly provides that the safety-valves shall be upon the boilers.

The Surveyors are instructed that in all *new boilers*, and whenever *alterations can be easily made*, the valve chest should be placed directly on the boiler ; and the neck, or part between the chest and the flange which is bolted on to the boiler, should be as short as possible, and be cast in one with the chest.

The Surveyors should note that it is not intended by this instruction that vessels with old boilers which have been previously passed with such an arrangement should be detained for the alterations to be carried out.

Of course, in any case in which a Surveyor is of opinion that it is positively dangerous to have a length of pipe between the boilers and the safety valve chest, it is his duty at once to insist on the requisite alterations being made before granting a declaration.

If any person place an undue weight on the safety-valve of *any* steamship, or in the case of steamships surveyed under the Act, increase such weight beyond the limits fixed by the Engineer Surveyor, he shall, in addition to any other liabilities he may incur by so doing, incur a penalty not exceeding one hundred pounds.

124. The area per square foot of fire grate surface of the locked-up safety-valves should not be less than that given in the following Tables opposite the boiler pressure intended, but in no case should the valves be less than two inches in diameter. This applies to new vessels or vessels which have not received a passenger certificate.

When, however, the valves are of the common description, and are made in accordance with the Tables, it will be necessary to fit them with springs having great elasticity, or to provide other means to keep the accumulation within moderate limits ; and as boilers with forced draught may require valves considerably larger than those found by the Tables, the design of the valves proposed for such boilers, together with the estimated coal consumption per square foot of fire-grate, should be submitted to the Board for consideration.

In ascertaining the fire-grate area, the length of the grate should be measured from the inner edge of the dead plate to the front of the bridge, and the width from side to side of the furnace on the top of the bars at the middle of their length.

In the case of vessels that have not had a passenger certificate, if there is only one safety-valve on any boiler, the Surveyor should not grant a declaration without first referring the case to the Board for special instructions.

Safety-Valve Areas (Board of Trade).

Boiler Pressure.	Area of Valve per square foot of Fire-grate.	Boiler Pressure.	Area of Valve per square foot of Fire-grate.	Boiler Pressure.	Area of Valve per square foot of Fire-grate.
15	1·250	54	·543	93	·347
16	1·209	55	·535	94	·344
17	1·171	56	·528	95	·340
18	1·136	57	·520	96	·337
19	1·102	58	·513	97	·334
20	1·071	59	·506	98	·331
21	1·041	60	·500	99	·328
22	1·013	61	·493	100	·326
23	·986	62	·487	101	·323
24	·961	63	·480	102	·320
25	·937	64	·474	103	·317
26	·914	65	·468	104	·315
27	·892	66	·462	105	·312
28	·872	67	·457	106	·309
29	·852	68	·451	107	·307
30	·833	69	·446	108	·304
31	·815	70	·441	109	·302
32	·797	71	·436	110	·300
33	·781	72	·431	111	·297
34	·765	73	·426	112	·295
35	·750	74	·421	113	·292
36	·735	75	·416	114	·290
37	·721	76	·412	115	·288
38	·707	77	·407	116	·286
39	·694	78	·403	117	·284
40	·681	79	·398	118	·281
41	·669	80	·394	119	·279
42	·657	81	·390	120	·277
43	·646	82	·386	121	·275
44	·635	83	·382	122	·273
45	·625	84	·378	123	·271
46	·614	85	·375	124	·269
47	·604	86	·371	125	·267
48	·595	87	·367	126	·265
49	·585	88	·364	127	·264
50	·576	89	·360	128	·262
51	·568	90	·357	129	·260
52	·559	91	·353	130	·258
53	·551	92	·350	131	·256

Saety-Valve Areas (Board of Trade)—*continued.*

Boiler Pressure.	Area of Valve per square foot of Fire-grate.	Boiler Pressure.	Area of Valve per square foot of Fire-grate.	Boiler Pressure.	Area of Valve per square foot of Fire-grate.
132	·255				
133	·253				
134	·251	155	·220	178	·194
135	·250	156	·219	179	·193
136	·248	157	·218	180	·192
137	·246	158	·216	181	·191
138	·245	159	·215	182	·190
139	·243	160	·214	183	·189
140	·241	161	·213	184	·188
141	·240	162	·211	185	·187
142	·238	163	·210	186	·186
143	·237	164	·209	187	·185
144	·235	165	·208	188	·184
145	·234	166	·207	189	·183
146	·232	167	·206	190	·182
147	·231	168	·204	191	·181
148	·230	169	·203	192	·181
149	·228	170	·202	193	·180
150	·227	171	·201	194	·179
151	·225	172	·200	195	·178
152	·224	173	·199	196	·177
153	·223	174	·198	197	·176
154	·221	175	·197	198	·176
		176	·196	199	·175
		177	·195	200	·174

125. The Surveyor, in his examination of the machinery and boilers is particularly to direct his attention to the safety-valves, and wheneve he considers it necessary, he is to satisfy himself as to the pressure o the boiler by actual trial.

The Surveyor is to fix the limits of the weight to be placed on th safety-valves, and the responsibility of issuing a declaration before h is fully satisfied on the point is very grave. The law places on th Surveyors the responsibility of "declaring" that the boilers are in hi judgment sufficient with the weights he states.

The Surveyor is to examine the whole of the valves, weights, an springs at every survey.

The responsibility of seeing to the efficiency of the mode by whicl the valves are fitted so as to be out of the control of the enginee when steam is up rests with the Surveyor, as long as it is efficient and the method adopted is approved of by the Board of Trade.

The safety-valves should be fitted with lifting gear, so arranged that the two or more valves on any one boiler can at all times be eased together, without interfering with the valves on any other boiler. The lifting gear should in all cases be arranged so that it can be worked by hand either from the engine-room or stoke-hole.

Care should be taken that the safety-valves have a lift equal to at least one-fourth their diameter; that the openings for the passage of steam to and from the valves, including the waste-steam pipe, should each have an area not less than the area of valves required by clause 124; and that each valve box has a drain pipe fitted at its lowest part. In the case of lever-valves, if the lever is not bushed with brass, the pin must be of brass; iron and iron working together must not be passed. Too much care cannot be devoted to seeing that there is proper lift, and free-means of escape of waste steam, as it is obvious that unless the lift and means for escape of waste steam are ample, the effect is the same as reducing the area of the valves or putting on an extra load. The valve seats should be secured by studs and nuts.

The Surveyors are, as far as in their power, to make the opinion of the Board on these points generally known to the owners of passenger steamers.

126. When the Surveyor has determined the amount of pressure he is to see the valves weighted accordingly, and the weights or springs fixed in such a manner as to preclude the possibility of their shifting or in any way increasing the pressure. The limits of the weight on the valves is to be inserted in the declaration, and should it at any time come to a Surveyor's knowledge that the weights or the loading of the valves have been shifted, or otherwise altered, or that the valves have been in any way interfered with, so as to increase the pressure, without the sanction of the Board of Trade, he is at once to report the facts to the Board of Trade.

127. If the following conditions are complied with the Surveyor need raise no question as to the substitution of spring loaded valves for dead weighted valves :—

(1.) That at least two valves are fitted to each boiler.

(2.) That the valves are of the proper size, as by clause 124.

(3.) That the springs and valves are so cased in that they cannot be tampered with.

(4.) That provision is made to prevent the valves flying off in case of the springs breaking.

(5.) That the requisite safety-valve area is cased in and locked up in the usual manner of the Government valves.

(6.) That screw lifting gear is provided to ease all the valves, as by clause 125.

(7.) That the size of the steel of which the springs are made is in accordance with that found by the following formula :

$$\sqrt[3]{\frac{s \times D}{c}} = d :$$

s = the load on the spring in lbs.

D = the diameter of the spring (from centre to centre of wire) in inches.

d = the diameter, or side of square, of the wire in inches.

c = 8000 for round steel.

c = 11,000 for square steel.

(8.) That the springs are protected from the steam and impurities issuing from the valves.

(9.) That when valves are loaded by direct springs, the compressing screws abut against metal stops or washers, when the loads sanctioned by the Surveyor are on the valves.

(10.) That the springs have a sufficient number of coils to allow a compression under the working load of at least one quarter the diameter of the valve.

128. In no case is the Surveyor to give a declaration for spring-loaded valves, unless he has examined them and is acquainted with the details of their construction, and unless he has tried them under full steam, and full firing, for at least 20 minutes with the feed-water shut off and stop-valve closed, and is fully satisfied with the result of the test. In special cases, or when the valves are of novel design, the results of the tests under full steam should be reported to the Board, but if the Surveyors are fully satisfied with the results of the tests they need not delay the granting of the declaration for the vessel subject to approval of the Board. If the accumulation of pressure exceed 10 per cent. of the loaded pressure, he should not give his declaration without first reporting the case to the Board of Trade, accompanied by a sketch, and full particulars of the trial and the strength pressure of the boilers.

129. In the case of valves, of which the principle and details have already been passed by the Board of Trade, the Surveyor need not require plans to be submitted so long as the details are unaltered, of which he must fully satisfy himself ; but in any new arrangement of valves, or in any case in which any detail of approved valves is altered, he should, before assuming the responsibility of passing them, report particulars, with a drawing to scale, to the Board of Trade. He can make this drawing himself from the actual parts of the valves fitted, but in order to save time, and to facilitate the survey, the owners or makers of engines may prefer to send in tracings of their own, before the valves are placed on the boiler. If they do this the survey can be more readily made, and delay and expense may be saved to owners, as the Surveyor will not then have to spend his time, and delay the ship, in preparing drawings and comparing them with the valves.

The tracings of new safety-valve designs should, if possible, be transmitted to the Board of Trade for consideration before the construction of the safety-valves is commenced.

In some spring valves the accumulation of pressure has reached cent. per cent., and therefore if the Surveyor had not required a trial, he would have passed valves which would have caused a pressure on the boiler double that intended by him. And in some cases in which the increase of pressure has not been great, defects that would have rendered the valves highly dangerous have been discovered on an examination of drawings.

The Surveyors should arrange with manufacturers so that the Surveyors may have the designs of valves which the manufacturers intend to use. An easy method of facilitating this matter is for the manufacturer to leave in the local Surveyor's office a plan or plans of his valve or valves when once agreed to, and then afterwards to inform the Surveyor that the valves fitted are according to drawing A, B, or C, as the case may be. By this means, when once a design has been agreed upon, and is adhered to, all subsequent questions and delays will be prevented.

131. It is clearly the duty of the masters and engineers of vessels to see, in the intervals between the surveys, that the locked-up safety-valves, as well as the other safety-valves and the rest of the machinery, are in proper working order. There is no provision in the Merchant Shipping Act, 1854, exempting the owner of any vessel, on the ground that she has been surveyed by the Board of Trade Surveyors, from any liability, civil or criminal, to which he would otherwise be subject. The Act of Parliament requires the Government safety-valves to be out of the control of the engineer when the steam is up; this enactment, far from implying that he is not to have access to them, and to see to their working, at proper intervals when the vessel is in port rather implies the contrary; and the master should take care that the engineer has access to them for that purpose. Substantial locks that cannot be easily tampered with, and as far as possible weather-proof, should be used for locking up the safety-valve boxes.

132. In witnessing the hydraulic tests of boilers, &c., and in witnessing all safety-valve tests for accumulation of pressure, the Surveyors are to use the pressure gauges supplied by the Board of Trade for the purpose. The steam gauge should not be used without a syphon filled with water between it and the boiler, and in all cases in which the Surveyors have to adjust the safety-valves of passenger steamships they should state in the Remarks column of their declarations which of the Board's gauges was used in making the adjustment of each set of valves.

The rules relating to feed check valves and feed pipes have already been given at the end of the section on "Feed pumps, &c." (page 143), and those relating to blow-off cocks and pipes at end of section on "Sea valves" (page 192).

Lloyd's rules relating to boiler mountings, &c., are as follows:—

17. Two safety-valves are to be fitted to each boiler, and loaded to the working pressure in the presence of a Surveyor. In the case of boilers of greater working pressure than 60 lbs. per square inch, the safety-valves may be loaded to 5 lbs. above the working pressure. If common valves are used, their combined areas are to be at least half a square inch to each square foot of grate surface. If improved valves are used, they are to be tested under steam in the presence of the Surveyor; the accumulation is in no case to exceed 10 per cent. of the working pressure.

18. An approved safety-valve is also to be fitted to the super-heater.

19. In winch boilers one safety-valve will be allowed, provided its area is not less than half a square inch per square foot of grate surface.

20. Each valve is to be arranged so that no extra load can be added when steam is up, and must be fitted with easing gear which lifts the valve itself. All safety-valve spindles are to extend through the covers, and are to be fitted with sockets and cross handles, so that the valves can be lifted and turned round in their seats, and their efficiency tested at any time.

21. Stop valves are to be fitted so that each boiler can be worked separately.

22. Each boiler is to be fitted with a separate steam gauge, to accurately indicate the pressure.

23. Each boiler is to be fitted with a blow-off cock, independent of that on the vessel's outside plating.

FURNACE FITTINGS.

Furnace fronts, &c.—Furnace fronts and fire doors should always be of wrought-iron or steel,—never of cast-iron; but the internal protecting or baffle plates are better of cast-iron, since it burns away less rapidly than the wrought material. The baffle plate should be in several small pieces, free to expand in all directions, rather than in one large piece, which would probably very soon crack and get adrift.

The size of fire door, in the clear, may vary from 12 inches high by 18 inches wide, to 16 inches high by 24 inches wide,—the top of the opening being arched, and struck with a radius equal to the *width* of the opening.

A very good arrangement is that in which the otherwise useless corners to right and left of, and above the fire door, inside the furnace, are filled in by curved cast-iron plates, perforated with small holes to allow the air to enter the furnace; the air is thus heated to a certain extent in these boxes or chambers before coming into contact with the fuel, and the riveted joint round the furnace mouth is protected from the fire.

Furnaces over 3 feet 6 inches diameter are perhaps better with two half doors hinged right and left, and meeting in the centre, only one of which need be opened at once ; less cold air is then admitted, and one side of the fire can be attended to at a time.

Means should be provided for holding the doors open in a sea-way.

Fire-bars, &c.—When chimney draught only, or chimney and forced draught not exceeding ·5 in. of water, is used, the bars are better in one length,—up to 5 feet, or even to 5 feet 6 inches long ; when a greater air pressure is used, or when the grate is longer than 5 feet 6 inches, two lengths of bars may be used.

When one long bar is used, it should be hooked to the inner edge of the dead plate, and free to expand inwards, and slide on the bridge-plate ; when two lengths are used, they should be hooked to the central bearer-bar, and free to slide on both the bridge-plate and the dead-plate.

In either case care should be taken that the dead-plate is formed so as to hold or support the bars with their faces or upper surfaces flush with its own upper surface.

The thickness of bar and width of air space between bars must depend on the class of coal that will generally be used, and on the air pressure with which it is intended to work.

The following are good average dimensions for fire-bars :—

Thickness on face, ...	1	inch.
" at bottom edge, ...	⅝	"
" on face, over distance pieces, .	1½	"
" at bottom edge, ...	1¼	"
Depth at centre, ...	·6$\sqrt{}$ length in inches.	
" near ends, ...	2	"
Width of air spaces, ...	½	inch.

The slope of the grate surface should never be less than 1 inch per foot of length, and is better 1½ inch, or even more, when possible.

When corrugated or ribbed furnaces are used, the two side bars should be made to templates from the furnaces, and should fit as closely as possible into all the recesses.

Fire-bars should, of course, be made of the most refractory iron obtainable ; fine grey irons are quite unsuitable.

Bridges, &c.—In return-tube boilers the grate should never be so long that the front face of the bridge is less than 9 inches from the face of the back tube-plate.

The height of bridge should be such that the clear area above it may be from ⅙th to ⅛th of the area of grate ; this proportion is obtained approximately when the clear height above the bridge at its centre is ⅓rd of the diameter of furnace.

In the case of corrugated or grooved furnaces, the ash-pits require to be fitted with thin lining plates to enable the rake to be used. ʼn Naval and other ships in which forced draught is used to any considerable extent, shallow pans which can be kept full of water are necessary.

Each ash-pit should also be fitted with a good stout pricker bar and with a damper; when the closed stoke-hole system of forced draught is used, the dampers are sometimes balanced and made to open inwards only,—closing against any pressure that may come from the furnace side.

LADDERS AND PLATFORMS, &c.

The following Table gives ordinary dimensions of ladders and gratings :—

Table LXXVI.—Ladders and Gratings.

	Ladders.				"Spill" Gratings.			
Width of ladder.	Size of side bars.		Width of cast-iron steps.	Bar steps, No., size, and pitch of bars.	Width of grating.	Size of side bars.	Diameter of "spills."	Pitch of "spills."
	Cast-iron steps.	Round bar-iron steps.						
in.	in. in.	in. in.	in.	in.	in. in.	in. in.	in.	in.
12	...	$2\frac{1}{4} \times \frac{3}{8}$	...	one $\frac{5}{8}$	15 & 18	$2\frac{3}{4} \times \frac{3}{8}$	$\frac{5}{8}$	$2\frac{1}{4}$
15	...	$3\frac{1}{4} \times \frac{3}{8}$	...	{ two $\frac{5}{8}$ / $1\frac{3}{8}$ pitch }	21	$3 \times \frac{7}{16}$	$\frac{3}{4}$	$2\frac{3}{8}$
18	$3\frac{1}{2} \times \frac{3}{8}$	$3\frac{1}{2} \times \frac{3}{8}$	$4\frac{1}{4}$	{ two $\frac{5}{8}$ / $1\frac{1}{2}$ pitch }	24	$3 \times \frac{7}{16}$	$\frac{3}{4}$	$2\frac{3}{8}$
21	$3\frac{3}{4} \times \frac{7}{16}$	...	$4\frac{3}{4}$	...	27	$3 \times \frac{7}{16}$	$\frac{3}{4}$	$2\frac{3}{8}$
24	$4 \times \frac{7}{16}$	...	$5\frac{1}{4}$	...	{ 30 and above }	$3 \times \frac{7}{16}$	{ $\frac{5}{8}$ with centre support }	$2\frac{1}{4}$

$\frac{5}{8}$-inch round iron "spills" should not be used for spans over 18 inches.

$\frac{3}{4}$-inch round iron "spills" should not be used for spans over 27 inches.

For wider spans, additional supporting bars, of the same section as the side bars, should be used.

Ladder steps should be from 9 to 10 inches apart (face to face).

The corners of cast-iron steps should be rounded away, and they should be attached to the side bars of the ladder by two ½-inch bolts at each end.

The main engine-room ladder should be 21 inches wide where possible, and in large ships, where there is plenty of room, 24 inches looks better.   Stoke-hole ladders need not exceed 18 inches in width.

Stoke-hole ladders are not usually fitted with cast-iron steps.

The inclination of a ladder to the vertical may be almost anything, and depends on the space available, and purpose for which ladder is fitted; the main engine-room ladder should be 1 in 2¼ when possible.

Handrails should be of solid wrought-iron, 1 inch in diameter.

Stanchions, when 3 feet high, may taper from 1 inch diameter at the top to 1¼-inch at the bottom ; when short, for ladders, ⅞-inch at top to 1 inch at bottom is enough.   The ball through which the rail passes may be 2 inches in diameter.

Engine and boiler room floors are best laid with chequered wrought-iron plates,—engine-room, $\frac{5}{16}$-inch thick, and boiler-room, $\frac{3}{8}$-inch, exclusive of the raised pattern.   In the Navy, ¼-inch and $\frac{5}{16}$-inch respectively are the usual thicknesses, but the numerous supporting bars necessary are supplied by the shipbuilders, and their weight is therefore not included under weight of machinery.

It is very desirable that floor plates should be secured to the bearers where possible, as cases have occurred in which an accumulation of water surging from side to side in the stoke-hole, has lifted the plates and driven them against the sea valves and pipes, and thus caused the loss of the vessel.

ENGINE AND BOILER SEATINGS, &c.

It is extremely difficult to lay down any general rules that will be of service in designing engine and boiler seatings, because so much depends on the type and structure of ship, strength and stiffness of engine framing or bedplate, type of boilers and position in which they are to be placed, &c., but the following hints may be of some use.

Seatings for vertical engines.—When the seating must be built upon the top edges of the ship's floors, it generally consists of two box girders, one under each side of the bedplate, parallel to the shaft axis. Whether additional cross girders should be fitted under each main bearing depends on the strength of the framing or bedplate.

Such girders usually have side plates varying from ½-inch to ¾-inch in thickness and top plates from ¾-inch to 1 inch thick, and the riveting varies from ¾-inch rivets at 4 inch pitch to ⅞-inch rivets at 4¾ or 5 inches pitch.

One of the chief difficulties with this type of structure is to get a sufficiently good attachment to the floors ; double reverse bars should always be fitted under the engines and boilers, and for large and heavy engines the attachment of the vertical side plates to the reverse bars should also be by means of double angle bars, so that there may be four rivets at every crossing point. The second bar is sometimes represented by a separate short piece at each frame. In very lightly built ships it is often advisable to carry the vertical plates down between the floors, and to attach them to the skin plating as well as to the reverse bars.

In all cases where rigid girders are added to the structure of the ship for the purpose of properly distributing weights or strains, care should be taken that they do not stop abruptly at any point, bulkhead or otherwise, but gradually decrease in section, or taper down for three or four frame spaces, as otherwise serious results may ensue from the localization of the flexure or spring of the structure.

Very large and heavy engines are generally so constructed as to require only a plain flat surface of the same length and breadth as the bedplate for a seating, and where the ship is built with open floors this should be obtained by placing a number of longitudinal girders across the tops of the floors and plating them over, using only so many athwartship girders as may be absolutely necessary.

When the vessel is of cellular construction and has an inner and outer skin, box girders may be placed upon the inner bottom,—the vertical plates being arranged to coincide with the longitudinals in the double bottom,—or the seating may be constructed by simply strengthening the longitudinal and athwartship girders already in the double bottom, putting an additional short length in here and there, and increasing the thickness of the inner bottom plating.

In this latter case, the holding down bolts should each be screwed through the top plating from below so as to make a water-tight joint.

In either of the cases, if the shipbuilders are communicated with in time, there will not generally be any difficulty in modifying the spacing of the longitudinals slightly to suit the engines; a few inches in height may also sometimes be gained by forming troughs or recesses in the inner bottom under the cranks.

Where box girders are placed upon the inner skin as described above, the Admiralty rule is that they shall be made water-tight, in order to prevent internal corrosion.

Seatings for horizontal engines.—These are usually plain flat surfaces constructed as described above for vertical engines and of sufficient extent to receive the whole of the cylinders and main-bearing frames. In Naval vessels the seating is usually formed by raising the inner bottom to the required height (thus increasing the depth of the double bottom) and increasing the scantlings as necessary.

Of course the engines are not always placed exactly horizontal, nor is the seating for the main bearing frames always the same height as that for the cylinder feet.

The seatings for paddle engines are usually constructed on one of the above described plans, or on a combination of them, and do not require any special description. The cases in which the framing of the engine is constructed of plates and angles and built into the ship are of too special a nature to be advantageously treated here; and the various methods of stiffening the sides of the ship and attaching the brackets for carrying the outer bearings also fall into the same category.

Thrust block seatings should have specially strong and well extended attachments; they should extend over at least four frame spaces in small vessels, and over six or more in large ones, and as many as possible of the longitudinals of the engine seating should be made continuous with those of the thrust seat.

Holding down bolts should be numerous and well distributed, and should have the points slightly burred over to prevent the slacking back of the nuts.

Staying of engines.—The cylinders, &c., of vertical engines should never under any circumstances be stayed to the decks or upper works of the ship; nor should the two sets of engines in a twin-screw vessel be stayed to one another; there is no difficulty in obtaining all necessary stiffness by the expenditure of a little care and thought in designing the framing, and the risks run by using stays are serious, both when the ship is in a sea-way and when she is on the blocks in dry dock. In paddle vessels it is not practically possible to avoid connection between the engines and the upper works of the ship, but even here, where a certain amount of spring is allowed for, cracked entablatures, or top frames, are by no means uncommon.

Boiler seatings.—When either single or double-ended boilers are placed with their axes athwartships, the best type of seating is that in which H section girders (10 or 12 inches deep) running the full length of the boiler room are used, their lower flanges being riveted to the reverse bars, or to the inner skin, and their upper flanges carrying the wedge shaped or roughly triangular chocks which hold the boilers in place. When the vessel is of cellular construction these girders should be arranged to coincide with the longitudinals in the double bottom.

When the boilers are placed with their axes fore and aft, and each cradle or bearer is built upon the top of a separate floor, the proper distribution of the weight is more difficult, but must be effected by putting in additional longitudinals, either intercostally or, preferably, on top of the floors. Care must be taken that these additional longitudinals do not prevent access to the underside of the boiler; if they are made of considerable depth manholes may be cut in them. When the construction of the vessel is cellular there is not usually any need to supplement the longitudinal connections.

Single ended boilers should have two cradles or pairs of chocks; double ended boilers of moderate size and weight, three; and very large and heavy double ended boilers, four.

The boilers should be prevented from moving end-ways by " toe " plates or brackets riveted to some convenient portion of the seating or of the vessel's structure.

The greatest care must also be taken so to secure the boilers in their seats that no probable movement of the vessel will throw any strain upon any of the pipes or connections. This is best done by riveting to the upper part of the shell of each boiler four plates, each of which has a "single eye" forged on it, and leading from each a rod or link to some convenient part of the vessel's structure (such as stringer or deck-beam), or to the neighbouring boiler. These eyes may be riveted in place when the riveting of the shell is done, and they then serve to sling the boiler by. Brackets of plate and angle-iron attached to the deck-beams and almost touching the top of the shell are also sometimes convenient as a means of fixing the boilers in their seats.

The present Admiralty method of securing the boilers is to rivet four eyes to the shell at about a foot above the top of the seatings and in the same planes with them, and four other similar eyes to the tops of the seatings themselves, and then connect each pair of eyes by a pair of flat links and pins, &c.

Lloyd's Rules relating to Seatings and to machinery spaces generally are as follows :—

Section 26 (Seatings, &c.)—1. In steam vessels care must be taken that the engine and boiler bearers are properly constructed, having efficient longitudinal ties ; and where the bearers may interfere with the longitudinal strength of the vessel, they must extend a sufficient distance beyond the bulkheads of the engine and boiler space to compensate for such interruption.

2. Where it is intended to fit engines of greater power than in ordinary cargo carrying steamers, the engine seating should be of proportionately greater strength, and be specially adapted with this object in view by being connected to the sides of the vessel ; and other means adopted to ensure greater rigidity and strength to withstand the extra vibration produced in this part of the vessel . . . . . .

3. As many upper, middle, and hold or lower-deck, beams of extra strength, having double angles at upper and lower edges, of sizes as per Table G 4, are to be introduced in the engine and boiler space as may be practicable ; all such beams to be pillared wherever practicable.

4. In the engine and boiler space, double reversed angles must be fitted to every floor, from bilge to bilge, and from margin plate to margin plate in vessels having double bottoms ; and in vessels where the number for plating is 15,000 and above (excepting in way of double bottoms), or the depth from the top of keel to top of hold beams is 17 ft. or above, they are to extend sufficiently high to admit of the bilge stringer angles being riveted to them, unless the bilges are otherwise additionally strengthened by web-frames, beyond the requirements of the Rules. . . . . . .

(Here follows a statement of the requirements as to strength, pitch, and attachments, &c., of web frames for vessels of various numerals.)

5. Where continuous bilge or side stringer bars pass through the web-frames, efficient compensation to be introduced in way of the same.

6. When hold beams are omitted in the engine and boiler space, the web-frames are to be closer spaced than above described.

18

7. Where it is desired to adopt other plans than the foregoing for maintaining the necessary rigidity in the engine and boiler space, sketches of the same must be submitted for the approval of the Committee.

(Paragraph 8 states the requirements as to strength and stiffening of shaft tunnels and provides that a water-tight sluice door, capable of being worked from the upper deck, is to be fitted on the engine-room bulkhead, at the entrance to the tunnel.)

Section 27 (Valves, etc.).—2. The shut-off valves or cocks of all openings for the inlet or outlet of water, in connection with the engines and boilers, are to be fitted close to the vessel's sides, and are to be accessible at all times.

4. Where soil pipes are attached to the outside plating below the load water-line, the lower length must be of steel or iron of substantial thickness, and be secured to the plating with a proper faced joint, and extended for some distance above the load water-line.

5. If the remainder of the pipe be of lead, care must be taken that it is of substantial thickness, and that it is properly protected externally with zinc or iron, to the satisfaction of the Society's Surveyors.

Section 29 (Openings in Decks).—1. The engine and boiler openings of the weather-deck of steam vessels are to be properly framed for a height of not less than 18 inches above the deck, the coaming plates to extend to the lower edge of the beams, and iron trunk bulkheads connected to the coamings should be fitted to a height of about 7 feet above the deck . . . . . . (here follows a statement of the thicknesses required and a reference to the illustrative sketches given at the end of the Rules).

2. The engine and boiler openings in the 'tween decks of all vessels are also to be enclosed by trunk bulkheads efficiently stiffened by angle bars 30 inches apart, and extending to the weather-deck beams, to which they are to be secured.

3. Strong iron doors will be allowed in these trunk bulkheads, provided their lower parts are at least 18 inches above the deck, and efficient arrangements made for their security.

4. When a poop, or bridge-house, covers the engine and boiler space, the coamings of the engine and boiler openings should not be less than 2 feet above such deck, unless these openings are constructed as provided for in the first paragraph of this section.

5. It is considered that in all cases the engine and boiler openings should be made as small as practicable, and be subdivided by athwart-ship iron divisional casings to secure the maximum safety of the vessel. The two sides of the casing should in all instances be efficiently connected by angle beams within them at the upper part.

6. The engine-room skylights are to be in all cases substantially constructed and to be securely bolted or riveted to the coamings, and where the skylight top is not solid with bull's-eyes fitted in the same, efficient dead-lights of metal or wood must be provided. The grating openings over the stoke-hole must also be protected by plates, fitted with hinges, or otherwise in a manner satisfactory to the Surveyors.

7. Where either of the openings exceeds 15 ft., or the combined length exceeds 30 ft., the beams in way of the same are to be plated over from the stringer to the tie-plates, the plating extending two beam spaces beyond the openings, and tapered from thence towards the stringer plate for a distance not less than the breadth of the plating required to be fitted; the thickness of this plating to be the same as given in Table G for iron decks.

8. Where large openings are adjacent to each other, the intervening space between the hatchways to be plated over.

Section 30 (Bunker Hatches, etc.).—Coal bunker pipes, where practicable, are to be formed so as to be at least 12 inches above the upper deck, fitted with lids having studs to fit in openings made in the pipes, for their security; the pipes to be so formed that tarpaulin may be securely lashed over them. When there are coal bunker hatches in the weather deck they must be properly framed with coaming plates of suitable height, having solid hatches secured by an iron bar or other approved fastening.

Section 35 (Cement).—1. The frames and plating of the bottom of all vessels to the upper parts of the bilges to be thickly and efficiently covered with Portland or other approved cement, which may be mixed with sand or other suitable substance. Care to be taken to have a proper substance of cement at its termination and to keep the watercourses clear all fore and aft.

Section 38 (Pumps, etc.).—1. In steam vessels the pumping arrangements according to the division of holds, &c., to be as follows :—

2. **Holds with double bottoms.**—In the double bottom of each compartment of the hold, and of engine and boiler space, a steam pump suction is to be fitted at the middle line, and one on each side to clear the tanks of water when the vessel has a heavy list. Where there is considerable rise of floor towards the ends of the vessels, the middle line suction only will be required. A steam pump suction and a hand pump are also to be fitted to each bilge in each hold where there is no well. When there is a well, one or three steam pump suctions are to be fitted in the same, according as there is considerable or little rise of floor, and hand pumps fitted at the bilges.

3. **Holds without double bottoms.**—Where there is considerable rise of floor, one steam pump suction and one hand pump are to be fitted in each hold. In vessels with little rise of floor, two or three steam pump suctions and at least one hand pump to be fitted to each hold.

4. **Engine and Boiler space.**—Where a double bottom extends the whole length of engine and boiler space, two steam pump suctions are to be fitted to the bilge on each side. Where there is a well one steam pump suction should be fitted in each bilge and one in the well. Where there is no double bottom in the machinery space, centre and wing steam pump suctions should be fitted. The rose box of the bilge injection is to be fitted where easily accessible, and is to be used for bilge water only. The main and donkey pumps to draw from all compartments and the donkey to have also a separate bilge suction in the engine-room.

5. Fore and After Peaks.—If the peaks are fitted as water ballast tanks a separate steam pump suction is to be led to each. If not used for water ballast, an efficient pump is to be fitted to the fore peak.

6. Tunnel.—The tunnel well is to be cleared by a steam pump suction.

7. All Hand Pumps to be capable of being worked from the upper or main decks above the deep load water-line.

8. No Sluice or other Valves or cocks are to be fitted which are not at all times accessible.

9. Sounding pipes to be fitted on each side of holds and ballast tanks, and a doubling plate is to be fitted under each.

10. Air pipes to be fitted to each ballast tank as required.

13 and 14. The pipes for bilge or ballast suctions are to be fitted with flanged joints, in convenient lengths, so that they may be easily disconnected for clearing. Those to fore and after peaks and to the tunnel well should not be less than $2\frac{1}{4}$ inches inside diameter, except in vessels of less than 500 tons under deck, in which case they may be made 2 inches.

15. The Bilge injection should not be less than two-thirds of the diameter of the sea inlet to the circulating pump.

The inside diameter of other bilge suction pipes should not be less than given in the following Table :—

Table LXVI. — Sizes of Bilge Suction Pipes by Lloyd's Rules.

Tonnage under upper deck.	Engine room and hold centre suctions, and separate donkey suction in engine room.	Wing suctions in holds when no centre suctions fitted, and wing suctions in engine room.	Wing suctions in holds when centre suctions are also fitted.
	Inches.	Inches.	Inches.
In vessels under 500 tons,	2	2	2
In vessels 500 tons but under 1000 tons, .	$2\frac{1}{4}$	2	2
In vessels 1000 tons but under 1500 tons, .	$2\frac{1}{2}$	$2\frac{1}{4}$	2
In vessels 1500 tons but under 2000 tons, .	3	$2\frac{3}{4}$	$2\frac{1}{4}$
In vessels 2000 tons but under 3000 tons, .	$3\frac{1}{2}$	3	$2\frac{1}{2}$
In vessels above 3000 tons, . . .	$3\frac{1}{2}$	$3\frac{1}{2}$	$2\frac{3}{4}$

In cases where two or more of the suctions from any one compartment are connected to the pumps by a single pipe, this pipe should ot be of less size than the centre suction.

LLOYD'S SURVEYS.

Lloyd's ordinary survey is made on the following items at the periods named :—

(*a*) On the different parts of the engines during erection.

(*b*) On the sea connections while being fitted to the vessel.

(*c*) On the boiler plates when they are bent, flauged and holed, ready for riveting, and on stays, &c., while being fitted.

(*d*) Testing the boilers by hydraulic pressure.

(*e*) When engines and boilers are being fixed on board the vessel.

(*f*) At the setting and testing of safety valves, and trying the machinery under steam.

Lloyd's special survey.—The additional regulations with regard to special surveys are as follows :—

(*g*) In steam vessels built under special survey, the machinery and boilers must also be constructed under special survey.

(*h*) In cases of machinery or new boilers being built under special survey, the distinguishing mark ✠ will be noted in red, thus : " ✠ LMC," or " ✠ NE & B," or " ✠ NB."

(*j*) In order to facilitate this inspection, the plans of the machinery and boilers are to be examined, and from them the working pressure fixed.

(*k*) The Surveyors are to examine the materials and workmanship from the commencement of the work until the final test of the machinery under steam ; any defects, &c., to be pointed out as early as possible.

(*l*) The Surveyors may also, if desired, compare the work as it progresses with the requirements of the specification agreed upon by the parties concerned, and certify to the conditions thereof, as far as can be seen, being satisfactorily complied with.

For Lloyd's requirements as to steel castings see under Cast Steel in section on " Materials " (page 292).

LLOYD'S RULES RELATING TO SPARE GEAR.

Lloyd's requirements as to spare gear are as follows :—

The articles of spare gear mentioned in the following list will be required to be carried in all steam vessels classed in the Society's Register Book, viz.:—

2 connecting rod, or piston rod, top end bolts and nuts.

2   ,,   bottom end bolts and nuts.

2 main bearing bolts.

1 set of coupling bolts.

1 set of feed and bilge pump valves.

1 set of piston springs (where common springs are used).

A quantity of assorted bolts and nuts.

Iron of various sizes.

In addition to the foregoing, the following articles are recommended to be carried with a view to expedite repairs and lessen delay in distant ports, viz.:—

Crank shaft.	1 set of link brasses.
Propeller shaft.	1 eccentric strap complete.
Propeller, or a full set of blades.	Air-pump rod.
Stern-bush, or lignum-vitæ lining for bush.	Circulating-pump rod.
1 pair connecting-rod brasses.	H. P. valve spindle.
1 pair crosshead brasses.	L. P. ,,
1 set of check valves.	2 dozen boiler tubes.
6 cylinder cover bolts.	3 ,, condenser tubes.
6 junk-ring bolts.	1 cylinder escape-valve and spring.
4 valve-chest cover bolts.	1 set of safety-valve springs.

ADDITIONAL BOARD OF TRADE RULES.

The following Board of Trade rules concern the engineer, but cannot properly be placed under any of the preceding section headings:—

63. There should be for each compartment a pump of sufficient size which can be worked from the upper deck. A rose or perforated box of sufficient size should be fitted at the end of the suction pipe of each pump, and means provided for clearing the end of the suction pipe.

It is very desirable that the steam winches (if any) be so fitted and arranged that the deck pumps can be worked by them as well as by hand.

Deck pumps should be provided with suitable handles, and those of the smaller size should have handles long enough for at least two men to work at them.

64. Sounding pipes should be fitted from the upper deck for ascertaining the depth of water in each compartment. The collision bulkhead should not have a valve or cock or any opening in it, nor should pipes be led through it.

65. Pipes connected with pumps worked by the engines should be carried through the bulkheads into all the compartments fore and aft of the engine-room, except the compartment in front of the collision bulkhead, so that each compartment can be pumped out separately by the engines as well as by the deck pump. The pipes should be well secured where they pass through the bulkheads, and it is very desirable that cocks or valves be fitted between these suction pipes and the bulkheads, capable of being opened or shut from the upper or main deck.

66. A spare tiller, properly fitted to the rudder head, relieving tackle, &c., should, in all foreign-going and home trade steamers, be kept near the after steering gear ready for immediate service. The steering gear, including chains, should be thoroughly overhauled at every survey, and taken to pieces and thoroughly examined at least once a year. The chains and blocks that are liable to interfere with or endanger the passengers or crew should be guarded by portable but properly secured guards.

With the view of relieving, as far as practicable, the rudders of vessels from severe and sudden shocks, springs have in some cases been fitted to the quadrant, or to the rods or chains at each side of the vessel, and the Board think that such fittings, or other efficient means, should be adopted, more particularly in the case of new vessels.

The Surveyors should note that the steam and exhaust pipes of steering engines in all new passenger steamships should be at least the same internal diameter respectively as the steam and exhaust connections on the cylinders, which they should see are sufficient for the purpose, and they should be so arranged that water cannot lodge in them. Right-angled bends should be avoided as much as possible, and the pipes should be used exclusively for the steering engines. In cases where this is not complied with, full particulars and sketches should be submitted to the Board for consideration.

Attention is also directed to a description of steam steering gear, in which a part of the shaft by which the helmsman actuates the controlling valve, passes through another shaft that is liable to be thrown out of line by the reaction of the spur gearing, and, consequently, liable to jam the inside shaft to such an extent as to deprive the helmsman of the control of the steering gear. All steam steering gear should be carefully examined, and if any be found constructed in the manner described above, their use should be discouraged, and should not be approved, unless they have been tested from midship to hard over in both directions, and found satisfactory when the vessel is running at full speed.

It is very desirable that the man at the helm should be so placed that he has a clear lookout ahead, more especially in steamers that frequent crowded harbours or rivers.

67. In passing a helm indicator, the Surveyor should ascertain by actual trial that whenever the pointer moves to the word "port" on the dial or plate, wherever that word may be, it shows that the helm is ported, and whenever the pointer points to the word "starboard," wherever that word may be, it means that the helm is starboarded.

In the foregoing directions it is assumed that "port helm" means that the helm is so moved as to turn the ship's head to the right, and "starboard helm" means that the helm is so moved as to turn the ship's head to the left.

77. Passenger steamers going to sea should be provided with a hose adapted for the purpose of extinguishing fire in any part of the ship, and capable of being connected with the engines of the ship, or with the donkey engine, if it can be worked from the main boiler. The Surveyor must take care that it answers the required purpose.

The fire hose should be connected and stretched, to judge of its length, and thoroughly examined at every survey, and at least once a year (and at any other time that the Surveyor thinks it necessary) tested with the conductor in its place by pumping water through it by the main or donkey engines at full speed. A proper conductor and metal bend or goose neck form part of its equipment, and should be provided. Generally, leather hoses are the most durable, and should be supplied when a declaration for 12 months is required.

121. In all boilers in which the Surveyors find that cast-iron is employed in such a manner as to be subject to the pressure of steam or water, they are directed to report the circumstances to the Board of Trade, in order that they may receive instructions how to act. Cast-iron stand-pipes or cocks through which hot brine would have to pass should never be passed. Cast-iron should not be used for stays, and Surveyors should also discourage the use of cast-iron for chocks and saddles for boilers. Particular attention should be paid to the chocking of boilers, more especially when they are fired athwartships.

143. In the case of steamers coming in for survey under the Passengers Acts, and other steamers performing ocean voyages, no question as to gear need be raised if the following spare gear and stores are supplied, or their equivalent which should be submitted to the Board for consideration. The heavier portions of this gear should have been fitted and tried in their places, and should be kept on board where access can at all times be had to them :—

1 pair of connecting rod brasses.
1 air pump bucket, and rod, with guide.
1 circulating pump bucket and rod.
1 air pump head valve, seat and guard.
1 set of india-rubber valves for air pumps.
1 circulating pump head valve, seat and guard.
1 set of india-rubber valves for circulating pumps.
2 main bearing bolts and nuts.
2 connecting rod bolts and nuts.
2 piston rod bolts and nuts.
8 screw shaft coupling bolts and nuts.
1 set of piston springs suitable for the pistons.
3 sets, if of india-rubber, or 1 set if of metal, of feed pump valves and seats.
8 sets, if of india-rubber, or 1 set if of metal, of bilge pump valves and seats.
1 hydrometer.
Boiler tubes, 3 for each boiler.
100 iron assorted bolts, nuts, and washers, screwed, but need not be turned.
12 brass bolts and nuts, assorted, turned, and fitted.
50 iron   ,,   ,,   ,,   ,,
50 condenser tubes.
100 sets of packing for condenser tube ends, or an equivalent.
At least one spare spring of each size for escape valves.
1 set of water-gauge glasses.
$\frac{1}{10}$th of the total number of fire-bars necessary.
3 plates of iron, assorted.
6 bars of iron, assorted.
1 complete set of stocks, dies, and taps, suitable for the engines.

1 smith's anvil.
1 fitter's vice.
Ratchet-braces, and suitable drills.
1 copper or metal hammer.
Suitable blocks and tackling for lifting weights.
1 dozen files, assorted, and handles for the same.
1 set of drifts or expanders for boiler tubes.
1 set of safety-valve springs (if so fitted) for every four valves; if
    there are not four valves, then at least one set of springs
    must be carried.
1 screw jack.
And a set of engineer's tools suitable for the service, including
    hammers and chisels for vice and forge; solder and solder-
    ing iron; sheets of tin and copper; spelter; muriatic acid,
    or other equivalent, &c., &c.

145. The distilling apparatus of emigrant ships should be taken to
pieces every voyage, except in the cases of steamers holding passenger
certificates, which should be taken to pieces at least once every six
months, or oftener, if the Surveyor thinks it necessary, and the tubes
or coils tested to at least twice the load on the safety-valve on the
apparatus, or in cases where no safety-valve is fitted, to twice the
highest working pressure of the boiler from which the apparatus can
be worked, and the machinery and boilers thoroughly examined.
After the distilling apparatus is put together again, it should be tested
as to the quantity and quality of the water made.

The water should be cold, pure, and fit to drink immediately it is
drawn off from the filter. No distilling apparatus should be passed,
unless fitted with a suitable sized filter, charged with animal charcoal;
the charcoal should be taken out, cleansed, or renewed every voyage,
except in case of steamers holding passenger certificates, in which case
it need not be taken out, unless the Surveyor thinks it necessary,
oftener than every six months. In such cases (where passenger steamers
are coming frequently in for survey under the Passenger Acts, and
such complete examination is not made previous to each voyage), the
Surveyor will be held wholly responsible for the efficiency of the
apparatus; but the quantity and quality of the water must be tried
previous to every voyage. It therefore rests with him whether he orders
them to be taken to pieces every voyage or not. The boilers should be
at least equal in strength to the boilers of passenger steamers, and should
have the same fittings as are necessary for them in accordance with the
Board of Trade Regulations. The Surveyor must satisfy himself as to
the capability of the man who is to have charge of the apparatus.

The steam for working the apparatus is not to be taken from the
main boilers. No exhaust steam must be permitted to go into the
condenser if appliances for the introduction of lubricants be fitted to
the steam pipes, or steam cylinder of the pumping engine. The
boiler of the apparatus must not be filled or fed with water from the
surface condensers of the main engines, and must not be fitted with
cocks, &c., for the introduction of tallow or oil.

When the water is pumped into the condenser there should be an efficient escape-valve on the condenser, which cannot be readily tampered with, and if the condensing portion of the apparatus or the cooler and filter be unfit to bear the pressure on the boiler, an efficient safety-valve that cannot be readily overloaded should be fitted between the steam pipe and the apparatus.

146. It is advisable that the donkey engine for pumping water through the condenser be so fitted that it can be made available in case of emergency for extinguishing fire in any part of the ship; a leather hose, with suitable bends and conductors, should be supplied for this purpose.

147. The following list of tools and materials should be provided for distilling apparatus :—

    1 set of stoking tools.
    1 scaling tool.
    1 spanner for boiler doors.
    1 set of fire bars, suitable for boiler.
    1 14-inch flat bastard file.
    1 14-inch half-round file.
    1 10-inch round file.
    3 file handles.
    2 hand cold chisels.
    1 chipping hammer.
    1 pair patent gas tongs.
    1 soldering iron.
    10 lbs. of solder.
    2 lbs. of rosin.
    6 gauge glasses.
    24 india-rubber gauge glass washers.
    30 bolts and nuts, assorted.
    1 slide rod for donkey pump.
    5 lbs. spun yarn.
    10 lbs. cotton waste.
    1 deal box, with lock, complete.
    2 gallons machinery oil.
    Animal charcoal sufficient to charge the filter at least twice.
    1 can for machinery oil.
    1 oil feeder.
    1 small bench vice.
    1 ratchet brace.
    4 drills, assorted.
    1 set dies and taps suitable for the bolts.
    2 glass salinometers.
    1 hydrometer and pot.
    1 shifting spanner.
    1 lamp for engineer.
    And other articles that the particular distiller and boiler supplied may, in the Surveyor's judgment, require.

CHAINS AND ROPES.

The following Tables give the Admiralty requirements as to chains :—

Table LXXVIII.—Admiralty Tests, &c., of Stud-link Chain Cable.

Dia. of cable in inches.	Breaking strength in tons.	Proof load in tons.	Weight of 100 fathoms in cwts.	Dia. of cable in inches.	Breaking strength in tons.	Proof load in tons.	Weight of 100 fathoms in cwts.
$7/16$	4·90	3·5	9·25	$1\frac{1}{2}$	56·70	40·5	108·0
$\frac{1}{2}$	6·30	4·5	12·0	$1\frac{5}{8}$	66·50	47·5	126·75
$9/16$	7·70	5·5	15·25	$1\frac{3}{4}$	77·17	55·125	147·0
$\frac{5}{8}$	9·80	7·0	18·75	$1\frac{7}{8}$	88·55	63·25	168·75
$11/16$	11·90	8·5	22·75	2	100·80	72·0	192·0
$\frac{3}{4}$	14·17	10·125	27·0	$2\frac{1}{8}$	113·75	81·25	216·75
$\frac{7}{8}$	19·25	13·75	36·75	$2\frac{1}{4}$	127·57	91·125	243·0
1	25·20	18·0	48·0	$2\frac{3}{8}$	142·10	101·5	270·75
$1\frac{1}{8}$	31·85	22·75	60·75	$2\frac{1}{2}$	157·50	112·5	300·0
$1\frac{1}{4}$	39·37	28·125	75·0	$2\frac{3}{4}$	181·02	129·3	363·0
$1\frac{3}{8}$	47·60	·0	90·75	3	204·12	145·8	432·0

Up to and including 2½ inches diameter the above proof loads are equal to 630 lbs. per circular ⅛ inch of section of one side of link, and are the same as those required by the Chain Cables and Anchors Acts; but for 2¾ inch diameter the load is reduced to 598·5 lbs. per circular ⅛ inch, and for 3 inch diameter to 567 lbs.   The formula for proof loads of chains up to 2½ inches diameter may also be written,—

Proof load in tons = 18 × (diameter in inches)².

The breaking strengths are placed at 40 per cent. above the proof loads.

Table LXXVIIIa.—Admiralty Tests, &c., of Short-link Chain.

Dia. of chain in inches.	Breaking strength in tons.	Proof load in tons.	Weight per fathom in pounds.	Dia. of chain in inches.	Breaking strength in tons.	Proof load in tons.	Weight per fathom in pounds.
$\frac{1}{4}$	1·87	·75	3·0	$15/16$	26·37	10·55	49·0
$5/16$	2·93	1·17	5·5	4	30·00	12·00	56·0
$\frac{3}{8}$	4·22	1·69	8·0	$1\frac{1}{16}$	33·87	13·54	63·0
$7/16$	5·74	2·30	10·5	$1\frac{1}{8}$	37·97	15·18	71·0
$\frac{1}{2}$	7·50	3·00	14·0	$1\frac{3}{16}$	42·30	16·92	79·0
$9/16$	9·49	3·80	18·0	$1\frac{1}{4}$	46·87	18·75	87·0
$\frac{5}{8}$	11·72	4·69	22·0	$1\frac{5}{16}$	51·68	20·67	96·0
$11/16$	14·18	5·67	27·0	$1\frac{3}{8}$	56·72	22·68	106·0
$\frac{3}{4}$	16·87	6·75	32·0	$1\frac{7}{16}$	62·00	24·80	116·0
$13/16$	19·80	7·92	37·0	$1\frac{1}{2}$	67·50	27·00	127·0
$\frac{7}{8}$	22·97	9·19	43·0	...	...	...	...

The proof loads for short-link chains are ⅔rds those for stud-link chains.

The rule may be written,—**Proof load in tons = 12 × (diameter in inches)².** The breaking strengths of short-link chains are placed at 2½ times the proof loads.

Lloyd's requirements as to chain cables, &c., are given in Table LXXIX.; the proof loads and breaking strengths are those required by the Act of Parliament; the breaking strengths for chains above 1½ inch diameter are the same as required by the Admiralty, but for chains of 1½ inch diameter and under, Lloyd's require a slightly higher breaking strength.

Table LXXIX.—Lloyd's Tests of Stud-link Chain Cables.

Diameter of chain in inches.	Proof load (Statutory) in tons.	Breaking strength (Statutory) in tons.	Diameter of chain in inches.	Proof load (Statutory) in tons.	Breaking strength (Statutory) in tons.
11/16	8·5	12·75	1⅝	47·5	66·5
¾	10·125	15·125	1 11/16	51·25	71·75
13/16	11·875	17·8	1¾	55·125	77·125
⅞	13·75	20·625	1 13/16	59·125	82·75
15/16	15·8	23·7	1⅞	63·25	88·5
1	18·0	27·0	1 15/16	67·5	94·5
1 1/16	20·3	30·4	2	72·0	100·8
1⅛	22·75	34·125	2 1/16	76·5	107·1
1 3/16	25·375	38·0	2⅛	81·25	113·75
1¼	28·125	42·125	2 3/16	86·125	120·5
1 5/16	31·0	46·5	2¼	91·125	127·5
1⅜	34·0	51·0	2 5/16	96·25	134·75
1 7/16	37·125	55·625	2⅜	101·5	142·1
1½	40·5	58·7	2 7/16	107·0	149·7
1 9/16	43·9	61·4	2½	112·5	157·5

Note.—Unstudded close-link chains will be admitted as cables if proved to two-thirds the load required for stud-link chains, and if the breaking strength is not less than twice such proof load.

In some recent tests of the chain cables of large steamers a 2⅝-inch cable gave an average ultimate strength of 212 tons, and a 2 9/16-inch cable gave 223 tons as the lowest and 229 tons as the highest of seven tests.

The safe working load on chains should not be taken higher than half the proof load; if this proportion be adopted the formula becomes :—

Working load in tons = $\begin{cases} 9 \times \text{(diameter in inches)}^2 \text{ for stud-link.} \\ 6 \times \text{(diameter in inches)}^2 \text{ for close-link.} \end{cases}$

For ordinary crane chains and slings 4 × (diameter in inches)² is igh enough.

Table LXXX.—Admiralty Flexible Steel Wire Ropes.

Size of rope (circumference) in inches.	Number of wires in each strand.	Weight per fathom in lbs.	Minimum breaking st'gth in tons.	Torsion test; No. of twists each wire must stand.	Ductility tests, &c.
8	30	53	148	9	Each rope to consist of 6 strands. Wires to be of best crucible steel galvanized with pure zinc.
7	30	41	113	11	
6½	30	35	98	14	The rope is to be laid up evenly and uniformly as regards size and angle, and is to contain a proper sized hemp core.
6	30	31	84	15	
5½	24	28	71	16	
5	24	23	59	17	A latitude not exceeding 5 per cent. over or under the prescribed weights will be allowed.
4½	12	14	39	15	
4	12	12	31	17	Elongation will be assumed to commence when one-sixth of the breaking-load has been applied.
3½	12	9	24	18	
3	12	7	17	22	Test A (torsion).—Each wire to stand being twisted through the number of revolutions stated in column 5, in a length of 8 inches.
2¾	12	5½	14½	25	
2½	12	4½	11¾	26	
2¼	12	3¾	9	28	Test B (bending).—Each wire to stand coiling around itself eight turns and back again.
2	12	2¾	7	33	
1¾	12	2	5½	36	If, in being tested, one wire in seven fail, but the other six give fair and uniform results, the average being up to the standard, the rope will be considered satisfactory in that respect.
1½	12	1¾	4	41	
1¼	12	1¼	2⅞	47	
1	12	¾	1¾	60	

Proportions of links of chains.—The standard proportions of the links of chains, in terms of the diameter of the bar from which they are made, are as follows :—

	Overall length	Overall breadth
Stud-link	6 diameters	3·6 diameters
Close-link	5 ,,	3·5 ,,

The stud has usually a diameter at the centre of ·6 × diameter of chain and at the ends 1 × diameter of chain.

Weight of chains.—The weight of stud-link chain cables is given very nearly by the rule, —

$$W = 55d^2$$

where W = weight per fathom in pounds, and d = diameter of bar from which chain is made.

For close-link chain the rule becomes,—

$$W = 58d^2$$

The above rules give weights about mid-way between those required by the Admiralty and those required by Lloyd's. The weight is of course much affected by the length of link used.

Steel Wire Ropes.

The Admiralty requirements with regard to flexible steel wire ropes are given in Table LXXX. on previous page.

Steel wire ropes for standing rigging are required to be made of fewer wires of larger diameter, are rather heavier, and must be of rather greater ultimate strength.

The following Table shows the breaking strengths that steel wire hawsers, &c. must show in order to be accepted by Lloyd's : —

Table LXXXI.—Breaking strengths of Steel Wire Ropes by Lloyd's Rules.

Size (circumference) in inches.	Breaking strength in tons.	Size (circumference) in inches.	Breaking strength in tons.
2	7	3¾	29
2¼	9½	4	33
2½	...	4¼	35
2¾	15½	4½	39
3	18	4¾	47
3¼	22	5	59
3½	26	5¼	71

A short length of each of the wires composing the hawser, &c. will also be required, after being galvanised, to show a tensile strength equivalent to that given in the above Table, and the aggregate strength of the wires must not be less than 10 per cent. in excess of that strength.

Each wire must also be capable of being twisted around itself not less than eight times, and of being untwisted and straightened again without breaking.

The strength of steel wire ropes, relatively to their girth, depends not only on the quality of the wire used, but also on the amount of hemp core used ; usually there is a central hemp core, and sometimes each strand has also a similar core, but sometimes there are no hemp cores at all.   There would be no particular difficulty in obtaining ropes to stand twice the tests given in the above Table, but such ropes would probably be wanting in flexibility, and would require drums of very large diameter if they were to work satisfactorily and to last any time.

Hemp Ropes.

Hemp is laid up right-handed in yarns ; and yarns are laid up left-handed into strands.

A hawser is composed of three strands laid up right-handed.

A cable is composed of three hawsers laid up left-handed.

Shroud-laid rope has a core surrounded by four strands.

The strength of hemp ropes depends on the quality of the hemp used, on the type or make of the rope, and on its condition (*i.e.,* wet or dry, tarred or untarred).

The twist diminishes the strength, but increases the solidity and durability, and the strength therefore depends to some extent on the twist.

When a rope is wet or tarred its strength is reduced by about one-fourth.

The working strength is commonly taken as ⅛th of the breaking strength.

The Admiralty requirements as to various sizes of hawser-laid cordage in common use are given in Table LXXXII. ; the 11-inch rope is specified to be of tarred Petersburg hemp, but all the smaller sizes are to be of tarred Riga hemp ; all are to be three strand.

Table LII.—Admiralty Tarred Hemp Cordage.

Size of rope (circumference) in inches.	Size of yarn.	Number of threads in the rope.	Standard breaking strength.		
			Tons.	cwt.	qrs.
½	40	6	0	3	0
¾	,,	12	0	6	0
1	,,	15	0	8	0
1½	,,	33	0	15	0
1¾	,,	42	1	0	0
2	,,	54	1	7	0
2½	,,	84	2	0	0
3	,,	120	3	0	0
..½	30	123	3	18	0
	,,	159	5	0	0
4½	,,	201	6	9	0
5	,,	249	7	18	0
11	25	1008	36	10	0

All sizes are specified to be formed at an angle of 27°, hardened at 37°, and finished at 42°.

If the smallest size be excluded, the above Table gives breaking strengths which approximate very closely to those given by the formula,—

Breaking strength in cwts. = $(6\cdot2 \times \text{girth in inches}^2) + 2$

The weight in pounds per fathom of hawser-laid hemp ropes is given approximately by the rule,—

Weight in lbs. per fathom = $\cdot17 \times (\text{girth in inches})^2$ for dry ropes.

,, ,, = $\cdot21 \times (\text{girth in inches})^2$ for wet or tarred ropes.

STRENGTH, &c., OF MATERIALS.

Cast-Iron.

The strength that an iron casting may be expected to possess depends on the quality of the iron, on the number of times it has been melted, on the design or form of the casting, and on the skill and care exercised in the foundry to ensure soundness and freedom from contraction stresses. The quality of the iron depends on its chemical constitution, and on the method of manufacture. The effects of variations in chemical composition are indicated in the following table :—

Table LXXXIII.—Compositions and qualities of Cast-Iron.

Quality.	Composition.		
	Combined carbon per cent.	Graphitic carbon per cent.	Silicon per cent.
Very soft, . . .	$\cdot15$	3·1	2·5
Very hard, . . .	...	...	under ·8
Great general strength, .	·5	2·8	1·42
Great tensile strength, .	...	...	1·8
Great crushing strength,	over 1·0	under 2·6	about ·8

Pig irons are as a rule divided into seven classes, each of which is known by a number. No. 1 contains the most free or graphitic carbon, and, when broken, exhibits a very coarsely granular fracture with dark grey scales of considerable size; when melted, "it runs very thin," or is of extreme fluidity, and is therefore used mostly for fine ornamental castings, and for mixing with other numbers when increased fluidity is required.

No. 2 pig is neither so soft nor so fluid as No. 1, but is not sufficiently close grained for general use.

No. 3 pig is that usually employed for marine engine castings ; by adding No. 1 a mixture suitable for complicated castings is obtained, and the addition of No. 4 gives a harder and closer-grained metal.

No. 4 is not much used in the foundry except for such mixing purposes ; it still shows a grey fracture, but the grain is finer and more crystalline, and there is an absence of the graphitic scales so marked in Nos. 1, 2, and 3.

Nos. 5 and 6 are not used in the foundry at all, but are made for conversion into wrought-iron, &c.

No. 7 pig shows a silvery white and crystalline fracture, contains practically no free carbon, and is extremely hard. Like Nos. 5 and 6 it is a "forge iron," and is sometimes called "white forge," whilst Nos. 5 and 6 are called "grey forge."

Iron from different districts, and made from different classes of ore, of course varies considerably in composition and quality, but quality is also considerably affected by the method of reduction,—cold-blast iron being usually stronger, tougher, and closer grained than hot-blast, and therefore often used for mixing with other irons where exceptional strength and toughness is required.

Iron mixtures.—As all cast-irons are improved by re-melting, no important casting should be made entirely of new pig, and if maximum strength is required the whole of the material should be re-melted.

For cylinders a strong, tough, and close-grained metal is required, and a metal of this character may be obtained by using equal proportions of picked scrap, best Scotch No. 3 pig, and Blaenavon (cold blast). If the cylinder is to have liners and false faces, most of the hardening elements may be omitted.

Cylinder liners and false faces require a fair amount of strength and great hardness, and may be made of No. 3 Scotch iron, Blaenavon, and selected hard scrap, or if that is not obtainable its place may be taken by some No. 4 pig.

Castings of simple form, such as propeller blades and bosses, may be increased in hardness, strength, and closeness of grain by the addition of steel boiler plate scrap to the extent of even 10 per cent. Hæmatite is used for the same purpose.

The contraction of iron castings in cooling varies considerably with the form and proportions of the casting, but is, on an average, $^1/_{10}$ to ⅛ inch per foot.

The cooling of large and intricate castings should be as gradual as possible, as the internal stresses are then somewhat relieved, and the risk of cracks diminished.

19

The weight of cast-iron also varies very considerably,—the difference between the heaviest and lightest kinds being nearly 40 lbs. per cubic foot; but a fair average value, and one easily remembered, is about 450 lbs. per cubic foot, or ·26 lbs. per cubic inch. A plate 1 foot square × 1 inch thick will then weigh 37·5 lbs. Ordinary marine castings are probably rather above than below this weight.

Strength of cast-iron.—Cast-iron suitable for ordinary marine castings should not have a lower ultimate tensile strength than 17,000 lbs. per square inch, and when weight is of importance, and scantlings are cut down, a strength of 20,000 to 22,000 lbs. per square inch should be aimed at.

The testing machine should be in constant use if there is to be any check on the foundry, or if any accurate knowledge as to the material being turned out is required.

The ultimate strength of cast-iron in compression is about 90,000 lbs. per square inch ; in ordinary construction it may carry three times as much as in tension.

The Admiralty requirements as to cast-iron are as follows :—

Test pieces to be taken from such castings as the inspecting officer may consider necessary. The minimum tensile strength to be 9 tons (20,160 lbs.) per square inch, taken on a length of not less than 2 inches.

The transverse breaking load for a bar one inch square, loaded at the middle between supports one foot apart, is not to be less than 2000 lbs.

Wrought Iron.

The quality of wrought iron,—provided it is free from such harmful ingredients as sulphur and phosphorus,—depends largely on the amount of work that has been done on it at the mill,—*i.e.*, on the extent to which it has been rolled down and the fibre developed.

The appearance of the fracture depends a good deal on the manner in which it is broken; if good bar iron is nicked at one side, and slowly broken or bent back, it should show a clear white silvery, and almost entirely fibrous, fracture, whereas if nicked on two sides, or quickly broken, the fracture will show more of a fine white crystalline grain, and less fibre.

A specimen from a good forging will show a clear and silvery grain, but larger than that of the bar iron, with about 20 to 30 per cent. of fibrous patches.

An inferior iron usually shows a coarse crystalline structure, or if fibre is present, it is dull and earthy looking.

Coarse crystals, or large shining plates, generally indicate a "cold-short" iron, and "red-shortness" is indicated by an earthy, dull, or dark fracture.

Merchant bar is the commonest quality generally used by engineers ; ladders, gratings, fire bars, bearer bars, &c., are made of it.

Best bar is the next quality; its tensile strength is about 24 tons (or say 54,000 lbs.) per square inch, and it may be used for all ordinary smithing purposes.

Best best bar is a higher quality again, and has an ultimate tenacity of 26 to 27 tons (or say 58,000 to 60,000 lbs.) per square inch, with an elongation of about 25 per cent. in 8 inches, and a contraction of area of about 50 per cent.; the fibre is uniform and silky in appearance, and the bar may be bent double, cold, without fracture.

Best Yorkshire boiler plates.—This at one time indispensable material is now rarely used by marine engineers. Its ultimate tensile strength, with the grain, is about 24 tons (54,000 lbs.), and across the grain about 22 tons (49,000 lbs.) per square inch, with an elongation of about 13 and 8 per cent. respectively. Its elastic strength, with the grain, is nearly 12 tons (26,000 lbs.) per square inch.

Staffordshire plates.—This quality of iron is also little used by marine engineers now; it was largely used for boiler shells, &c. Its ultimate tensile strength is about as follows:—

Plates under ¾ inch thick $\begin{cases} \text{with the grain—23 tons (51,500 lbs.).} \\ \text{across the grain—19 tons (42,500 lbs.).} \end{cases}$
Plates of ¾ inch thick and upwards $\begin{cases} \text{with the grain—22 tons (50,000 lbs.).} \\ \text{across the grain—18 tons (40,500 lbs.).} \end{cases}$

Iron forgings.—The strength of iron forgings depends both on the quality of the scrap from which they are made, and on the size of forging, or the amount of work that has been done on them under the hammer.

For a large forging, 20 tons (45,000 lbs.) per square inch, with an extension of 8 or 10 per cent. in 8 inches, would be a very fair test result; with a small forging, about 22 tons (50,000 lbs.) per square inch, and an extension of 12 or 15 per cent., should be reached.

Weight of wrought iron.—Wrought iron varies slightly in density according to the method of manufacture, the form into which it is put, and the amount of impurity it contains. A fair average value is 485 lbs. per cubic foot, or ·28 lb. per cubic inch; bars and Yorkshire plates give about this figure, but Staffordshire plates are rather lighter (about 480 lbs.), and large forgings lighter again (about 477 lbs.),—owing probably to the presence of cinder in a minutely divided condition. A square foot of plate, one inch thick, usually weighs 40 lbs.

Cast Steel.

Steel castings for the pistons, covers, framing, &c., of marine engines should have an ultimate tensile strength of 29 to 31 tons (65,000 to 69,000 lbs.) per square inch, with an extension of about 10 or 12 per cent. in 8 inches. For castings that are at all intricate, or thin in parts, it is necessary to use a rather stronger steel say up to

36 or 38 tons ultimate strength), as the milder and tougher steels are not sufficiently fluid when melted, but the extension obtainable will not then exceed about 8 per cent.

The contraction of cast steel in cooling is more variable than that of cast iron, but is on an average about $\frac{3}{16}$ inch per foot, or the same as that of brass. Special care should therefore be taken in designing large pieces to be cast in this metal, and forms that will interfere with the contraction, or cause the casting to "draw" in cooling, should be avoided ; when possible an open or H section should be preferred to a close or box section for framings, as the former is more likely to give a sound casting,—especially if plenty of small ribs are placed along in the angles (say ⅜-inch thick and 6 inches pitch,—for large framings,—dying away at 3 inches out from the angle in both directions) and a good radius is used.

Annealing, for the purpose of relieving internal stresses set up in cooling, is very necessary for large or intricate castings, and desirable for all.

Soft steels of this class will contain about ·3 per cent. of carbon ; when the proportion reaches about 1 per cent. it becomes possible to harden or temper the steel.

Good hard cast steel, with an ultimate strength of 50 or 55 tons, will probably contain 1·3 or 1·4 per cent. of carbon.

To weld properly, steel must not contain more than ·5 per cent. of carbon.

Weight of steel castings.—Soft steels, of 28 to 35 tons ultimate strength, weigh about 490 lbs. per cubic foot, or ·284 lb. per cubic inch : a plate 1 foot square and 1 inch thick will therefore weigh very nearly 41 lbs.

The Admiralty requirements as regards steel castings for machinery are as follows :—

Tensile strength to be not less than 28 tons (63,000 lbs.) per square inch, with an extension, in 2 inches of length, of, at least, 10 per cent. for intricate castings, and not less than 23 per cent. for simple ones.

Bars 1 inch square to bend cold without fracture, over 1⅝-inch radius :—for ordinary castings 90° for 28 ton steel, 60° for 35 ton steel, and other strengths in proportion ; and for intricate thin castings 20° degrees for 28 ton steel, 15° for 35 ton steel, and other strengths in proportion.

Test pieces are to be taken from each casting.

All steel castings are also to stand being dropped from a height of about 12 feet upon a hard road or floor.

Lloyd's requirements as to steel castings for machinery are as follows :—

For purposes for which **cast-iron** is ordinarily employed, such as propeller bosses and blades, bed-plates, engine framing and columns, ackets, weigh-shaft levers, pistons, cylinder covers, eccentric straps, bearing bushes, &c., the castings must be sound, and are to be subjected to such drop and hammering tests as are practicable.

For shafts or parts of shafts, and for purposes for which **forgings** are ordinarily employed, the material must also be subjected to the following tests:—

A tensile test is to be made of a piece taken from each casting. The tensile strength is not to exceed 30 tons (67,000 lbs.) per square inch, and the elongation is not to be less than 10 per cent. in a length of 8 inches, and a test piece turned to 1¼ inches diameter, or planed to 1¼ inches square, is to be capable of being bent, cold, through an angle of 90°, over a radius not exceeding 1¾ inches, without fracture.

All steel castings are to be thoroughly annealed.

Steel Bars and Plates.

This material has now almost entirely superseded bars and plates of wrought-iron; even for ventilators, uptakes, chimneys, &c., it is cheaper than iron of the quality that would be necessary to stand the working, rolling, &c.

The tests required by the Admiralty, by the Board of Trade, and by Lloyd's, are fully given in the section on "Steel Boilers" (pages 229 and 230).

In steel boiler plates rolled from ingots, the strength of test pieces is found to be practically independent of the direction in which they are cut from the plate, *i.e.*, whether the material is pulled asunder in the direction of rolling, or across that direction.

The percentage of carbon in mild steel bars and plates (27 to 30 tons per square inch) is usually between ·15 and ·25 per cent.

Weight of steel bars and plates.—The average weight of steel of this description is about 490 lbs. per cubic foot, or ·284 lb. per cubic inch, and a plate 1 foot square and 1 inch thick weighs about 41 lbs.

Experience in handling mild steel has shown that work cannot safely be continued after the red colour of the heat has disappeared; when cold, the material will stand very severe treatment, and it is equally ductile at a full red heat (say 1500° to 1800°), but below 700° it seems to be in a critical and at times brittle and unreliable condition, and should not be handled again until it has cooled below 400°, or been re-heated.

Also, when a plate has been locally heated or worked, the internal stresses set up seem to be much more severe than in the case of wrought-iron, and to avoid all risk of cracks, the plate should be carefully annealed, as soon as possible.

Steel Forgings.

Steel forgings are similar in most respects to bars and plates, but, of course, not quite so dense or so uniform in texture and strength.

The Admiralty require that all important steel forgings for machinery shall be made from ingots, and test pieces from all important ingots and forgings must satisfy the following conditions :—

Ultimate tensile strength, not less than 28 tons, with an extension in 2 inches of length, of at least 30 per cent.

Bars 1 inch square should bend cold, without fracture, through an angle of 180°, over a radius not greater than ¼-inch.

The ultimate tensile strength of material for crank and propeller shafts is not to exceed 32 tons per square inch. 30% of top end of ingot to be removed before forging, and at least 3% from bottom end after forging. Sectional area of body of forging is not to exceed ⅙th original sectional area of ingot.

The weight of steel forgings may be taken as about 487 lbs. per cubic foot, or ·282 lb. per cubic inch.

Whitworth's fluid compressed steel weighs about 495 lbs. per cubic foot.

Copper.

Copper in its unalloyed condition is used mainly for pipes and for fire-box plates of loco type boilers.

The tenacity of sheet copper is about 13½ tons (30,000 lbs.) per square inch.

Annealed copper wire has a strength of about 18 tons (40,000 lbs.) per square inch.

Sheet copper has a weight equivalent to about 550 lbs. per cubic foot, or ·318 lb. per cubic inch.

For weights of copper pipes see page 318.

The Admiralty specify that strips cut from steam and other pipes, either longitudinally or transversely, are to have an ultimate tensile strength of not less than 13 tons when annealed in water, and are to elongate at least 35 per cent. in 2 inches, or 30 per cent. in 4 inches. They are also to bend through 180° until the two sides touch, and to stand hammering to a fine edge, when cold, without cracking.

Gun-metal.

Gun-metal is composed of copper and tin in various proportions, and a small percentage of zinc is usually added to ensure sound castings.

Its strength depends mainly on its composition, but is much affected by such circumstances as the size of the casting, the rate of cooling, and the skill of the founder in mixing the metals, ventilating the moulds, relieving the cores, &c.

In large castings which cool slowly there is a great tendency for the metals to separate from one another to some extent, and the average strength is therefore usually less than in small castings ; it is also a general rule that the more quickly the casting is cooled the stronger the metal is.

The metal sets very rapidly, and contracts nearly ¾₆ inch per foot on an average, and in large castings the cores must be very quickly relieved if the casting is not to be drawn and porous and of low tenacity.

With a mixture of 90 per cent. copper and 10 per cent. tin, a carefully made test bar may be got to show a strength of nearly 17 tons (38,000 lbs.) per square inch. With 84 per cent. copper and 16 per cent. tin, a much harder metal is obtained (the hardness of gun-metal varies almost directly with the percentage of tin in the mixture), with an ultimate strength of test piece of about 16 tons (35,000 lbs.) per square inch.

For heavy bearings, 79 per cent. copper and 21 per cent. tin is sometimes used ; the resulting metal is very hard, and test pieces show a strength of 13½ to 14 tons (30,000 to 31,000 lbs.) per square inch.

Admiralty gun-metal.—For all ordinary castings in connection with the machinery, the gun-metal used must contain *not less* than 8 per cent. of tin, and *not more than* 5 per cent. of zinc. For air-compressing machinery and torpedo fittings, &c. (where the working pressure is 1700 lbs. per square inch), the mixture specified is,—Copper, *not less than* 86 per cent.; Tin, *not less than* 10 *nor more than* 12 per cent.; Zinc, *not more than* 2 per cent. No gun-metal fittings subject to steam pressure are to contain more than 10 per cent. of tin.

The ultimate tensile strength of gun-metal is to be not less than 14 tons (31,360 lbs.) per square inch, and it must extend at least 7½ per cent. in 2 inches before breaking.

Fairly good ordinary gun-metal ought to show a strength of about 12 tons (27,000 lbs.) per square inch, and should extend 10 per cent. in a length of 2 inches before breaking.

Its weight is about 545 lbs. per cubic foot, or ·315 lb. per cubic inch.

Phosphor Bronze.

This metal is composed of copper and tin, with a small proportion (about ½ per cent.) of phosphorus. It is harder than ordinary gun-metal, very close grained, and of superior strength. The average ultimate strength is about 15¼ tons (35,000 lbs.) per square inch, while that of some grades of the metal is as high as 22 tons ; it is, however, very "red-short," and, when heated, is liable to crack. Great care is required in melting and running it, and repeated meltings very much reduce its virtue.

Manganese Bronze.

Manganese bronze consists of copper and tin, or copper, tin, and zinc (according to the grade of metal required), with the addition of a proportion of ferro-manganese. The quality generally used for propeller blades shows a strength of about 21 to 22 tons (47,000 to 49,000 lbs.) as ordinarily cast, but rather better results can be obtained by casting the blades on end, and giving a head of about 2 feet.

The weight of this material is about 535 lbs. per cubic foot, or ·31 lbs. per cubic inch.

Rolled rods of manganese bronze can be obtained of strengths varying from 28 to 32 tons (63,000 to 72,000 lbs.) per square inch, and showing an elongation of 40 per cent. to 15 per cent. in 2 inches.

Muntz Metal.

Muntz metal is composed of 60 per cent. of copper, and 40 per cent. of zinc. It is very ductile, and can be forged when hot, and has an ultimate tensile strength of about 22 tons (49,000 lbs.) per square inch. A cubic foot weighs about 512 lbs., and a cubic inch about ·296 lbs.

Naval Brass.

Naval brass is composed of 62 per cent. copper, 37 per cent. zinc, and 1 per cent. tin, and possesses qualities very similar to those of muntz metal; the small addition of tin is found to protect it against the corrosive action of sea water, which causes muntz metal to "perish" or decay. Its strength is rather higher than that of muntz metal, and may be taken as about 24 tons (54,000 lbs.) per square inch.

The Admiralty specify, for bars of ¾-inch diameter and under, an ultimate tensile strength of not less than 26 tons per square inch; and for square bars and round bars over ¾-inch diameter not less than 22 tons per square inch. Extension before fracture to be at least 10 per cent. in 2 inches. The bars are also to stand being hammered to a fine point when hot, and are to bend through an angle of 75°, over a radius equal to diameter or thickness of bar, when cold.

White Metals.

The following Table shows the compositions of some of the best known white metals :—

	Copper.	Tin.	Zinc.	Lead.	Antimony.
Parson's white brass,	1	68	30·5	·5	...
Babbit's white metal,	8·5	83	...	...	8·5
Fenton's white metal,	4·4	16·6	79	...	...
Admiralty ,,	5·5	86	...	...	8·5

A very good white metal is made by mixing 6 parts of tin with 1 of copper, and 6 parts of tin with 1 of antimony, and then adding the two mixtures together.

The exact Admiralty specification is,—at least 85 per cent. tin, not less than 8 per cent. antimony, and about 5 per cent. copper; zinc or lead not to be used.

TABLE LXXXIV.—STRENGTH, ETC., OF MATERIALS.   297

Table LXXXIV.—Strengths, &c., of Materials (Summary).

Material.	Ultimate tensile strength. lbs. per square inch.	Elastic strength. lbs. per square inch.	Elongation per cent., when broken by tensile stress.
Cast-iron (ordinary good)	18,000	11,000	...
" (Admiralty) .	{ not less than 20,160	...	...
Wrought-iron bars (ordinary good . .	54,000	29,000	15 % in 8 ins.
Yorkshire plate—			
With grain . .	54,000	26,000	20 % "
Across " . .	49,000	...	14 % "
Staffordshire plate—			
With grain . .	50,000	24,000	12 % "
Across " . .	41,000	...	8 % "
Iron forgings—			
Large . . .	45,000	...	9 % "
Small . . .	50,000	...	13 % "
Steel castings (ordinary good). . . .	67,000	35,000	10 % "
Steel castings (Admiralty)	{ not less than 63,000 }	...	{ not less than 10–23 % in 2 ins.
" (Lloyd's) .	{ not exc'd'g 67,000 }	...	{ not less than 10 % in 8 ins.
Steel boiler plate—			
(Ordinary good) .	65,000	36,000	20 % "
(Admiralty) internal .	51,520–58,240	...	{ 27 %–26 % in 8 ins.
" shell .	56,000–67,200	...	{ 25 %–20 % in 8 ins.
(B. of T.) internal .	58,240–67,200	...	
" shell .	60,480–71,680	...	18 % in 10 ins.
Lloyd's . . .	58,240–67,200	...	{ not less than 20 % in 8 ins.
Steel forgings (Admiralty)	62,720–71,680	...	30 % in 2 ins.
Sheet copper . . .	30,000	5,600	35 % in 8 ins.
" (Admiralty)	29,120	...	35 % in 2 ins.
Copper wire (annealed) .	40,000	...	...
Gun - metal (ordinary good) . . .	27,000	6,500	10 % in 2 ins.
Gun-metal (Admiralty) .	31,360	...	7½ % "
Phosphor bronze (cast) .	35,000	19,000	12 % "
Manganese bronze . .	55,000	...	10 % "
" (rolled) .	67,000	...	20 % "
Muntz metal . . .	50,000	30,000	30 % "
Naval brass . . .	49,280–58,240	...	{ at least 10 % in 2 ins.

BENDING STRAINS.

Table LXXXV.—Graphic Representation of Bending Moments and Shearing Strains in Simple Cases.

Bending Moment.	Shearing Strain.

f = greatest permissible stress on material; Z = modulus of section —*see* Table LXXXVI.; M = greatest bending moment; S = greatest shearing strain.

Beams fixed at one end.

$$M = Wl$$

Working load = $\dfrac{fz}{l}$

$$S = W$$

$M_1 = W_1 \, l_1$; and $M_2 = W_2 \, l_2$
$M_3 = W_1 \, l_1 + W_2 \, l_2$

$S_1 = W_1$; and $S_2 = W_2$
$S_3 = W_1 + W_2$

Beams fixed at one end—*continued*.

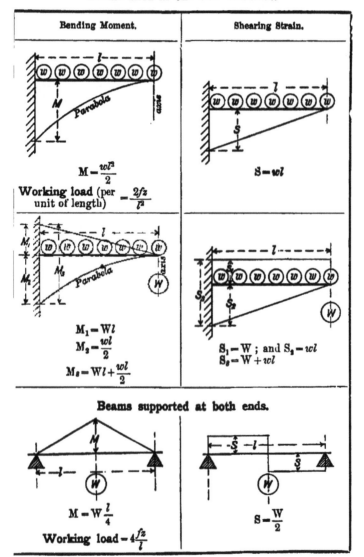

Bending Moment.	Shearing Strain.

$$M = \frac{wl^2}{2}$$

Working load (per unit of length) $= \frac{2fz}{l^2}$

$$S = wl$$

$$M_1 = Wl$$
$$M_2 = \frac{wl}{2}$$
$$M_3 = Wl + \frac{wl}{2}$$

$S_1 = W$; and $S_2 = wl$
$S_3 = W + wl$

Beams supported at both ends.

$$M = W\frac{l}{4}$$

Working load $= 4\frac{fz}{l}$

$$S = \frac{W}{2}$$

Beams supported at both ends—*continued*.

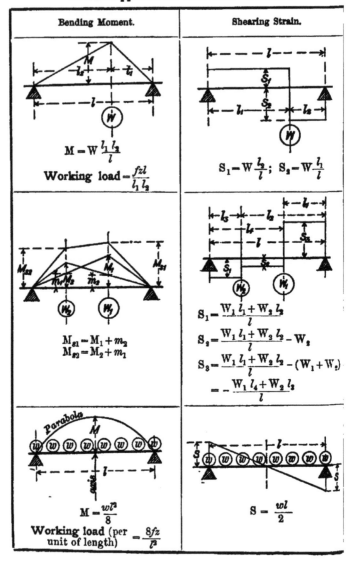

Bending Moment.	Shearing Strain.

$$M = W\frac{l_1 l_2}{l}$$

$$\text{Working load} = \frac{fzl}{l_1 l_2}$$

$$S_1 = W\frac{l_2}{l}; \quad S_2 = W\frac{l_1}{l}$$

$$M_{s1} = M_1 + m_2$$
$$M_{s2} = M_2 + m_1$$

$$S_1 = \frac{W_1 l_1 + W_2 l_2}{l}$$

$$S_2 = \frac{W_1 l_1 + W_3 l_3}{l} - W_2$$

$$S_3 = \frac{W_1 l_1 + W_2 l_2}{l} - (W_1 + W_2)$$
$$= -\frac{W_1 l_4 + W_2 l_3}{l}$$

$$M = \frac{wl^2}{8}$$

$$\text{Working load (per unit of length)} = \frac{8fz}{l^2}$$

$$S = \frac{wl}{2}$$

Beams fixed at both ends.

Bending Moment.

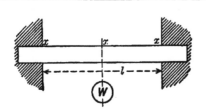

In this case the bending moments at x, x, and x are equal, and have the value,—

$$M = \frac{Wl}{8}$$

$$\text{Working load} = \frac{8fz}{l}$$

Shearing Strain.

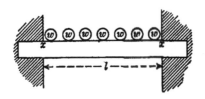

In this case the greatest bending moment occurs at x and x, and has the value,—

$$M = \frac{wl^2}{12}$$

$$\text{Working load (per unit of length)} = \frac{12fz}{l^2}$$

Table LXXXVI.—Moment of Inertia, Modulus, &c., of some Sections.

The plane of bending is supposed perpendicular to plane of paper, and parallel to side of page.

Form of section.	Area of section. A	Moment of inertia of section about axis through centre of gravity. I	Square of rads. of gyration of section. $\rho^2 = \dfrac{I}{A}$	"Modulus" of section. $Z = \dfrac{I}{y}$
	s^2	$\dfrac{s^4}{12}$	$\dfrac{s^2}{12}$	$\dfrac{s^3}{6}$
	bh	$\dfrac{bh^3}{12}$	$\dfrac{h^2}{12}$	$\dfrac{bh^2}{6}$
	$S^2 - s^2$	$\dfrac{b(S^4 - s^4)}{12}$	$\dfrac{S^2 + s^2}{12}$	$\tfrac{1}{6}\left(\dfrac{S^4 - s^4}{S}\right)$
	$b(H - h)$	$\dfrac{b(H^3 - h^3)}{12}$	$\dfrac{H^2 + hH + h^2}{12}$	$\dfrac{b(H^3 - h^3)}{6H}$
	$\cdot 7854\,d^2$	$\cdot 0491\,d^4$	$\dfrac{d^2}{16}$	$\cdot 0982\,d^3$
	$\cdot 7854\,(D^2 - d^2)$	$\cdot 0491\,(D^4 - d^4)$	$\dfrac{D^2 + d^2}{16}$	$\cdot 0982\left(\dfrac{D^4 - d^4}{D}\right)$

TABLE LXXXVI.—MOMENT OF INERTIA, MODULUS, ETC. 303

Form of section.	Area of section. A	Moment of inertia of section about axis through centre of gravity. I	Square of rads. of gyration of section. $\rho^2 = \dfrac{I}{A}$	"Modulus" of section. $Z = \dfrac{I}{y}$
	$BH - bh$	$\dfrac{BH^3 - bh^3}{12}$	$\dfrac{1}{12}\left(\dfrac{BH^3 - bh^3}{BH - bh}\right)$	$\dfrac{BH^3 - bh^3}{6H}$
	$Bh + bH$	$\dfrac{Bh^3 + bH^3}{12}$	...	$\dfrac{Bh^3 + bH^3}{6H}$
	$BH - bh$	$\dfrac{(BH^2 - bh^2)^2 - 4BHbh(H - h)^2}{12(BH - bh)}$	...	$\dfrac{(BH^3 - bh^2)^2 - 4BHbh(H - h)^2}{6(BH^2 + bh^2 - 2bHh)}$

Table LXXXVII.—Forms of Beams of Uniform Strength.

Breadth (b) uniform throughout.	
Elevation of arrangement.	**Equation for dimensions.**
	$y = \sqrt{\dfrac{6Wx}{bf}}$
	$y = \sqrt{\dfrac{3wx^2}{bf}}$
	$y_1 = \sqrt{\dfrac{6Wl_2x_1}{bfl}}$ $y_2 = \sqrt{\dfrac{6Wl_1x_2}{bfl}}$
	$y = \sqrt{\dfrac{3w(l^2 - 4x^2)}{4bf}}$

TABLE LXXXVII.—FORMS OF BEAMS, ETC.   305

Table LXXXVII.—Forms of Beams of Uniform Strength.—*cont.*

Depth (h) uniform throughout.	
Plan of arrangement.	**Equation for dimensions.**
	$z = \dfrac{6Wx}{h^2 f}$
	$z = \dfrac{3wx^2}{h^2 f}$
	$z_1 = \dfrac{6Wl_2 x_1}{h^2 fl}$ $z_2 = \dfrac{6Wl_1 x_2}{h^2 fl}$
	$z = \dfrac{3w\,(l^2 - 4x^2)}{4h^2 f}$

EXPANSION OF METALS BY HEAT.

The following table shows approximately the extent to which metals, &c., expand under the action of heat:—

Table LXXXVIII.—Expansion of Metals, &c., for rise of Temperature of 180° F.

Material	Change of length.	
	Fraction of total length.	Inches per foot. (C)
Cast-iron, ...	·00117	·0140
Wrought-iron, . . .	·00122	·0146
Steel, ...	·00120	·0146
Copper, ...	·00182	·0218
Gun-metal, . .	·00187	·0247
Fire-brick, . . .	·000423	·0050

If t_1 and t_2 be the highest and lowest temperatures to which the object is exposed, the *alteration* in length is given by,—

$$\text{Alteration in length (increase or diminution)} = C \times \left(\frac{t_1 - t_2}{180}\right) \times L$$

where L is length in feet, value of C is taken from above table, and alteration of length is given in inches.

The expansion of metals, per degree rise of temperature, increases slightly as higher temperatures are reached, but for all practical purposes it may be assumed to be constant.

Effect of temperature on strengths of metals.—At 400° F. almost all alloys of copper, tin, and zinc lose from 15 to 20 per cent. of their ultimate strength.
Copper is similarly affected, and loses 8 to 10 per cent. of its strength at 400° F.
Cast-iron is practically unaffected up to 400° F.
Wrought-iron and mild steel *gain* from 10 to 15 per cent. in ultimate strength at 400° F.
The changes seem to progress regularly with the changes of temperature.

MELTING POINTS OF METALS.

Table LXXXIX. gives, approximately, the melting points of some ordinary metals; it should be understood that all temperatures above 900° to 1000° F. are only roughly approximate, as the readings of pyrometers cannot be relied upon to a hundred degrees or so.

Table LXXXIX.—Melting Points of some Metals.

Wrought-iron melts at about	2700°–2900°
,, welds ,,	2700°
Steel (low carbon) melts,,	2600°
,, (high carbon) ,, ,,	2400°
Cast-iron (white) ,, ,,	2250°
,, (grey) ,, ,,	2050°
Copper ,, ,,	2000°
Wrought-iron—" cherry-red " heat .	1500°–1800°
Gun-metal melts at about	1700°
Red heat visible by daylight about .	1100°
Zinc melts at about	840°
Antimony ,,	800°
Red heat visible in dark about	750°– 800°
Lead melts at about	630°
Bismuth ,,	510°
Tin ,,	440

WEIGHTS OF MATERIALS.
Table XC.—Weights of Materials (Summary).

Material.	Weight of a cubic foot in lbs.	Weight of a cubic inch in lbs.
Air (32° F. and 14·7 lbs. pressure)	·0807	...
⎰ pure,	62·4	·036
⎱ river,	63	...
⎰ sea,	64	·037
Colza, linseed, or olive oil,	57–58	...
Mineral oils,	56	...
Tallow,	58	...
Waste (moderately pressed),	11	...
Elm, pine, or fir timber,*	30–40	...
Beech, ash, or birch, ,,	40–45	...
Oak or teak, ,,	45–55	...
Greenheart, ,,	about 65	...
Lignum-vitæ, ,,	about 80	...
Fire brick,	about 140	...
Wrought-iron bars or plates,	485	·281
Staffordshire plates,	480	·277
Iron forgings (large),	477	·276
Steel bars and plates,	490	·284
Steel forgings (large),	487	·282
Steel—Whitworth compressed,	496	·287
Cast-iron,	450	·260
Cast-steel (mild),	490	·284
Sheet copper,	550	·318
Gun-metal,	545	·315
Muntz-metal,	512	·296
Naval brass (rolled),	530	·307
White metal (Babbit's),	456	·263
Zinc ⎰ cast,	435	·252
⎱ rolled,	450	·260
Lead ⎰ cast,	708	·408
⎱ sheet,	711	·411

A plate of cast-iron 1 foot square and 1 inch thick weighs 37·5 lbs.
" 　wrought-iron　 " 　 " 　40 "
" 　cast-steel (mild)　 " 　 " 　41 "
" 　wrought-steel (mild)　 " 　 " 　41 "
" 　gun-metal　 " 　 " 　45 "
" 　rolled brass　 " 　 " 　44 "

The Admiralty reckon 40 cubic feet of bunker space as 1 ton, but the usual allowance is 45 cubic feet to the ton; the actual average bulk of a ton is about 43 cubic feet, but if taken at 45 cubic feet a fair allowance is made for the upper portions under the deck-beams, which cannot be filled.

* The following are the exact weights of the various kinds of pine wood in general use, as derived from a series of careful experiments recently made by Mr Seaton:—
Yellow pine, very dry, . . . 24 lbs. per cubic foot.
　　"　　 in planks, seasoned, . . 28 " 　 "
Baltic red pine, " 　 " 　 . . 30·5 " 　 "
Oregon pine, " 　 " 　 . . 35 " 　 "
Kauri pine, " 　 " 　 . . 42 " 　 "
Pitch pine, " 　 " 　 . . 45 " 　 "

TABLE XCI.—WEIGHT OF ROUND AND SQUARE IRON BARS. 309

Table XCI.—Table of the Weight of Round and Square Wrought-Iron Bars in lbs. per Lineal Foot.

Dia. or Side.	Weight in lbs.		Dia. or Side.	Weight in lbs.		Dia. or Side.	Weight in lbs.	
	Round.	Square.		Round.	Square.		Round.	Square.
3/16	·093	·117	3¾	36·812	46·875	8	167·53	213·33
1/4	·164	·208	7/8	39·306	50·052	1/8	172·81	220·05
5/16	·256	·326				1/4	178·17	226·88
3/8	·368	·469	4	41·884	53·333	3/8	183·61	233·80
7/16	·501	·638	1/8	44·542	56·719	1/2	189·13	240·83
1/2	·654	·833	1/4	47·283	60·208	5/8	194·73	247·97
9/16	·828	1·055	3/8	50·105	63·802	3/4	200·42	255·21
5/8	1·023	1·302	1/2	53·009	67·500	7/8	206·19	262·55
11/16	1·237	1·576	5/8	55·995	71·302			
3/4	1·473	1·875	3/4	59·062	75·208	9	212·04	270·00
13/16	1·728	2·201	7/8	62·212	79·219	1/8	217·97	277·55
7/8	2·004	2·552				1/4	223·98	285·21
15/16	2·300	2·930	5	65·443	83·333	3/8	230·07	292·97
			1/8	68·756	87·552	1/2	236·25	300·83
1	2·618	3·333	1/4	72·151	91·875	5/8	242·51	308·80
1/8	3·313	4·219	3/8	75·628	96·302	3/4	248·85	316·88
1/4	4·090	5·208	1/2	79·186	100·83	7/8	255·27	325·05
3/8	4·949	6·302	5/8	82·827	105·47			
1/2	5·890	7·500	3/4	86·549	110·21	10	261·77	333·33
5/8	6·912	8·802	7/8	90·353	115·05	1/8	268·36	341·72
3/4	8·017	10·208				1/4	275·03	350·21
7/8	9·203	11·719	6	94·238	120·00	3/8	281·77	358·80
			1/8	98·206	125·05	1/2	288·60	367·50
2	10·471	13·333	1/4	102·26	130·21	5/8	295·52	376·30
1/8	11·821	15·052	3/8	106·39	135·47	3/4	302·51	385·21
1/4	13·252	16·875	1/2	110·60	140·83	7/8	309·59	394·22
3/8	14·766	18·802	5/8	114·89	146·30			
1/2	16·361	20·833	3/4	119·27	151·88	11	316·75	403·33
5/8	18·038	22·969	7/8	123·73	157·55	1/8	323·99	412·55
3/4	19·797	25·208				1/4	331·31	421·88
7/8	21·637	27·552	7	128·27	163·33	3/8	338·71	431·30
			1/8	132·89	169·22	1/2	346·20	440·83
3	23·560	30·000	1/4	137·60	175·21	5/8	353·76	450·47
1/8	25·564	32·552	3/8	142·98	181·30	3/4	361·41	460·21
1/4	27·650	35·208	1/2	147·25	187·50	7/8	369·14	470·05
3/8	29·818	37·969	5/8	152·20	193·80			
1/2	32·067	40·833	3/4	157·23	200·21	12	376·95	480·00
5/8	34·399	43·802	7/8	162·34	206·72			

For larger sizes take weight of bar of half the diameter or side and multiply it by four.

Table XCII.—Table of the Weight of Round and Square Bar Steel in lbs. per Lineal Foot.

Dia. or Side.	Weight in lbs. Round.	Weight in lbs. Square.	Dia. or Side.	Weight in lbs. Round.	Weight in lbs. Square.	Dia. or Side.	Weight in lbs. Round.	Weight in lbs. Square.
⅛	·042	·053	3⅝	35·090	44·678	7⅞	165·60	210·85
³⁄₁₆	·094	·120	¾	37·552	47·813			
¼	·167	·213	⅞	40·097	51·053	8	170·90	217·60
⁵⁄₁₆	·261	·332				⅛	176·29	225·25
⅜	·375	·478	4	42·726	54·400	¼	181·75	231·41
⁷⁄₁₆	·511	·651	⅛	45·438	57·853	⅜	187·30	238·48
½	·667	·850	¼	48·238	61·413	½	192·93	245·65
⁹⁄₁₆	·845	1·076	⅜	51·112	65·078	⅝	198·65	252·93
⅝	1·043	1·328	½	54·075	68·850	¾	204·45	260·31
¹¹⁄₁₆	1·262	1·607	⅝	57·121	72·728	⅞	210·33	267·80
¾	1·502	1·913	¾	60·250	76·713			
¹³⁄₁₆	1·762	2·245	⅞	63·463	80·803	9	216·30	275·40
⅞	2·044	2·603				⅛	222·35	283·10
¹⁵⁄₁₆	2·347	2·988	5	66·759	85·000	¼	228·48	290·91
			⅛	70·139	89·303	⅜	234·70	298·83
1	2·670	3·400	¼	73·602	93·713	½	241·00	306·85
⅛	3·380	4·303	⅜	77·148	98·229	⅝	248·38	314·98
¼	4·172	5·313	½	80·778	102·85	¾	253·85	323·21
⅜	5·049	6·428	⅝	84·492	107·58	⅞	260·40	331·55
½	6·008	7·650	¾	88·288	112·41			
⅝	7·051	8·978	⅞	92·169	117·35	10	267·04	340·00
¾	8·178	10·413				⅛	273·75	348·55
⅞	9·388	11·953	6	96·133	122·40	¼	280·55	357·21
			⅛	100·18	127·55	⅜	287·44	365·98
2	10·681	13·600	¼	104·31	132·81	½	294·41	374·85
⅛	12·058	15·353	⅜	108·52	138·18	⅝	301·46	383·83
¼	13·519	17·213	½	112·82	143·65	¾	308·59	392·91
⅜	15·062	19·178	⅝	117·20	149·23	⅞	315·81	402·10
½	16·690	21·250	¾	121·67	154·91			
⅝	18·400	23·428	⅞	126·22	160·70	11	323·11	411·40
¾	20·195	25·713				⅛	330·50	420·80
⅞	22·072	28·103	7	130·85	166·60	¼	337·97	430·31
			⅛	135·56	172·60	⅜	345·52	439·93
3	24·033	30·600	¼	140·36	178·71	½	353·15	449·65
⅛	26·078	33·203	⅜	145·24	184·93	⅝	360·87	459·48
¼	28·206	35·913	½	150·21	191·25	¾	368·68	469·41
⅜	30·417	38·728	⅝	155·26	197·68	⅞	376·56	479·45
½	32·712	41·650	¾	160·39	204·21	12	384·53	489·60

For larger sizes take weight of bar of half the diameter or side and ·ltiply it by four.

TABLE XCIII.—WEIGHT OF FLAT WROUGHT-IRON BARS. 311

Table XCIII.—Table of the Weight of Flat Wrought-Iron Bars in lbs. per Lineal Foot.

Breadth of bar (ins.)	1/4	5/16	3/8	7/16	1/2	9/16	5/8	11/16	3/4	13/16	7/8	15/16	1 in.	Breadth of bar (ins.)
1	·83	1·04	1·25	1·46	1·67	1·88	2·08	2·29	2·50	2·71	2·92	3·13	3·33	1
1¼	1·04	1·30	1·56	1·82	2·08	2·34	2·60	2·86	3·13	3·39	3·65	3·91	4·16	1¼
1½	1·25	1·56	1·88	2·19	2·50	2·81	3·13	3·44	3·75	4·06	4·38	4·69	5·00	1½
1¾	1·46	1·82	2·19	2·55	2·92	3·28	3·65	4·01	4·38	4·74	5·10	5·47	5·83	1¾
2	1·67	2·08	2·50	2·92	3·33	3·75	4·17	4·58	5·00	5·42	5·83	6·25	6·67	2
2¼	1·88	2·34	2·81	3·28	3·75	4·22	4·69	5·16	5·63	6·09	6·56	7·03	7·50	2¼
2½	2·08	2·60	3·13	3·65	4·17	4·69	5·21	5·73	6·25	6·77	7·29	7·81	8·33	2½
2¾	2·29	2·86	3·44	4·01	4·58	5·16	5·73	6·30	6·88	7·45	8·02	8·59	9·17	2¾
3	2·50	3·13	3·75	4·38	5·00	5·63	6·25	6·88	7·50	8·13	8·75	9·38	10·00	3
3¼	2·71	3·39	4·06	4·74	5·42	6·09	6·77	7·45	8·13	8·80	9·48	10·16	10·83	3¼
3½	2·92	3·65	4·38	5·10	5·83	6·56	7·29	8·02	8·75	9·48	10·21	10·94	11·67	3½
3¾	3·13	3·91	4·69	5·47	6·25	7·03	7·81	8·59	9·38	10·16	10·94	11·72	12·50	3¾
4	3·33	4·17	5·00	5·83	6·67	7·50	8·33	9·17	10·00	10·83	11·67	12·50	13·33	4
4¼	3·54	4·43	5·31	6·20	7·08	7·97	8·85	9·74	10·63	11·51	12·40	13·28	14·17	4¼
4½	3·75	4·69	5·63	6·56	7·50	8·44	9·38	10·31	11·25	12·19	13·13	14·06	15·00	4½
4¾	3·96	4·95	5·94	6·93	7·92	8·91	9·90	10·89	11·88	12·86	13·85	14·84	15·83	4¾
5	4·17	5·21	6·25	7·29	8·33	9·38	10·42	11·46	12·50	13·54	14·58	15·63	16·67	5

Table XCIV.—Table of the Weight of Flat Rolled Steel Bars in lbs. per Lineal Foot.

Breadth of bar (ins.)	Thickness in fractions of an inch													Breadth of bar (ins.)
	¼	5/16	⅜	7/16	½	9/16	⅝	11/16	¾	13/16	⅞	15/16	1 in.	
1	·85	1·06	1·28	1·49	1·70	1·91	2·13	2·34	2·55	2·76	2·98	3·19	3·40	1
1¼	1·06	1·33	1·59	1·86	2·13	2·39	2·66	2·92	3·19	3·45	3·72	3·98	4·25	1¼
1½	1·28	1·59	1·91	2·23	2·55	2·87	3·19	3·51	3·83	4·14	4·46	4·78	5·10	1½
1¾	1·49	1·86	2·23	2·60	2·98	3·35	3·72	4·09	4·46	4·83	5·21	5·58	5·95	1¾
2	1·70	2·13	2·55	2·98	3·40	3·83	4·25	4·68	5·10	5·53	5·95	6·38	6·80	2
2¼	1·91	2·39	2·87	3·35	3·83	4·30	4·78	5·26	5·74	6·22	6·69	7·17	7·65	2¼
2½	2·13	2·66	3·19	3·72	4·25	4·78	5·31	5·84	6·38	6·91	7·44	7·97	8·50	2½
2¾	2·34	2·92	3·51	4·09	4·68	5·26	5·84	6·43	7·01	7·60	8·18	8·77	9·35	2¾
3	2·55	3·19	3·83	4·46	5·10	5·74	6·38	7·01	7·65	8·29	8·93	9·56	10·20	3
3¼	2·76	3·45	4·14	4·83	5·53	6·22	6·91	7·60	8·29	8·98	9·67	10·36	11·05	3¼
3½	2·98	3·72	4·46	5·21	5·95	6·69	7·44	8·18	8·93	9·67	10·41	11·16	11·90	3½
3¾	3·19	3·98	4·78	5·58	6·38	7·17	7·97	8·77	9·56	10·36	11·16	11·95	12·75	3¾
4	3·40	4·25	5·10	5·95	6·80	7·65	8·50	9·35	10·20	11·05	11·90	12·75	13·60	4
4¼	3·61	4·52	5·42	6·32	7·23	8·13	9·03	9·93	10·84	11·74	12·64	13·55	14·45	4¼
4½	3·83	4·78	5·74	6·69	7·65	8·61	9·56	10·52	11·48	12·43	13·39	14·34	15·30	4½
4¾	4·04	5·05	6·06	7·07	8·08	9·08	10·09	11·10	12·11	13·12	14·13	15·14	16·15	4¾
5	4·25	5·31	6·38	7·44	8·50	9·56	10·63	11·69	12·75	13·81	14·88	15·94	17·00	5

TABLE XCV.—WEIGHT OF IRON ANGLE-BARS. 313

Table XCV.—Table of the Weight of Iron Angle-Bars in lbs. per Lineal Foot.

Sum of flanges (ins.)	Thickness in fractions of an inch.										
	³⁄₁₆	¼	⁵⁄₁₆	⅜	⁷⁄₁₆	½	⁹⁄₁₆	⅝	¹¹⁄₁₆	¾	¹³⁄₁₆
2½	1·45	1·88	2·28	2·66	...	...	...	...	...	...	...
2¾	1·60	2·08	2·54	2·97	...	...	...	...	...	...	...
3	1·76	2·29	2·80	3·28	...	...	...	...	...	...	...
3¼	1·91	2·50	3·06	3·59	4·10	...	...	...	...	...	...
3½	2·07	2·71	3·32	3·91	4·47	...	...	...	...	...	...
3¾	2·23	2·92	3·58	4·22	4·83	...	...	...	...	...	...
4	2·38	3·13	3·84	4·53	5·20	5·83	...	...	...	...	...
4¼	2·54	3·33	4·10	4·84	5·56	6·25	...	...	...	...	...
4½	2·70	3·54	4·36	5·16	5·92	6·67	...	...	...	...	...
4¾	2·85	3·75	4·62	5·47	6·29	7·08	7·85	...	...	...	...
5	...	3·96	4·88	5·78	6·65	7·50	8·32	...	...	...	...
5¼	...	4·17	5·14	6·09	7·02	7·92	8·79	...	...	...	...
5½	...	4·38	5·40	6·41	7·38	8·33	9·26	10·16	...	...	...
5¾	...	...	5·66	6·72	7·75	8·75	9·73	10·68	...	...	...
6	...	...	5·92	7·03	8·11	9·17	10·20	11·20	...	...	...
6¼	...	...	6·18	7·34	8·48	9·58	10·66	11·72	12·75	...	...
6½	...	...	...	7·66	8·84	10·00	11·13	12·24	13·32	...	...
6¾	...	...	...	7·97	9·21	10·42	11·60	12·76	13·89	...	...
7	...	...	...	8·28	9·57	10·83	12·07	13·28	14·46	15·63	...
7¼	...	...	...	...	9·93	11·25	12·54	13·80	15·04	16·25	...
7½	...	...	...	...	10·30	11·67	13·01	14·32	15·61	16·88	...
7¾	...	...	...	...	10·66	12·08	13·48	14·84	16·18	17·50	18·79
8	...	...	...	...	...	12·50	13·95	15·36	16·76	18·13	19·47
8¼	...	...	...	...	...	12·92	14·44	15·89	17·33	18·75	20·14
8½	...	...	...	...	...	13·33	14·88	16·41	17·90	19·38	20·82
8¾	...	...	...	...	...	...	15·35	16·93	18·48	20·00	21·50
9	...	...	...	...	...	...	15·82	17·45	19·05	20·63	22·17
9¼	...	...	...	...	...	...	16·29	17·97	19·62	21·25	22·85
9½	...	...	...	...	...	...	...	18·49	20·20	21·88	23·53
9¾	...	...	...	...	...	...	...	19·01	20·77	22·50	24·21
10	...	...	...	...	...	...	...	19·53	21·34	23·13	24·88

Table XCVI.—Table of the Weight of Steel Angle-Bars in lbs. per Lineal Foot.

Sum of flanges (ins.)	Thickness in fractions of an inch.										
	3/16	1/4	5/16	3/8	7/16	1/2	9/16	5/8	11/16	3/4	13/16
2½	1·47	1·91	2·32	2·71	...	...	...	...	...	...	...
2¾	1·63	2·12	2·59	3·03	...	...	...	...	...	...	...
3	1·79	2·34	2·86	3·35	...	...	...	...	...	...	...
3¼	1·95	2·55	3·12	3·67	4·18	...	...	...	...	...	...
3½	2·11	2·76	3·39	3·98	4·56	...	...	...	...	...	...
3¾	2·27	2·97	3·65	4·30	4·93	...	...	...	...	...	...
4	2·43	3·19	3·92	4·62	5·30	5·95	...	...	...	...	...
4¼	2·59	3·40	4·18	4·94	5·67	6·38	...	...	...	...	...
4½	2·75	3·61	4·45	5·26	6·04	6·80	...	...	...	...	...
4¾	2·91	3·82	4·71	5·58	6·41	7·23	8·01	...	...	...	...
5	...	4·04	4·98	5·90	6·79	7·65	8·49	...	...	...	...
5¼	...	4·25	5·25	6·22	7·16	8·08	8·96	...	...	...	...
5½	...	4·46	5·51	6·53	7·53	8·50	9·44	10·36	...	...	...
5¾	...	...	5·78	6·85	7·90	8·93	9·92	10·89	...	...	...
6	...	...	6·04	7·17	8·27	9·35	10·40	11·42	...	...	...
6¼	...	...	6·31	7·49	8·65	9·78	10·88	11·95	13·00	...	...
6½	...	...	...	7·81	9·02	10·20	11·36	12·48	13·59	...	...
6¾	...	...	...	8·13	9·39	10·63	11·83	13·02	14·17	...	...
7	...	...	...	8·45	9·76	11·05	12·31	13·55	14·76	15·94	...
7¼	...	...	...	...	10·13	11·48	12·79	14·08	15·34	16·58	...
7½	...	...	...	...	10·51	11·90	13·27	14·61	15·92	17·21	...
7¾	...	...	...	...	10·88	12·33	13·75	15·14	16·51	17·85	19·16
8	...	...	...	...	...	12·75	14·22	15·67	17·09	18·49	19·85
8¼	...	...	...	...	...	13·18	14·70	16·20	17·68	19·13	20·55
8½	...	...	...	...	...	13·60	15·18	16·73	18·26	19·76	21·24
8¾	...	...	...	...	...	...	15·66	17·27	18·85	20·40	21·93
9	...	...	...	...	...	...	16·14	17·80	19·43	21·04	22·62
9¼	...	...	...	...	...	...	16·61	18·33	20·01	21·68	23·31
9½	...	...	...	...	...	...	...	18·86	20·60	22·31	24·00
9¾	...	...	...	...	...	...	...	19·39	21·18	22·95	24·69
10	...	...	...	...	...	...	...	19·92	21·77	23·59	25·38

TABLE XCVII.—WHITWORTH GAS THREADS. 315

Table XCVII.—Whitworth Gas Threads.

Diameter of pipe (inside).	Diameter over threads.	Diameter at bottom of thread.	Number of threads per inch.
⅛″	·3825	·3367	28
¼	·5180	·4506	19
⅜	·6563	·5889	19
½	·8257	·7342	14
⅝	·9022	·8107	14
¾	1·041	·9495	14
⅞	1·189	1·0975	14
1	1·309	1·1925	11
1⅛	1·492	1·3755	11
1¼	1·65	1·5335	11
1⅜	1·745	1·6285	11
1½	1·8825	1·766	11
1⅝	2·021	1·9045	11
1¾	2·047	1·9305	11
1⅞	2·245	2·1285	11
2	2·347	2·2305	11
2¼	2·5875	2·471	11
2½	3·0013	2·8848	11
2¾	3·247	3·1305	11
3	3·485	3·3685	11
3¼	3·6985	3·582	11
3½	3·912	3·7955	11
3¾	4·1225	4·009	11
4	4·339	4·2225	11

Table XCVIII.—Weight of Brass Condenser Tubes.

L.S.G.		14	15	16	17	18	19	20
Inches.		·080	·072	·064	·056	·048	·040	·036
Millimetres.		2·032	1·829	1·626	1·422	1·219	1·016	·914
Ext. Diameter.		Weight of a Lineal Foot in pounds.						
Ins.	mm.							
⅝	15·9	·51	·46	·42	·37	·32	·27	·25
¾	19·0	·62	·57	·51	·45	·39	·33	·30
⅞	22·2	·74	·67	·60	·53	·46	·39	·35
1	25·4	·86	·78	·70	·62	·53	·45	·40

Note.—The above weights are for tubes containing 70 per cent. copper.

Table XCIX.—Weights of

	THICKNESS											
External Diam. Ins. / Usual L.S.G. and Weight			16	15	14	13	12	11	10	9	8	7	6	5
L.S.G. Inches			·064	·072	·080	·092	·104	·116	·128	·144	·160	·176	·192	·215
Millimetres			1·626	1·829	2·082	2·337	2·642	2·946	3·251	3·658	4·064	4·470	4·877	5·385
Ins.	L.S.G.	Weight	WEIGHT PER											
1	14	0·771	0·627	0·700	0·771	0·875	0·976	1·074	1·169	1·291	1·407	1·519	1·624	1·749
1⅛	14	0·875	0·711	0·794	0·875	0·995	1·112	1·226	1·336	1·479	1·617	1·749	1·876	2·027
1¼	13	1·116		0·888	0·980	1·116	1·248	1·377	1·504	1·668	1·826	1·979	2·127	2·304
1⅜	13	1·236		0·982	1·085	1·236	1·384	1·529	1·671	1·856	2·036	2·210	2·378	2·582
1½	13	1·356		1·077	1·190	1·356	1·520	1·681	1·889	2·045	2·245	2·440	2·680	2·859
1⅝	13	1·477		1·171	1·294	1·477	1·656	1·838	2·007	2·233	2·455	2·671	2·881	3·137
1¾	13	1·597		1·265	1·399	1·597	1·798	1·985	2·174	2·422	2·664	2·901	3·132	3·414
1⅞	13	1·718		1·359	1·504	1·718	1·929	2·137	2·342	2·610	2·878	3·131	3·384	3·602
2	12	2·065			1·608	1·888	2·065	2·288	2·509	2·799	3·083	3·362	3·685	3·969
2⅛	12	2·201			1·718	1·959	2·201	2·440	2·677	2·987	3·292	3·592	3·896	4·247
2¼	12	2·337			1·818	2·079	2·337	2·592	2·844	3·176	3·502	3·822	4·138	4·524
2⅜	11	2·744				2·199	2·478	2·744	3·012	3·364	3·711	4·053	4·389	4·803
2½	11	2·896				2·320	2·609	2·896	3·179	3·553	3·921	4·283	4·640	5·079
2⅝	11	3·048				2·440	2·745	3·048	3·347	3·741	4·130	4·514	4·892	5·357
2¾	11	3·200				2·561	2·882	3·200	3·514	3·930	4·339	4·744	5·143	5·634
2⅞	11	3·351				2·681	3·018	3·351	3·682	4·118	4·549	4·974	5·394	5·912
3	11	3·503				2·802	3·154	3·503	3·850	4·307	4·758	5·205	5·646	6·189
3⅛	10	4·017					3·290	3·655	4·017	4·496	4·968	5·435	5·897	6·467
3¼	10	4·185					3·426	3·807	4·185	4·684	5·177	5·665	6·148	6·744
3⅜	10	4·352					3·562	3·959	4·352	4·872	5·387	5·896	6·400	7·022
3½	10	4·520					3·698	4·111	4·520	5·061	5·596	6·126	6·651	7·299
3⅝	10	4·687					3·835	4·262	4·687	5·249	5·806	6·357	6·902	7·577
3¾	10	4·855					3·971	4·414	4·855	5·433	6·015	6·587	7·154	7·854
3⅞	9	5·626						4·566	5·022	5·626	6·224	6·817	7·405	8·133
4	9	5·815						4·718	5·190	5·815	6·434	7·048	7·656	8·410

This table of weights is based on the specific gravity of Iron at 7·698. Actual weights may differ slightly.

For other metals—
Steel, ...   multiply by 1·012.
Brass, ...   „   1·080.
Copper, ...   „   1·135.

TABLE XCIX.—WEIGHTS OF IRON BOILER TUBES.   317

Iron Boiler Tubes.

OF IRON.											External Diam.
4	3	2	1	$\frac{1}{8}''$	$\frac{3}{16}''$	$\frac{1}{4}''$	$\frac{5}{16}''$	$\frac{3}{8}''$	$\frac{7}{16}''$	$\frac{1}{2}''$	
·232	·252	·276	·300	·125	·187	·250	·313	·375	·437	·500	
5·893	6·401	7·010	7·620	3·175	4·726	6·350	7·937	9·525	11·112	12·700	
FOOT IN POUNDS.											Ins.
1·866	1·974	2·092	2·199	1·145	1·595	1·968	2·250	2·454	2·577	2·618	1
2·169	2·304	2·454	2·592	1·309	1·841	2·291	2·659	2·945	3·150	3·272	1⅛
2·473	2·634	2·815	2·984	1·473	2·086	2·618	3·068	3·436	3·723	3·927	1¼
2·777	2·968	3·176	3·377	1·686	2·332	2·945	3·477	3·927	4·295	4·581	1⅜
3·081	3·293	3·538	3·770	1·800	2·577	3·272	3·886	4·418	4·868	5·236	1½
3·384	3·623	3·899	4·162	1·963	2·822	3·600	4·295	4·909	5·440	5·890	1⅝
3·688	3·953	4·260	4·555	2·127	3·068	3·927	4·704	5·400	6·013	6·545	1¾
3·992	4·283	4·621	4·948	2·291	3·313	4·254	5·113	5·890	6·586	7·199	1⅞
4·295	4·613	4·983	5·341	2·454	3·559	4·581	5·522	6·381	7·159	7·854	2
4·599	4·943	5·344	5·733	2·618	3·804	4·909	5·931	6·872	7·731	8·508	2⅛
4·903	5·273	5·705	6·126	2·782	4·050	5·236	6·340	7·363	8·304	9·163	2¼
5·206	5·602	6·067	6·519	2·945	4·295	5·563	6·749	7·854	8·877	9·817	2⅜
5·510	5·932	6·428	6·911	3·109	4·541	5·890	7·159	8·345	9·449	10·472	2½
5·814	6·262	6·789	7·304	3·272	4·786	6·218	7·568	8·836	10·022	11·126	2⅝
6·117	6·592	7·150	7·697	3·436	5·031	6·545	7·977	9·327	10·595	11·781	2¾
6·421	6·922	7·512	8·090	3·600	5·277	6·872	8·386	9·818	11·167	12·435	2⅞
6·725	7·252	7·873	8·482	3·763	5·522	7·200	8·795	10·308	11·740	13·090	3
7·029	7·582	8·234	8·875	3·927	5·768	7·527	9·204	10·799	12·313	13·744	3⅛
7·332	7·911	8·596	9·268	4·091	6·013	7·854	9·613	11·290	12·885	14·399	3¼
7·636	8·241	8·957	9·660	4·254	6·259	8·181	10·022	11·781	13·458	15·053	3⅜
7·940	8·571	9·318	10·053	4·418	6·504	8·508	10·431	12·272	14·031	15·708	3½
8·243	8·901	9·679	10·446	4·581	6·750	8·836	10·840	12·763	14·604	16·362	3⅝
8·547	9·231	10·041	10·838	4·745	6·996	9·163	11·249	13·254	15·176	17·017	3¾
8·851	9·561	10·402	11·231	4·908	7·240	9·490	11·658	13·745	15·749	17·671	3⅞
9·154	9·891	10·763	11·624	5·072	7·486	9·817	12·067	14·235	16·322	18·326	4

Table C.—Weights of Seamless

L. S. G.	4/0	3/0	2/0	0	1	2	3	4	5
Inches	·400	·372	·348	·324	·300	·276	·252	·232	·212
Millimetres	10·160	9·449	8·839	8·229	7·620	7·010	6·401	5·893	5·385

Internal Diameter.		Weight of a lineal								
Inches	mm.									
¾	19·0	...	...	...	...	...	...	...	...	...
1	25·4	...	...	...	...	...	...	...	...	...
1¼	31·7	...	...	...	...	...	...	...	...	...
1½	38·1	...	...	...	...	...	...	...	...	...
1¾	44·4	...	...	...	...	...	...	...	...	...
2	50·8	...	...	...	...	...	...	...	...	...
2¼	57·1	...	...	...	...	...	...	...	...	...
2½	63·5	...	...	...	...	...	...	...	...	...
2¾	69·8	...	...	...	...	...	...	...	...	7·60
3	76·2	...	...	...	...	...	...	...	9·07	8·24
3¼	82·5	...	...	...	...	...	...	...	9·77	8·88
3½	88·9	...	...	...	...	...	...	11·44	10·47	9·52
3¾	95·2	...	...	...	...	...	13·44	12·20	11·18	10·16
4	101·6	...	...	...	...	...	14·28	12·96	11·88	10·80
4½	114·3	...	...	...	...	17·42	15·95	14·49	13·28	12·08
5	127·0	...	...	...	20·87	19·23	17·62	16·01	14·68	13·37
5½	139·7	...	...	...	22·83	21·05	19·29	17·54	16·09	14·65
6	152·4	...	...	26·72	24·79	22·86	20·95	19·06	17·49	15·93
6½	165·1	...	30·93	28·83	26·75	24·68	22·62	20·58	18·89	17·21
7	177·8	...	33·18	30·93	28·71	26·49	24·29	22·11	20·30	18·50
7½	190·5	38·23	35·43	33·04	30·67	28·31	25·96	23·63	21·70	19·78
8	203·2	40·65	37·68	35·14	32·63	30·12	27·63	25·16	23·10	21·06
8½	215·9	43·07	39·93	37·25	34·59	31·94	29·30	26·68	24·51	22·34
9	228·6	45·49	42·18	39·35	36·55	33·75	30·97	28·21	25·91	23·63
9½	241·3	47·91	44·43	41·46	38·51	35·57	32·64	29·73	27·31	24·91
10	254·0	50·33	46·68	43·56	40·47	37·38	34·31	31·25	28·72	26·19
10½	266·7	52·75	48·93	45·67	42·43	39·20	35·98	32·78	30·12	27·47
11	279·4	55·16	51·18	47·77	44·39	41·01	37·65	34·30	31·52	28·75
11½	292·1	57·58	53·43	49·88	46·35	42·83	39·32	35·83	32·93	30·04
12	304·8	60·00	55·68	51·98	48·31	44·64	40·99	37·35	34·33	31·32
12½	317·5	62·42	57·93	54·09	50·26	46·45	42·66	38·88	35·73	32·60
13	330·2	64·84	60·18	56·19	52·22	48·27	44·33	40·40	37·14	33·88
13½	342·9	67·26	62·43	58·30	54·18	50·08	46·00	41·92	38·54	35·17
14	355·6	69·68	64·68	60·40	56·14	51·90	47·67	43·45	39·94	36·45

Mandril drawn Brazed Tubes weigh the
Ordinary Brazed Copper Tubes weigh rather more than Seamless Tubes, the

TABLE C.—WEIGHTS OF SEAMLESS COPPER TUBES.   319

Copper Tubes.

6	7	8	9	10	11	12	13	14	15	16
·192	·176	·160	·144	·128	·116	·104	·092	·080	·072	·064
4·877	4·470	4·064	3·658	3·251	2·946	2·642	2·377	2·032	1·829	1·626

foot in pounds.

6	7	8	9	10	11	12	13	14	15	16
...	...	...	...	1·36	1·21	1·07	·94	·80	·72	·63
...	...	...	...	1·75	1·57	1·39	1·21	1·04	·93	·82
...	...	...	2·43	2·13	1·92	1·70	1·49	1·29	1·15	1·02
...	...	3·21	2·86	2·52	2·27	2·02	1·77	1·53	1·37	1·21
...	...	3·70	3·30	2·91	2·62	2·33	2·05	1·77	1·59	1·40
...	4·63	4·18	3·73	3·29	2·97	2·65	2·33	2·01	1·80	1·60
5·67	5·16	4·66	4·17	3·68	3·32	2·96	2·61	2·25	2·02	1·79
6·25	5·70	5·15	4·61	4·07	3·67	3·28	2·88	2·50	2·24	1·98
6·83	6·23	5·63	5·04	4·46	4·02	3·59	3·16	2·74	2·46	2·18
7·41	6·76	6·12	5·48	4·84	4·37	3·90	3·44	2·98	2·68	2·37
7·99	7·29	6·60	5·91	5·23	4·72	4·22	3·72	3·22	2·89	2·57
8·58	7·83	7·08	6·35	5·62	5·07	4·53	4·00	3·46	3·11	2·76
9·16	8·36	7·57	6·78	6·00	5·42	4·85	4·28	3·71	3·33	2·95
9·74	8·89	8·05	7·22	6·39	5·78	5·16	4·55	3·95	3·55	3·15
10·90	9·96	9·02	8·09	7·17	6·48	5·79	5·11	4·43	3·98	3·53
12·06	11·02	9·99	8·96	7·94	7·18	6·42	5·67	4·92	4·42	3·92
13·22	12·08	10·96	9·83	8·71	7·88	7·05	6·22	5·40	4·85	4·31
14·38	13·15	11·92	10·70	9·49	8·58	7·68	6·78	5·88	5·29	4·69
15·54	14·21	12·89	11·57	10·26	9·28	8·31	7·34	6·37	5·72	5·08
16·70	15·28	13·86	12·44	11·04	9·99	8·94	7·89	6·85	6·16	5·47
17·87	16·34	14·83	13·32	11·81	10·69	9·57	8·45	7·34	6·60	5·86
19·03	17·41	15·79	14·19	12·59	11·39	10·20	9·01	7·82	7·03	6·24
20·19	18·47	16·76	15·06	13·36	12·09	10·82	9·56	8·30	7·47	6·63
21·35	19·54	17·73	15·93	14·13	12·79	11·45	10·12	8·79	7·90	7·02
22·51	20·60	18·70	16·80	14·91	13·49	12·08	10·68	9·27	8·34	...
23·67	21·67	19·67	17·67	15·68	14·20	12·71	11·23	9·76	8·77	...
24·83	22·73	20·63	18·54	16·46	14·90	13·34	11·79	10·24	...	...
26·00	23·80	21·60	19·41	17·23	15·60	13·97	12·34	10·72	...	...
27·16	24·86	22·57	20·28	18·01	16·30	14·60	12·90	...	...	...
28·32	25·92	23·54	21·16	18·78	17·00	15·23	13·46	...	...	...
29·48	26·99	24·50	22·03	19·55	17·70	15·86	...	...	...	...
30·64	28·05	25·47	22·90	20·33	18·41	16·49	...	...	...	...
31·80	29·12	26·44	23·77	21·10	19·11	...	...	...	...	...
32·96	30·18	27·41	24·64	21·88	19·81	...	...	...	...	...

Same as Seamless Tubes.
difference varying with the thickness, diameter, and class of joint.

Table CI.—Weight of Lead Pipes.

Thick-ness in inches.	Weight in pounds per foot run.										
	2″	2¼″	2½″	2¾″	3″	3¼″	3½″	3¾″	4″	4½″	5″
¼	8·7	9·7	10·6	11·6	12·6	13·5	14·5	15·5	16·4	18·4	20·3
5/16	...	...	13·6	14·8	16·0	17·2	18·4	19·7	20·9	23·3	25·7
3/8	...	...	...	...	...	21·0	22·5	23·9	25·4	28·3	31·2
7/16	...	...	...	...	...	...	...	...	...	33·4	36·8

Lead pipes of these sizes and weights are usually manufactured in 10 ft. lengths.

Table CII.—Weight of Sheet Metals.

Thickness L.S.G.	Weight in pounds per square foot.					
	Steel.	Iron.	Copper.	Brass.	Lead.	Zinc.
7/0	20·40	20·00	22·83	21·98	29·65	18·72
6/0	18·93	18·56	21·19	20·39	27·51	17·38
5/0	17·63	17·28	19·73	18·99	25·62	16·18
4/0	16·32	16·00	18·27	17·58	23·72	14·98
3/0	15·18	14·88	16·99	16·35	22·06	13·93
2/0	14·20	13·92	15·89	15·30	20·64	13·03
0	13·22	12·96	14·80	14·24	19·21	12·13
1	12·24	12·00	13·70	13·19	17·79	11·23
2	11·26	11·04	12·60	12·13	16·37	10·34
3	10·28	10·08	11·51	11·08	14·94	9·44
4	9·47	9·28	10·59	10·20	13·76	8·69
5	8·65	8·48	9·68	9·32	12·57	7·94
6	7·83	7·68	8·77	8·44	11·38	7·19
7	7·18	7·04	8·04	7·74	10·44	6·59
8	6·53	6·40	7·31	7·03	9·49	5·99
9	5·88	5·76	6·58	6·33	8·54	5·39
10	5·22	5·12	5·84	5·63	7·59	4·79
11	4·73	4·64	5·30	5·10	6·88	4·34
12	4·24	4·16	4·75	4·57	6·17	3·89
13	3·75	3·68	4·20	4·04	5·45	3·44
14	3·26	3·20	3·65	3·52	4·74	3·00
15	2·94	2·88	3·29	3·16	4·27	2·70
16	2·61	2·56	2·92	2·81	3·79	2·40
17	2·28	2·24	2·56	2·46	3·32	2·10
18	1·96	1·92	2·19	2·11	2·85	1·80
19	1·63	1·60	1·83	1·76	2·37	1·50
20	1·47	1·44	1·64	1·58	2·13	1·35
21	1·31	1·28	1·46	1·41	1·90	1·20
22	1·14	1·12	1·28	1·23	1·66	1·05

TABLE CIII.—WEIGHTS OF ENGINES AND BOILERS. 321

Table CIII.—Weights of Engines and Boilers, &c.

Description of Machinery, &c.	Engine room weights, including screw or wheels, &c. Lbs. per I.H.P.	Boiler room weights, lbs. per I.H.P.	Total weights. Lbs. and tons per I.H.P.
(a) Ordinary merchant steamer with triple engines and ordinary cylindrical boilers working at 150 or 160 lbs. pressure, with chimney draught only; cylinders about 7 to 1; single screw.	235	225	{ 460 ·205
(b) Passenger or mail boat with machinery similar to above, but larger, and worked with chimney draught + ·5 to ·75 inches of air pressure; cylinders, say 6·75 to 1; (1) single screw, and (2) twin screw.	(1) 215 (2) 230	200 200	{ 415 ·185 { 430 ·192
(c) Battle ship or first-class cruiser, with vertical triple engines and ordinary cylindrical boilers working at 150 lbs. pressure, with chimney draught + ·5 inches, and 2 inches of air pressure for N.D. and F.D. respectively; cylinders, 5 to 1; twin screws.	N.D. 115 F.D. 85	150 110	{ 265 ·118 { 195 ·087
(d) Torpedo gunboat with vertical triple engines and loco type boilers working at 150 lbs. pressure, with about 2·5 inches of air pressure; cylinders, 5 to 1; twin screw.	35	60	{ 95 ·043
(e) Paddle steamer with oscillating jet condensing engines and rectangular or box boilers working at 30 to 35 lbs. pressure, with slight assisted draught; (1) cylinders fore and aft, (2) cylinders side by side.	(1) 140 (2) 160	170 170	{ 310 ·138 { 330 ·147
(f) Paddle steamer with compound oscillating engines, surface condensers, and ordinary cylindrical boilers, working at about 100 lbs. pressure, with slight assisted draught: cylinders, 3·25 to 1.	205	175	{ 380 ·170
(g) Paddle steamer with compound diagonal engines, surface condensers, and ordinary cylindrical boilers, working at about 120 lbs. pressure, with slight assisted draught; cylinders, 3·25 or 3·5 to 1; (1) cylinders one on top of other, (2) cylinders side by side.	(1) 170 (2) 185	180 180	{ 350 ·156 { 365 ·163
(h) Vertical compound two-crank engines and ordinary cylindrical boilers working at 80 to 90 lbs. pressure, as formerly fitted in merchant steamers.	220	240	{ 460 ·205
(k) Vertical compound single-crank engines and cylindrical boilers working at 80 to 90 lbs. pressure, as formerly fitted in some merchant steamers.	270	245	{ 515 ·230
(l) Twin-screw compound engines, as fitted in some fast cargo vessels; ordinary cylindrical boilers; 80 lbs. pressure.	220	230	{ 450 ·201

21

Between (c) and (d) come various classes of Naval vessels (second and third-class cruisers, &c.), the machinery for which varies from 150 to 200 lbs. per I.H.P.

The machinery of torpedo boats weighs about 30 to 35 lbs. per I.H.P. in the engine room, and about 45 to 50 lbs. in the boiler room, or say ·036 ton per I.H.P. for both engine and boiler room. When water tube boilers are used, the boiler room weights are reduced to about 33 lbs. per I.H.P.

The machinery of Naval pinnaces weighs from 90 to 115 lbs., or say ·045 ton per I.H.P., and that of Naval cutters from 150 to 190 lbs., or say ·076 ton per I.H.P.,—the smaller sizes weighing rather heavier than the larger.

The weights of paddle engines depend to some extent on the service for which the vessel is intended, since for cross channel work the wheels will be much heavier than for comparatively smooth water work.

CENTRE OF GRAVITY OF MACHINERY, &c.

When the common centre of gravity of a number of detached bodies, —such as the various portions of the machinery of a steamship,—is required, it is obtained as follows:—

1. **Longitudinal position of centre of gravity.**—On the longitudinal section drawing of the machinery, near where it is supposed the longitudinal position of the centre of gravity will be, draw a vertical line, to represent the edge of a plane cutting the ship transversely. Then, take the weight of each portion of the machinery, and multiply it by the distance of its centre of gravity from this transverse plane; distinguish the resulting quantities, or moments, by the signs + and −, according as they happen to be derived from the right or left side of the plane; and add together all the plus quantities, and all the minus quantities. Then obtain the resultant, by subtracting the less quantity from the greater, and place before it the sign of the greater.

If this resultant moment be now divided by the *total* weight of the whole system of bodies, on both sides of the plane, the quotient will be the distance of the longitudinal position of the centre of gravity from the plane; and it will lie to the right, or to the left, of the plane, according as the sign of the quotient is plus or minus.

2. **Transverse position of centre of gravity.**—Assume another plane, perpendicular to the first, and represent it by a line on the transverse section drawings, and then proceed exactly as described above.

3. **Vertical position of centre of gravity.**—Assume a third plane, perpendicular to both those previously used (parallel to the water-line, that is), represent it by a line or lines as before, and again calculate position as described above.

MISCELLANEOUS TABLES.

Table CIV.—Surface of Tubes in square feet per foot run.

Diam. in inches.	0	⅛	¼	⅜	½	⅝	¾	⅞
0	...	·0327	·0654	·0982	·1309	·1636	·1963	·2291
1	·2618	·2945	·3272	·3600	·3927	·4254	·4581	·4909
2	·5236	·5563	·5890	·6218	·6545	·6872	·7200	·7527
3	·7854	·8181	·8508	·8836	·9163	·9490	·9817	1·0145
4	1·0472	1·0799	1·1126	1·1454	1·1781	1·2108	1·2435	1·2763
5	1·3090	1·3417	1·3744	1·4072	1·4399	1·4726	1·5053	1·5381

Bulk and Weight &c. of Water.

Table CV.—Fresh Water.

Cubic inches.	Cubic feet.	Pounds.	Gallons.	Ton.
1	...	·036	...	...
27·7	·016	1·0	·1	...
277	·160	10·0	1·0	·00446
1728	1·0	62·4	6·24	·0279
...	35·9	2240	224	1·0

Table CVa.—Salt Water.

Cubic inches.	Cubic feet.	Pounds.	Gallons.	Ton.
1	...	·037	...	...
27	·0156	1·0	·097	...
277	·160	10·26	1·0	·00458
1728	1·0	64·0	6·24	·0285
...	35·0	2240	218·3	1·0

The above figures are very nearly correct for 62° F.: for bulk of water at other temperatures see page 259.

Table CVI.—Boiling Points &c. of Sea Water.

Proportion of salt in the water.	Temperature, Fahrenheit, at which it boils.	Specific Gravity.
$\frac{1}{32}$	213·2°	1·029
$\frac{2}{32}$	214·4°	1·058
$\frac{3}{32}$	215·5°	1·087
$\frac{4}{32}$	216·7°	1·116
$\frac{5}{32}$	217·9°	1·145
$\frac{6}{32}$	219·1°	1·174
$\frac{7}{32}$	220·3°	1·203
$\frac{8}{32}$	221·5°	1·232
$\frac{9}{32}$	222·7°	1·261
$\frac{10}{32}$	223·8°	1·290
$\frac{11}{32}$	225·0°	1·319
$\frac{12}{32}$	226·1°	1·348

Table CVII.—Litre and Cubic Metre.

Litres.	Gallons.	Cubic Metre.	Cubic feet.
1	·22	·001	·0353
4·543	1·0	·00454	·1605
1000	220	1·0	35·316

TABLE CVIII.—PRESSURE OF WATER, ETC. 325

Table CVIII.—Pressure of Water due to various Heads.

Depth of Water.	Pressure in		Depth of Water.	Pressure in		Depth of Water.	Pressure in	
	Pounds per sq. in.	Kilos. per sq. c/m.		Pounds per sq. in.	Kilos. per sq. c/m.		Pounds per sq. in.	Kilos. per sq. c/m.
1 in.	·03608	·002537	27 ft.	11·691	·82196	64 ft.	27·712	1·94836
2	·07216	·005074	28	12·124	·85240	65	28·145	1·97880
3	·10824	·007611	29	12·557	·88284	66	28·578	2·00925
4	·14432	·010148	30	12·990	·91329	67	29·011	2·03969
5	·18040	·012685	31	13·423	·94373	68	29·444	2·07013
6	·21648	·015222	32	13·856	·97417	69	29·877	2·10057
7	·25256	·017759	33	14·289	1·00462	70	30·310	2·13102
8	·28864	·020296	34	14·722	1·03406	71	30·743	2·16146
9	·32472	·022833	35	15·155	1·06450	72	31·176	2·19190
10	·36080	·025370	36	15·588	1·09495	73	31·609	2·22235
11	·39688	·027907	37	16·021	1·12539	74	32·042	2·25279
1 ft.	·433	·030443	38	16·454	1·15583	75	32·475	2·28323
2	·866	·060886	39	16·887	1·18627	76	32·908	2·31368
3	1·299	·091329	40	17·320	1·21773	77	33·341	2·34412
4	1·732	·121773	41	17·753	1·24817	78	33·774	2·37456
5	2·165	·152216	42	18·186	1·27861	79	34·207	2·40500
6	2·598	·182659	43	18·619	1·30906	80	34·640	2·43545
7	3·031	·213102	44	19·052	1·33950	81	35·073	2·46589
8	3·464	·243545	45	19·485	1·36994	82	35·506	2·49633
9	3·897	·273989	46	19·918	1·40039	83	35·939	2·52678
10	4·330	·30443	47	20·351	1·43083	84	36·372	2·55722
11	4·763	·33487	48	20·784	1·46127	85	36·805	2·58766
12	5·196	·36531	49	21·217	1·49171	86	37·238	2·61811
13	5·629	·39576	50	21·650	1·52216	87	37·671	2·64855
14	6·062	·42620	51	22·083	1·55260	88	38·104	2·67899
15	6·495	·45664	52	22·516	1·58804	89	38·537	2·70943
16	6·928	·48709	53	22·949	1·61349	90	38·970	2·73989
17	7·361	·51753	54	23·382	1·64393	91	39·403	2·77033
18	7·794	·54797	55	23·815	1·67437	92	39·836	2·80077
19	8·227	·57841	56	24·248	1·70482	93	40·269	2·83122
20	8·660	·60886	57	24·681	1·73526	94	40·702	2·86166
21	9·093	·63930	58	25·114	1·76570	95	41·135	2·89210
22	9·526	·66974	59	25·557	1·79614	96	41·568	2·92255
23	9·959	·70019	60	25·980	1·82659	97	42·001	2·95299
24	10·392	·73063	61	26·413	1·85703	98	42·434	2·98343
25	10·825	·76107	62	26·846	1·88747	99	42·867	3·01887
26	11·258	·79152	63	27·279	1·91792	100	43·300	3·04432

The above Table is calculated for fresh water at a temperature of 62° F

Table CIX.—Comparison of Thermometers.

Fahrenheit degs.	Centigrade degs.	Reaumur degs.	Fahrenheit degs.	Centigrade degs.	Reaumur degs.	Fahrenheit degs.	Centigrade degs.	Reaumur degs.
-461·20	-274·00	-219·22	+	-	-	+	+	+
+	-	-	21·20	6·	4·80	44·60	7·	5·60
0·	17·78	14·22	22·	5·56	4·44	45·	7·23	5·77
0·50	17·50	14·	23·	5·	4·	45·50	7·50	6·
1·	17·23	13·77	24·	4·45	3·55	46·	7·78	6·22
1·40	17·	13·60	24·80	4·	3·20	46·40	8·	6·40
2·	16·67	13·33	25·	3·89	3·11	47·	8·34	6·66
2·75	16·25	13·	25·25	3·75	3·	47·75	8·75	7·
3·	16·11	12·88	26·	3·34	2·66	48·	8·89	7·11
3·20	16·	12·80	26·60	3·	2·40	48·20	9·	7·20
4·	15·56	12·44	27·	2·78	2·22	49·	9·45	7·55
5·	15·	12·	27·50	2·50	2·	50·	10·	8·
6·	14·45	11·55	28·	2·23	1·77	51·	10·56	8·44
6·80	14·	11·20	28·40	2·	1·60	51·80	11·	8·80
7·	13·89	11·11	29·	1·67	1·33	52·	11·11	8·88
7·25	13·75	11·	29·75	1·25	1·	52·25	11·25	9·
8·	13·34	10·66	30·	1·11	0·88	53·	11·67	9·33
8·60	13·	10·40	30·20	1·	0·80	53·60	12·	9·60
9·	12·78	10·22	31·	0·56	0·44	54·	12·23	9·77
9·50	12·50	10·	32·	0·	0·	54·50	12·50	10·
10·	12·23	9·77	+33·	+0·56	+0·44	55·	12·78	10·22
10·40	12·	9·60	33·80	1·	0·80	55·40	13·	10·40
11·	11·67	9·33	34·	1·11	0·88	56·	13·34	10·66
11·75	11·25	9·	34·25	1·25	1·	56·75	13·75	11·
12·	11·11	8·88	35·	1·67	1·33	57·	13·89	11·11
12·20	11·	8·80	35·60	2·	1·60	57·20	14·	11·20
13·	10·56	8·44	36·	2·23	1·77	58·	14·45	11·55
14·	10·	8·	36·50	2·50	2·	59·	15·	12·
15·	9·45	7·55	37·	2·78	2·22	60·	15·56	12·44
15·80	9·	7·20	37·40	3·	2·40	60·80	16·	12·80
16·	8·89	7·11	38·	3·34	2·66	61·	16·11	12·88
16·25	8·75	7·	38·75	3·75	3·	61·25	16·25	13·
17·	8·34	6·66	39·	3·89	3·11	62·	16·67	13·33
17·60	8·	6·40	39·20	4·	3·20	62·60	17·	13·60
18·	7·78	6·22	40·	4·45	3·55	63·	17·23	13·77
18·50	7·50	6·	41·	5·	4·	63·50	17·50	14·
19·	7·23	5·77	42·	5·56	4·44	64·	17·78	14·22
19·40	7·	5·60	42·80	6·	4·80	64·40	18·	14·40
20·	6·67	5·33	43·	6·11	4·88	65·	18·34	14·66
20·75	6·25	5·	43·25	6·25	5·	65·75	18·75	15·
21·	6·11	4·88	44·	6·67	5·33	66·	18·89	15·11

TABLE CIX.—COMPARISON OF THERMOMETERS. 327

Table CIX.—Comparison of Thermometers—*continued.*

Fahren-heit degs.	Centi-grade degs.	Reau-mur degs.	Fahren-heit degs.	Centi-grade degs.	Reau-mur degs.	Fahren-heit degs.	Centi-grade degs.	Reau-mur degs.
+	+	+	+	+	+	+	+	+
66·20	19·	15·20	89·60	32·	25·60	111·20	44·	35·20
67·	19·45	15·55	90·	32·23	25·77	112·	44·45	35·55
68·	20·	16·	90·50	32·50	26·	113·	45·	36·
69·	20·56	16·44	91·	32·78	26·22	114·	45·56	36·44
69·80	21·	16·80	91·40	33·	26·40	114·80	46·	36·80
70·	21·11	16·88	92·	33·34	26·66	115·	46·11	36·88
70·25	21·25	17·	92·75	33·75	27·	115·25	46·25	37·
71·	21·67	17·33	93·	33·89	27·11	116·	46·67	37·33
71·60	22·	17·60	93·20	34·	27·20	116·60	47·	37·60
72·	22·23	17·77	94·	34·45	27·55	117·	47·23	37·77
72·50	22·50	18·	95·	35·	28·	117·50	47·50	38·
73·	22·78	18·22	96·	35·56	28·44	118·	47·78	38·22
73·40	23·	18·40	96·80	36·	28·80	118·40	48·	38·40
74·	23·34	18·66	97·	36·11	28·88	119·	48·34	38·66
74·75	23·75	19·	97·25	36·25	29·	119·75	48·75	39·
75·	23·89	19·11	98·	36·67	29·33	120·	48·89	39·11
75·20	24·	19·20	98·60	37·	29·60	120·20	49·	39·20
76·	24·45	19·55	99·	37·23	29·77	121·	49·45	39·55
77·	25·	20·	99·50	37·50	30·	122·	50·	40·
78·	25·56	20·44	100·	37·78	30·22	123·	50·56	40·44
78·80	26·	20·80	100·40	38·	30·40	123·80	51·	40·80
79·	26·11	20·88	101·	38·34	30·66	124·	51·11	40·88
79·25	26·25	21·	101·75	38·75	31·	124·25	51·25	41·
80·	26·67	21·33	102·	38·89	31·11	125·	51·67	41·33
80·60	27·	21·60	102·20	39·	31·20	125·60	52·	41·60
81·	27·23	21·77	103·	39·45	31·55	126·	52·23	41·77
81·50	27·50	22·	104·	40·	32·	126·50	52·50	42·
82·	27·78	22·22	105·	40·56	32·44	127·	52·78	42·22
82·40	28·	22·40	105·80	41·	32·80	127·40	53·	42·40
83·	28·34	22·66	106·	41·11	32·88	128·	53·34	42·66
83·75	28·75	23·	106·25	41·25	33·	128·75	53·75	43·
84·	28·89	23·11	107·	41·67	33·33	129·	53·89	43·11
84·20	29·	23·20	107·60	42·	33·60	129·20	54·	43·20
85·	29·45	23·55	108·	42·23	33·77	130·	54·45	43·55
86·	30·	24·	108·50	42·50	34·	131·	55·	44·
87·	30·56	24·44	109·	42·78	34·22	132·	55·56	44·44
87·80	31·	24·80	109·40	43·	34·40	132·80	56·	44·80
88·	31·11	24·88	110·	43·34	34·66	133·	56·11	44·88
88·25	31·25	25·	110·75	43·75	35·	133·25	56·25	45·
89·	31·67	25·33	111·	43·89	35·11	134·	56·67	45·33

Table CIX.—Comparison of Thermometers—*continued.*

Fahren-heit degs.	Centi-grade degs.	Reau-mur degs.	Fahren-heit degs.	Centi-grade degs.	Reau-mur degs.	Fahren-heit degs.	Centi-grade degs.	Reau-mur degs.
+	+	+	+	+	+	+	+	+
134·60	57·	45·60	156·20	69·	55·20	179·60	82·	65·60
135·	57·23	45·77	157·	69·45	55·55	180·	82·23	65·77
135·50	57·50	46·	158·	70·	56·	180·50	82·50	66·
136·	57·78	46·22	159·	70·56	56·44	181·	82·78	66·22
136·40	58·	46·40	159·80	71·	56·80	181·40	83·	66·40
137·	58·34	46·66	160·	71·11	56·88	182·	83·34	66·66
137·75	58·75	47·	160·25	71·25	57·	182·75	83·75	67·
138·	58·89	47·11	161·	71·67	57·33	183·	83·89	67·11
138·20	59·	47·20	161·60	72·	57·60	183·20	84·	67·20
139·	59·45	47·55	162·	72·23	57·77	184·	84·45	67·55
140·	60·	48·	162·50	72·50	58·	185·	85·	68·
141·	60·56	48·44	163·	72·78	58·22	186·	85·56	68·44
141·80	61·	48·80	163·40	73·	58·40	186·80	86·	68·80
142·	61·11	48·88	164·	73·34	58·66	187·	86·11	68·88
142·25	61·25	49·	164·75	73·75	59·	187·25	86·25	69·
143·	61·67	49·33	165·	73·89	59·11	188·	86·67	69·33
143·60	62·	49·60	165·20	74·	59·20	188·60	87·	69·60
144·	62·23	49·77	166·	74·45	59·55	189·	87·23	69·77
144·50	62·50	50·	167·	75·	60·	189·50	87·50	70·
145·	62·78	50·22	168·	75·56	60·44	190·	87·78	70·22
145·40	63·	50·40	168·80	76·	60·80	190·40	88·	70·40
146·	63·34	50·66	169·	76·11	60·88	191·	88·34	70·66
146·75	63·75	51·	169·25	76·25	61·	191·75	88·75	71·
147·	63·89	51·11	170·	76·67	61·33	192·	88·89	71·11
147·20	64·	51·20	170·60	77·	61·60	192·20	89·	71·20
148·	64·45	51·55	171·	77·23	61·77	193·	89·45	71·55
149·	65·	52·	171·50	77·50	62·	194·	90·	72·
150·	65·56	52·44	172·	77·78	62·22	195·	90·56	72·44
150·80	66·	52·80	172·40	78·	62·40	195·80	91·	72·80
151·	66·11	52·88	173·	78·34	62·66	196·	91·11	72·88
151·25	66·25	53·	173·75	78·75	63·	196·25	91·25	73·
152·	66·67	53·33	174·	78·89	63·11	197·	91·67	73·33
152·60	67·	53·60	174·20	79·	63·20	197·60	92·	73·60
153·	67·23	53·77	175·	79·45	63·55	198·	92·23	73·77
153·50	67·50	54·	176·	80·	64·	198·50	92·50	74·
154·	67·78	54·22	177·	80·56	64·44	199·	92·78	74·22
154·40	68·	54·40	177·80	81·	64·80	199·40	93·	74·40
155·	68·34	54·66	178·	81·11	64·88	200·	93·34	74·66
155·75	68·75	55·	178·25	81·25	65·	200·75	93·75	75·
156·	68·89	55·11	179·	81·67	65·33	201·	93·89	75·11

TABLE CIX.—COMPARISON OF THERMOMETERS. 329

Table CIX.—Comparison of Thermometers—*continued*.

Fahren-heit degs.	Centi-grade degs.	Reau-mur degs.	Fahren-heit degs.	Centi-grade degs.	Reau-mur degs.	Fahren-heit degs.	Centi-grade degs.	Reau-mur degs.
+	+	+	+	+	+	+	+	+
201·20	94·	75·20	224·60	107·	85·60	246·20	119·	95·20
202·	94·45	75·55	225·	107·23	85·77	247·	119·45	95·55
203·	95·	76·	225·50	107·50	86·	248·	120·	96·
204·	95·56	76·44	226·	107·78	86·22	249·	120·56	96·44
204·80	96·	76·80	226·40	108·	86·40	249·80	121·	96·80
205·	96·11	76·88	227·	108·34	86·66	250·	121·11	96·88
205·25	96·25	77·	227·75	108·75	87·	250·25	121·25	97·
206·	96·67	77·33	228·	108·89	87·11	251·	121·67	97·33
206·60	97·	77·60	228·20	109·	87·20	251·60	122·	97·60
207·	97·23	77·77	229·	109·45	87·55	252·	122·23	97·77
207·50	97·50	78·	230·	110·	88·	252·50	122·50	98·
208·	97·78	78·22	231·	110·56	88·44	253·	122·78	98·22
208·40	98·	78·40	231·80	111·	88·80	253·40	123·	98·40
209·	98·34	78·66	232·	111·11	88·88	254·	123·34	98·66
209·75	98·75	79·	232·25	111·25	89·	254·75	123·75	99·
210·	98·89	79·11	233·	111·67	89·33	255·	123·89	99·11
210·20	99·	79·20	233·60	112·	89·60	255·20	124·	99·20
211·	99·45	79·55	234·	112·23	89·77	256·	124·45	99·55
212·	100·	80·	234·50	112·50	90·	257·	125·	100·
213·	100·56	80·44	235·	112·78	90·22	258·	125·56	100·44
213·80	101·	80·80	235·40	113·	90·40	258·80	126·	100·80
214·	101·11	80·88	236·	113·34	90·66	259·	126·11	100·88
214·25	101·25	81·	236·75	113·75	91·	259·25	126·25	101·
215·	101·67	81·33	237·	113·89	91·11	260·	126·67	101·33
215·60	102·	81·60	237·20	114·	91·20	260·60	127·	101·60
216·	102·23	81·77	238·	114·45	91·55	261·	127·23	101·77
216·50	102·50	82·	239·	115·	92·	261·50	127·50	102·
217·	102·78	82·22	240·	115·56	92·44	262·	127·78	102·22
217·40	103·	82·40	240·80	116·	92·80	262·40	128·	102·40
218·	103·34	82·66	241·	116·11	92·88	263·	128·34	102·66
218·75	103·75	83·	241·25	116·25	93·	263·75	128·75	103·
219·	103·89	83·11	242·	116·67	93·33	264·	128·89	103·11
219·20	104·	83·20	242·60	117·	93·60	264·20	129·	103·20
220·	104·45	83·55	243·	117·23	93·77	265·	129·45	103·55
221·	105·	84·	243·50	117·50	94·	266·	130·	104·
222·	105·56	84·44	244·	117·78	94·22	267·	130·26	104·44
222·80	106·	84·80	244·40	118·	94·40	267·80	131·	104·80
223·	106·11	84·88	245·	118·34	94·66	268·	131·11	104·88
223·25	106·25	85·	245·75	118·75	95·	268·25	131·25	105·
224·	106·67	85·33	246·	118·89	95·11	269·	131·67	105·33

Table CIX.—Comparison of Thermometers—*continued*.

Fahrenheit degs.	Centigrade degs.	Reaumur degs.	Fahrenheit degs.	Centigrade degs.	Reaumur degs.	Fahrenheit degs.	Centigrade degs.	Reaumur degs.
+	+	+	+	+	+	+	+	+
269·60	132·	105·60	291·20	144·	115·20	314·60	157·	125·60
270·	132·23	105·77	292·	144·45	115·55	315·	157·23	125·77
270·50	132·50	106·	293·	145·	116·	315·50	157·50	126·
271·	132·78	106·22	294·	145·56	116·44	316·	157·78	126·22
271·40	133·	106·40	294·80	146·	116·80	316·40	158·	126·40
272·	133·34	106·66	295·	146·11	116·88	317·	158·34	126·66
272·75	133·75	107·	295·25	146·25	117·	317·75	158·75	127·
273·	133·89	107·11	296·	146·67	117·33	318·	158·89	127·11
273·20	134·	107·20	296·60	147·	117·60	318·20	159·	127·20
274·	134·45	107·55	297·	147·23	117·77	319·	159·45	127·55
275·	135·	108·	297·50	147·50	118·	320·	160·	128·
276·	135·56	108·44	298·	147·78	118·22	321·	160·56	128·44
276·80	136·	108·80	298·40	148·	118·40	321·80	161·	128·80
277·	136·11	108·88	299·	148·34	118·66	322·	161·11	128·88
277·25	136·25	109·	299·75	148·75	119·	322·25	161·25	129·
278·	136·67	109·33	300·	148·89	119·11	323·	161·67	129·33
278·60	137·	109·60	300·20	149·	119·20	323·60	162·	129·60
279·	137·23	109·77	301·	149·45	119·55	324·	162·23	129·77
279·50	137·50	110·	302·	150·	120·	324·50	162·50	130·
280·	137·78	110·22	303·	150·56	120·44	325·	162·78	130·22
280·40	138·	110·40	303·80	151·	120·80	325·40	163·	130·40
281·	138·34	110·66	304·	151·11	120·88	326·	163·34	130·66
281·75	138·75	111·	304·25	151·25	121·	326·75	163·75	131·
282·	138·89	111·11	305·	151·67	121·33	327·	163·89	131·11
282·20	139·	111·20	305·60	152·	121·60	327·20	164·	131·20
283·	139·45	111·55	306·	152·23	121·77	328·	164·45	131·55
284·	140·	112·	306·50	152·50	122·	329·	165·	132·
285·	140·56	112·44	307·	152·78	122·22	330·	165·56	132·44
285·80	141·	112·80	307·40	153·	122·40	330·80	166·	132·80
286·	141·11	112·88	308·	153·34	122·66	331·	166·11	132·88
286·25	141·25	113·	308·75	153·75	123·	331·25	166·25	133·
287·	141·67	113·33	309·	153·89	123·11	332·	166·67	133·33
287·60	142·	113·60	309·20	154·	123·20	332·60	167·	133·60
288·	142·23	113·77	310·	154·45	123·55	333·	167·23	133·77
288·50	142·50	114·	311·	155·	124·	333·50	167·50	134·
289·	142·78	114·22	312·	155·56	124·44	334·	167·78	134·22
289·40	143·	114·40	312·80	156·	124·80	334·40	168·	134·40
290·	143·34	114·66	313·	156·11	124·88	335·	168·34	134·66
290·75	143·75	115·	313·25	156·25	125·	335·75	168·75	135·
291·	143·89	115·11	314·	156·67	125·33	336·	168·89	135·11

TABLE CIX.—COMPARISON OF THERMOMETERS.   331

Table CIX.—Comparison of Thermometers—*continued*.

Fahrenheit degs.	Centigrade degs.	Reaumur degs.	Fahrenheit degs.	Centigrade degs.	Reaumur degs.	Fahrenheit degs.	Centigrade degs.	Reaumur degs.
+	+	+	+	+	+	+	+	+
336·20	169·	135·20	359·60	182·	145·60	381·20	194·	155·20
337·	169·45	135·55	360·	182·23	145·77	382·	194·45	155·55
338·	170·	136·	360·50	182·50	146·	383·	195·	156·
339·	170·56	136·44	361·	182·78	146·22	384·	195·56	156·44
339·80	171·	136·80	361·40	183·	146·40	384·80	196·	156·80
340·	171·11	136·88	362·	183·34	146·66	385·	196·11	156·88
340·25	171·25	137·	362·75	183·75	147·	385·25	196·25	157·
341·	171·67	137·33	363·	183·89	147·11	386·	196·67	157·38
341·60	172·	137·60	363·20	184·	147·20	386·60	197·	157·60
342·	172·23	137·77	364·	184·45	147·55	387·	197·23	157·77
342·50	172·50	138·	365·	185·	148·	387·50	197·50	158·
343·	172·78	138·22	366·	185·56	148·44	388·	197·78	158·22
343·40	173·	138·40	366·80	186·	148·80	388·40	198·	158·40
344·	173·34	138·66	367·	186·11	148·88	389·	198·34	158·66
344·75	173·75	139·	367·25	186·25	149·	389·75	198·75	159·
345·	173·89	139·11	368·	186·67	149·33	390·	198·89	159·11
345·20	174·	139·20	368·60	187·	149·60	390·20	199·	159·20
346·	174·45	139·55	369·	187·23	149·77	391·	199·45	159·55
347·	175·	140·	369·50	187·50	150·	392·	200·	160·
348·	175·56	140·44	370·	187·78	150·22	393·	200·56	160·44
348·80	176·	140·80	370·40	188·	150·40	393·80	201·	160·80
349·	176·11	140·88	371·	188·34	150·66	394·	201·11	160·88
349·25	176·25	141·	371·75	188·75	151·	394·25	201·25	161·
350·	176·67	141·33	372·	188·89	151·11	395·	201·67	161·33
350·60	177·	141·60	372·20	189·	151·20	395·60	202·	161·60
351·	177·23	141·77	373·	189·45	151·55	396·	202·23	161·77
351·50	177·50	142·	374·	190·	152·	396·50	202·50	162·
352·	177·78	142·22	375·	190·56	152·44	397·	202·78	162·22
352·40	178·	142·40	375·80	191·	152·80	397·40	203·	162·40
353·	178·34	142·66	376·	191·11	152·88	398·	203·34	162·66
353·75	178·75	143·	376·25	191·25	153·	398·75	203·75	163·
354·	178·89	143·11	377·	191·67	153·33	399·	203·89	163·11
354·20	179·	143·20	377·60	192·	153·60	399·20	204·	163·20
355·	179·45	143·55	378·	192·23	153·77	400·	204·45	163·55
356·	180·	144·	378·50	192·50	154·	401·	205·	164·
357·	180·56	144·44	379·	192·78	154·22	402·	205·56	164·44
357·80	181·	144·80	379·40	193·	154·40	402·80	206·	164·80
358·	181·11	144·88	380·	193·34	154·66	403·	206·11	164·88
358·25	181·25	145·	380·75	193·75	155·	403·25	206·25	165·
359·	181·67	145·33	381·	193·89	155·11	404·	206·67	165·33

Table CIX.—Comparison of Thermometers—*continued.*

Fahrenheit degs.	Centigrade degs.	Reaumur degs.	Fahrenheit degs.	Centigrade degs.	Reaumur degs.	Fahrenheit degs.	Centigrade degs.	Reaumur degs.
+	+	+	+	+	+	+	+	+
404·60	207·	165·60	422·	216·67	173·33	439·25	226·25	181·
405·	207·23	165·77	422·60	217·	173·60	440·	226·67	181·33
405·50	207·50	166·	423·	217·23	173·77	440·60	227·	181·60
406·	208·75	166·22	423·50	217·50	174·	441·	227·23	181·77
406·40	208·	166·40	424·	217·78	174·22	441·50	227·50	182·
407·	208·34	166·66	424·40	218·	174·40	442·	227·78	182·22
407·75	208·75	1€7·	425·	218·34	174·66	442·40	228·	182·40
408·	208·89	167·11	425·75	218·75	175·	443·	228·34	182·66
408·20	209·	167·20	426·	218·89	175·11	443·75	228·75	183·
409·	209·45	167·55	426·20	219·	175·20	444·	228·89	183·11
410·	210·	168·	427·	219·45	175·55	444·20	229·	183·20
411·	210·56	168·44	428·	220·	176	445·	229·45	183·55
411·80	211·	168·80	429·	220·56	176·44	446·	230·	184·
412·	211·11	168·88	429·80	221·	176·80	447·	230·56	184·44
412·25	211·25	169·	430·	221·11	176·88	447·80	231·	184·80
413·	211·67	169·33	430·25	221·25	177·	448·	231·11	184·88
413·60	212·	169·60	431·	221·67	177·33	448·25	231·25	185·
414·	212·23	169·77	431·60	222·	177·60	449·	231·67	185·33
414·50	212·50	170·	432·	222·23	177·77	449·60	232·	185·60
415·	212·78	170·22	432·50	222·50	178·	450·	232·23	185·77
415·40	213·	170·40	433·	222·78	178·22	450·50	232·50	186·
416·	213·34	170·66	433·40	223·	178·40	451·	232·78	186·22
416·75	213·75	171·	434·	223·34	178·66	451·40	233·	186·40
417·	213·89	171·11	434·75	223·75	179·	452·	233·34	186·66
417·20	214·	171·20	435·	223·89	179·11	452·75	233·75	187·
418·	214·45	171·55	435·20	224·	179·20	453·	233·89	187·11
419·	215·	172·	436·	224·45	179·55	453·20	234·	187·20
420·	215·56	172·44	437·	225·	180·	454·	234·45	187·55
420·80	216·	172·80	438·	225·56	180·44	455·	235·	188·
421·	216·11	172·88	438·80	226·	180·80			
421·25	216·25	173·	439·	226·11	180·88			

TABLE CX.—PROPERTIES OF SATURATED STEAM.   333

Table CX.—Properties of Saturated Steam.

Pressure per Square Inch from Mean Atmospheric Pressure.	Temperature in Fahrenheit Degrees.	Specific or Relative Volume of the Steam.	Density or Weight of 1 Cubic Foot of the Steam.	Cubic Feet of the Steam per lb.	Latent Heat of Evaporation in Thermal Units per lb. of the Steam.	Total Heat in Thermal Units from 32° Fahrenheit per lb. of the Steam.	Absolute Pressure per Square Inch.
lbs.			lb.				lbs.
− 14	90·4	28740	·002170	460·7	1051·1	1109·5	·7
− 13	120·3	12480	·004998	200·1	1030·3	1118·6	1·7
− 12	137·5	8080	·007720	129·5	1018·3	1123·9	2·7
− 11	149·8	6009	·01038	96·32	1009·7	1127·6	3·7
− 10	159·7	4799	·01300	76·92	1002·8	1130·7	4·7
− 9	167·9	4003	·01558	64·17	997·0	1133·2	5·7
− 8	174·9	3439	·01814	55·12	992·1	1135·3	6·7
− 7	181·1	3010	·02072	48·25	987·8	1137·2	7·7
− 6	186·7	2690	·02319	43·12	983·9	1138·9	8·7
− 5	191·8	2428	·02568	38·93	980·3	1140·4	9·7
− 4	196·4	2215	·02817	35·50	977·1	1141·9	10·7
− 3	200·7	2036	·03063	32·64	974·0	1143·2	11·7
− 2	204·7	1886	·03307	30·24	971·2	1144·4	12·7
− 1	208·5	1755	·03553	28·14	968·5	1145·5	13·7
Atmosphere.	212·0	1643	·03797	26·34	966·1	1146·6	14·7
1	215·3	1544	·04039	24·76	963·8	1147·6	15·7
2	218·5	1457	·04280	23·36	961·5	1148·6	16·7
3	221·5	1380	·04521	22·12	959·4	1149·5	17·7
4	224·4	1310	·04761	21·0	957·3	1150·4	18·7
5	227·1	1248	·05000	20·0	955·4	1151·2	19·7
6	229·8	1191	·05238	19·09	953·5	1152·0	20·7
7	232·3	1139	·05475	18·26	951·8	1152·8	21·7
8	234·7	1092	·05712	17·51	950·1	1153·5	22·7
9	237·1	1049	·05949	16·81	948·4	1154·3	23·7
10	239·4	1009	·06184	16·17	946·7	1155·0	24·7
11	241·6	971·8	·06419	15·58	945·2	1155·6	25·7
12	243·7	937·5	·06654	15·03	943·7	1156·3	26·7
13	245·8	905·7	·06888	14·52	942·2	1156·9	27·7
14	247·8	876·0	·07122	14·04	940·8	1157·5	28·7
15	249·7	818·2	·07355	13·60	939·4	1158·1	29·7
16	251·6	822·2	·07587	13·18	938·1	1158·7	30·7

Table CX.—Properties of Saturated Steam—*continued.*

Pressure per Square Inch from Mean Atmospheric Pressure. lbs.	Temperature in Fahrenheit Degrees.	Specific or Relative Volume of the Steam.	Density or Weight of 1 Cubic Foot of the Steam. lb.	Cubic Feet of the Steam per lb.	Latent Heat of Evaporation in Thermal Units per lb. of the Steam.	Total Heat in Thermal Units from 32° Fahrenheit per lb. of the Steam.	Absolute Pressure per Square Inch. lbs.
17	253·5	797·8	·07819	12·79	936·8	1159·3	31·7
18	255·3	774·9	·08050	12·42	935·5	1159·8	32·7
19	257·0	753·2	·08282	12·07	934·3	1160·3	33·7
20	258·7	732·8	·08513	11·75	933·1	1160·9	34·7
21	260·4	713·5	·08743	11·44	931·9	1161·4	35·7
22	262·0	695·2	·08973	11·14	930·7	1161·9	36·7
23	263·6	677·9	·09203	10·87	929·6	1162·3	37·7
24	265·2	661·4	·09433	10·60	928·4	1162·8	38·7
25	266·7	645·7	·09661	10·35	927·4	1163·3	39·7
26	268·2	630·8	·09890	10·11	926·3	1163·8	40·7
27	269·7	616·6	·10М	9·883	925·2	1164·2	41·7
28	271·1	603·0	·1034	9·666	924·3	1164·6	42·7
29	272·6	590·0	·1057	9·458	923·2	1165·1	43·7
30	273·9	577·6	·1080	9·259	922·3	1165·5	44·7
31	275·3	565·7	·1102	9·068	921·3	1165·9	45·7
32	276·7	554·3	·1125	8·885	920·3	1166·3	46·7
33	278·0	543·4	·1148	8·710	919·3	1166·7	47·7
34	279·3	532·9	·1170	8·541	918·4	1167·1	48·7
35	280·5	522·8	·1193	8·380	917·6	1167·5	49·7
36	281·8	513·1	·1215	8·225	916·6	1167·9	50·7
37	283·0	503·8	·1238	8·075	915·8	1168·3	51·7
38	284·2	494·8	·1260	7·931	914·9	1168·6	52·7
39	285·4	486·1	·1283	7·792	914·1	1169·0	53·7
40	286·6	477·7	·1305	7·658	913·2	1169·4	54·7
41	287·8	469·7	·1328	7·529	912·3	1169·7	55·7
42	288·9	461·9	·1350	7·404	911·6	1170·1	56·7
43	290·1	454·4	·1373	7·283	910·7	1170·4	57·7
44	291·2	447·1	·1395	7·167	909·9	1170·8	58·7
45	292·3	440·0	·1417	7·054	909·1	1171·1	59·7
46	293·4	433·2	·1440	6·944	908·3	1171·4	60·7
47	294·4	426·6	·1462	6·839	907·6	1171·7	61·7

TABLE CX.—PROPERTIES OF SATURATED STEAM.   335

Table CX.—Properties of Saturated Steam—*continued.*

Pressure per Square Inch from Mean Atmospheric Pressure. lbs.	Temperature in Fahrenheit Degrees.	Specific or Relative Volume of the Steam.	Density or Weight of 1 Cubic Foot of the Steam. lb.	Cubic Feet of the Steam per lb.	Latent Heat of Evaporation in Thermal Units per lb. of the Steam.	Total Heat in Thermal Units from 32° Fahrenheit per lb. of the Steam.	Absolute Pressure per Square Inch. lbs.
48	295·5	420·2	·1484	6·736	906·8	1172·1	62·7
49	296·5	414·1	·1506	6·636	906·1	1172·4	63·7
50	297·5	408·0	·1529	6·540	905·4	1172·7	64·7
51	298·6	402·2	·1551	6·446	904·6	1173·0	65·7
52	299·6	396·5	·1573	6·356	903·9	1173·3	66·7
53	300·6	391·0	·1595	6·267	903·2	1173·6	67·7
54	301·5	385·6	·1617	6·181	902·5	1173·9	68·7
55	302·5	380·4	·1639	6·098	901·8	1174·2	69·7
56	303·5	375·4	·1661	6·017	901·1	1174·5	70·7
57	304·4	370·4	·1684	5·938	900·5	1174·8	71·7
58	305·3	365·6	·1706	5·861	899·8	1175·1	72·7
59	306·3	361·0	·1728	5·786	899·1	1175·4	73·7
60	307·2	356·4	·1750	5·714	898·5	1175·6	74·7
61	308·1	352·0	·1772	5·643	897·8	1175·9	75·7
62	309·0	347·7	·1794	5·573	897·2	1176·2	76·7
63	309·9	343·5	·1816	5·506	896·5	1176·5	77·7
64	310·8	339·4	·1838	5·440	895·9	1176·7	78·7
65	311·6	335·4	·1860	5·376	895·3	1177·0	79·7
66	312·5	331·5	·1882	5·313	894·7	1177·3	80·7
67	313·3	327·7	·1904	5·252	894·1	1177·5	81·7
68	314·2	323·9	·1926	5·192	893·4	1177·8	82·7
69	315·0	320·3	·1947	5·134	892·9	1178·0	83·7
70	315·8	316·7	·1969	5·077	892·3	1178·3	84·7
71	316·7	313·3	·1991	5·021	891·6	1178·5	85·7
72	317·5	309·9	·2013	4·967	891·1	1178·8	86·7
73	318·3	306·5	·2035	4·914	890·5	1179·0	87·7
74	319·1	303·3	·2056	4·862	889·9	1179·3	88·7
75	319·9	300·1	·2078	4·811	889·3	1179·5	89·7
76	320·7	297·0	·2100	4·761	888·8	1179·8	90·7
77	321·4	294·0	·2122	4·712	888·3	1180·0	91·7
78	322·2	291·0	·2144	4·664	887·7	1180·2	92·7

Table CX.—Properties of Saturated Steam—*continued.*

Pressure per Square Inch from Mean Atmospheric Pressure.	Tempera- ture in Fahren- heit Degrees.	Specific or Rela- tive Volume of the Steam.	Densi'y or Weight of 1 Cubic Foot of the Steam.	Cubic Feet of the Steam per lb.	Latent Heat of Evapora- tion in Thermal Units per lb. of the Steam.	Total Heat in Ther- mal Units from 32° Fahrenheit per lb. of the Steam.	Absolute Pressure per Square Inch.
lbs.			lb.				lbs.
79	323·0	288·0	·2165	4·617	887·1	1180·5	93·7
80	323·8	285·2	·2187	4·572	886·5	1180·7	94·7
81	324·5	282·4	·2209	4·527	886·0	1180·9	95·7
82	325·2	279·6	·2230	4·483	885·5	1181·1	96·7
83	326·0	276·9	·2252	4·439	885·0	1181·4	97·7
84	326·7	274·3	·2274	4·397	884·4	1181·6	98·7
85	327·4	271·7	·2295	4·356	883·9	1181·8	99·7
86	328·1	269·2	·2317	4·315	883·4	1182·0	100·7
87	328·9	266·7	·2339	4·275	882·9	1182·3	101·7
88	329·6	264·3	·2360	4·236	882·4	1182·5	102·7
89	330·3	261·9	·2382	4·198	881·9	1182·7	103·7
90	331·0	259·5	·2404	4·160	881·3	1182·9	104·7
91	331·7	257·2	·2425	4·123	880·8	1183·1	105·7
92	332·3	254·9	·2447	4·087	880·4	1183·3	106·7
93	333·0	252·5	·2468	4·051	879·9	1183·5	107·7
94	333·7	250·5	·2490	4·016	879·4	1183·7	108·7
95	334·4	248·4	·2511	3·981	878·9	1183·9	109·7
96	335·1	246·3	·2533	3·948	878·4	1184·2	110·7
97	335·7	244·2	·2554	3·914	878·0	1184·3	111·7
98	336·4	242·2	·2576	3·882	877·4	1184·6	112·7
99	337·0	240·1	·2598	3·849	877·0	1184·7	113·7
100	337·7	238·2	·2619	3·818	976·5	1184·9	114·7
101	338·3	236·2	·2640	3·787	876·1	1185·1	115·7
102	339·0	234·3	·2662	3·757	875·6	1185·3	116·7
103	339·6	232·5	·2683	3·727	875·1	1185·5	117·7
104	340·2	230·6	·2704	3·697	874·7	1185·7	118·7
105	340·9	228·8	·2726	3·668	874·2	1185·9	119·7
106	341·5	227·0	·2747	3·639	873·8	1186·1	120·7
107	342·1	225·1	·2769	3·611	873·3	1186·3	121·7
108	342·7	223·6	·2790	3·584	872·9	1186·5	122·7
109	343·3	221·9	·2811	3·556	872·5	1186·7	123·7

TABLE CX.—PROPERTIES OF SATURATED STEAM.' 337

Table CX.—Properties of Saturated Steam—*continued.*

Pressure per Square Inch from Mean Atmospheric Pressure. lbs.	Temperature in Fahrenheit Degrees.	Specific or Relative Volume of the Steam.	Density or Weight of 1 Cubic Foot of the Steam. lb.	Cubic Feet of the Steam per lb.	Latent Heat of Evaporation in Thermal Units per lb. of the Steam.	Total Heat in Thermal Units from 32° Fahrenheit per lb. of the Steam.	Absolute Pressure per Square Inch. lbs.
110	343·9	220·1	·2833	3·530	872·0	1186·8	124·7
111	344·5	218·5	·2854	3·503	871·6	1187·0	125·7
112	345·1	216·9	·2875	3·477	871·2	1187·2	126·7
113	345·7	215·3	·2897	3·452	870·7	1187·4	127·7
114	346·3	213·7	·2918	3·426	870·3	1187·6	128·7
115	346·9	212·2	·2939	3·401	869·9	1187·8	129·7
116	347·5	210·7	·2961	3·377	869·4	1187·9	130·7
117	348·1	209·2	·2982	3·353	869·0	1188·1	131·7
118	348·7	207·7	·3003	3·329	868·6	1188·3	132·7
119	349·2	206·2	·3025	3·306	868·2	1188·5	133·7
120	349·8	204·8	·3046	3·283	867·8	1188·6	134·7
121	350·4	203·4	·3067	3·260	867·3	1188·8	135·7
122	351·0	202·0	·3088	3·237	866·9	1189·0	136·7
123	351·5	200·6	·3110	3·215	866·5	1189·2	137·7
124	352·1	199·2	·3131	3·194	866·1	1189·3	138·7
125	352·6	197·9	·3152	3·172	865·7	1189·5	139·7
126	353·2	196·6	·3173	3·151	865·3	1189·7	140·7
127	353·7	195·3	·3194	3·130	864·9	1189·8	141·7
128	354·3	194·0	·3216	3·109	864·5	1190·0	142·7
129	354·8	192·7	·3237	3·089	864·1	1190·2	143·7
130	355·4	191·5	·3258	3·069	863·7	1190·3	144·7
131	355·9	190·2	·3279	3·049	863·3	1190·5	145·7
132	356·4	189·0	·3300	3·030	863·0	1190·7	146·7
133	357·0	187·8	·3321	3·010	862·5	1190·8	147·7
134	357·5	186·6	·3343	2·991	862·2	1191·0	148·7
135	358·0	185·4	·3364	2·978	861·8	1191·1	149·7
136	358·5	184·3	·3385	2·954	861·4	1191·3	150·7
137	359·1	182·9	·3406	2·936	861·0	1191·5	151·7
138	359·6	182·0	·3427	2·918	860·6	1191·6	152·7
139	360·1	180·9	·3448	2·900	860·3	1191·8	153·7
140	360·6	179·8	·3469	2·882	859·9	1191·9	154·7

22

Table CX.—Properties of Saturated Steam—*continued.*

Pressure per Square Inch from Mean Atmospheric Pressure.	Temperature in Fahrenheit Degrees.	Specific or Relative Volume of the Steam.	Density or Weight of 1 Cubic Foot of the Steam.	Cubic Feet of the Steam per lb.	Latent Heat of Evaporation in Thermal Units per lb. of the Steam.	Total Heat in Thermal Units from 32° Fahrenheit per lb. of the Steam.	Absolute Pressure per Square Inch.
Ibs.			Ib.				Ibs.
141	361·1	178·7	·3490	2·865	859·6	1192·1	155·7
142	361·6	177·6	·3511	2·848	859·2	1192·2	156·7
143	362·1	176·6	·3532	2·831	858·8	1192·4	157·7
144	362·6	175·5	·3554	2·814	858·5	1192·5	158·7
145	363·1	174·5	·3575	2·797	858·1	1192·7.	159·7
146	363·6	173·5	·3596	2·781	857·7	1192·8	160·7
147	364·1	172·5	·3617	2·765	857·4	1193·0	161·7
148	364·6	171·5	·3638	2·749	857·0	1193·2	162·7
149	365·1	170·5	·3659	2·733	856·6	1193·3	163·7
150	365·6	169·5	·3680	2·717	856·3	1193·5	164·7
151	366·1	168·6	·3701	2·702	855·9	1193·6	165·7
152	366·6	167·6	·3722	2·687	855·5	1193·8	166·7
153	367·1	166·7	·3743	2·672	855·2	1193·9	167·7
154	367·5	165·7	·3764	2·657	854·9	1194·0	168·7
155	368·0	164·8	·3785	2·642	854·5	1194·2	169·7
156	368·5	163·9	·3806	2·627	854·2	1194·3	170·7
157	369·0	163·0	·3827	2·613	853·8	1194·5	171·7
158	369·4	162·1	·3847	2·599	853·5	1194·6	172·7
159	369·9	161·2	·3868	2·585	853·1	1194·8	173·7
160	370·4	160·4	·3889	2·571	852·8	1194·9	174·7
161	370·8	159·5	·3910	2·557	852·5	1195·0	175·7
162	371·3	158·7	·3931	2·543	852·1	1195·2	176·7
163	371·7	157·8	·3952	2·530	851·8	1195·3	177·7
164	372·2	157·0	·3973	2·517	851·5	1195·5	178·7
165	372·7	156·2	·3994	2·504	851·1	1195·6	179·7
166	373·1	155·4	·4015	2·491	850·8	1195·7	180·7
167	373·6	154·6	·4036	2·478	850·4	1195·9	181·7
168	374·0	153·8	·4057	2·465	850·2	1196·0	182·7
169	374·5	153·0	·4077	2·452	849·8	1196·2	183·7
170	374·9	152·2	·4098	2·440	849·5	1196·3	184·7
171	375·4	151·4	·4119	2·427	849·1	1196·4	185·7

TABLE CX.—PROPERTIES OF SATURATED STEAM. 339

Table CX.—Properties of Saturated Steam—*continued.*

Pressure per Square Inch from Mean Atmospheric Pressure.	Temperature in Fahrenheit Degrees.	Specific or Relative Volume of the Steam.	Density or Weight of 1 Cubic Foot of the Steam.	Cubic Feet of the Steam per lb.	Latent Heat of Evaporation in Thermal Units per lb. of the Steam.	Total Heat in Thermal Units from 32° Fahrenheit per lb. of the Steam.	Absolute Pressure per Square Inch.
lbs.			lb.				lbs.
172	375·8	150·7	·4140	2·415	848·8	1196·6	186·7
173	376·2	149·9	·4161	2·403	848·5	1196·7	187·7
174	376·7	149·2	·4182	2·391	848·2	1196·8	188·7
175	377·1	148·4	·4203	2·379	847·9	1197·0	189·7
176	377·5	147·7	·4223	2·368	847·6	1197·1	190·7
177	378·0	147·0	·4244	2·356	847·2	1197·2	191·7
178	378·4	146·3	·4265	2·344	846·9	1197·4	192·7
179	378·8	145·5	·4286	2·333	846·7	1197·5	193·7
180	379·3	144·8	·4307	2·322	846·3	1197·6	194·7
181	379·7	144·1	·4327	2·311	846·0	1197·8	195·7
182	380·1	143·5	·4348	2·300	845·7	1197·9	196·7
183	380·5	142·8	·4369	2·289	845·4	1198·0	197·7
184	381·0	142·1	·4390	2·278	845·0	1198·2	198·7
185	381·4	141·4	·4410	2·267	844·8	1198·3	199·7
186	381·8	140·8	·4431	2·257	844·5	1198·4	200·7
187	382·2	140·1	·4452	2·246	844·2	1198·5	201·7
188	382·6	139·5	·4473	2·236	843·9	1198·6	202·7
189	383·0	138·8	·4493	2·225	843·6	1198·8	203·7
190	383·5	138·2	·4514	2·215	843·2	1198·9	204·7
191	383·9	137·5	·4535	2·205	842·9	1199·0	205·7
192	384·3	136·9	·4556	2·195	842·6	1199·2	206·7
193	384·7	136·3	·4576	2·185	842·3	1199·3	207·7
194	385·1	135·7	·4597	2·175	842·0	1199·4	208·7
195	385·5	135·1	·4618	2·165	841·8	1199·5	209·7
196	385·9	134·5	·4639	2·156	841·5	1199·6	210·7
197	386·3	133·9	·4659	2·146	841·2	1199·8	211·7
198	386·7	133·3	·4680	2·137	840·9	1199·9	212·7
199	387·1	132·7	·4701	2·127	840·6	1200·0	213·7
200	387·5	132·1	·4721	2·118	840·3	1200·1	214·7

Table CXI.—Knots, Miles, Kilometres, &c.

Admiralty Knots.	Statute Miles.	Kilometres.	Feet per minute.	Feet per second.
4	4·6061	7·413	405·8	6·75
4·25	4·8939	7·876	430·7	7·17
4·5	5·1818	8·339	456	7·59
4·75	5·4697	8·802	481·8	8·02
5	5·7576	9·266	506·7	8·44
5·25	6·0454	9·729	532	8·86
5·5	6·3333	10·192	557·3	9·28
5·75	6·6212	10·656	582·7	9·71
6	6·9091	11·119	608	10·13
6·25	7·1970	11·582	633·3	10·55
6·5	7·4848	12·046	658·7	10·97
6·75	7·7727	12·509	684	11·40
7	8·0606	12·927	709·3	11·82
7·25	8·3485	13·435	734·7	12·24
7·5	8·6364	13·899	760	12·66
7·75	8·9242	14·362	785·3	13·09
8	9·2121	14·825	810·7	13·51
8·25	9·5000	15·289	836	13·93
8·5	9·7879	15·752	861·3	14·35
8·75	10·0757	16·215	886·7	14·78
9	10·3636	16·678	912	15·20
9·25	10·6515	17·142	937·3	15·62
9·5	10·9394	17·605	962·7	16·04
9·75	11·2273	18·068	988	16·47
10	11·5151	18·532	1013·3	16·89
10·25	11·8030	18·995	1038·7	17·31
10·5	12·0909	19·458	1064	17·73
10·75	12·3788	19·921	1089·3	18·15
11	12·6667	20·385	1114·7	18·58
11·25	12·9545	20·848	1140	19·00
11·5	13·2424	21·311	1165·3	19·42
11·75	13·5303	21·775	1190·7	19·85
12	13·8182	22·238	1216	20·27
12·25	14·1061	22·701	1241·3	20·69
12·5	14·3939	23·165	1266·7	21·11
12·75	14·6818	23·628	1292	21·54
13	14·9697	24·091	1317·3	21·95
13·25	15·2576	24·554	1342·7	22·38
13·5	15·5454	25·018	1368	22·80
13·75	15·8333	25·481	1393·3	23·22
14	16·1212	25·944	1418·7	23·64
14·25	16·4091	26·408	1444	24·07
14·5	16·6970	26·871	1469·3	24·49
14·75	16·9848	27·334	1494·7	24·91
15	17·2727	27·797	1520	25·33
15·25	17·5606	28·261	1545·3	25·76
15·5	17·8485	28·724	1570·7	26·18
15·75	18·1364	29·187	1596	26·60

TABLE CXI.—KNOTS, MILES, KILOMETRES, ETC. 341

	Statute Mile.			Kilometre.		Feet per min		Foot per sec.	
	18·4242	29·651	1621·3	27·02	23·25	26·7727	43·086	2356	39·27
16	18·7121	30·114	1646·7	27·45	23·5	27·0606	43·549	2381·3	39·69
16·25	19·0000	30·577	1672	27·87	23·75	27·3485	44·013	2406·7	40·11
16·5	19·2879	31·041	1697·3	28·29	24	27·6364	44·476	2432	40·53
16·75	19·5757	31·504	1722·7	28·71	24·25	27·9242	44·939	2457·3	40·95
17	19·8636	31·967	1748	29·13	24·5	28·2121	45·408	2482·7	41·38
17·25	20·1515	32·430	1773·3	29·55	24·75	28·5000	45·866	2508	41·80
17·5	20·4394	32·894	1798·7	29·98	25	28·7879	46·329	2533·3	42·22
17·75	20·7273	33·357	1824	30·40	25·25	29·0757	46·792	2558·7	42·64
18	21·0151	33·820	1849·3	30·82	25·5	29·3636	47·256	2584	43·07
18·25	21·3030	34·284	1874·7	31·24	25·75	29·6515	47·719	2609·3	43·49
18·5	21·5909	34·747	1900	31·67	26	29·9394	48·182	2634·7	43·91
18·75	21·8788	35·210	1925·3	32·09	26·25	30·2273	48·646	2660	44·33
19	22·1667	35·673	1950·7	32·51	26·5	30·5151	49·109	2685·3	44·75
19·25	22·4545	36·137	1976	32·93	26·75	30·8030	49·572	2710·7	45·18
19·5	22·7424	36·600	2001·3	33·35	27	31·0909	50·035	2736	45·60
19·75	23·0303	37·063	2026·7	33·78	27·25	31·3788	50·499	2761·3	46·02
20	23·3182	37·526	2053	34·20	27·5	31·6667	50·962	2786·7	46·44
20·25	23·6061	37·990	2077·3	34·62	27·75	31·9545	51·425	2812	46·87
20·5	23·8939	38·453	2102·7	35·04	28	32·2424	51·889	2837·3	47·29
20·75	24·1818	38·916	2128	35·47	28·25	32·5303	52·352	2862·7	47·71
21	24·4697	39·379	2153·3	35·89	28·5	32·8182	52·815	2888	48·13
21·25	24·7576	39·843	2178·7	36·31	28·75	33·1061	53·279	2913·3	48·55
21·5	25·0454	40·306	2204	36·73	29	33·3939	53·742	2938·7	48·98
21·75	25·3333	40·769	2229·3	37·15	29·25	33·6818	54·205	2964	49·40
22	25·6212	41·233	2254·7	37·58	29·5	33·9697	54·668	2989·3	49·82
22·25	25·9091	41·696	2280	38·00	29·75	34·2576	55·132	3014·7	50·24
22·5	26·1970	42·160	2305·3	38·42	30	34·5454	55·595	3040	50·67
22·75	26·4848	42·623	2330·7	38·84	30·25	34·8333	56·058	3065·3	51·09
23									

Statute Mile. Knot = ·11515, 1/10 = ·011515

Kilometre. ·1853, ·01853

Feet per min 10·133, 1·0133

Foot per sec. ·1688, ·01688

The English standard yard is the distance between two marks on a bronze bar, measured when the bar is at a temperature of 62° F.

The French standard metre is the length of a platinum bar, measured when the bar is at a temperature of 0° Cent. (32° F.).

The Standards Commission, in their Report of 1871–72, considered that a correction was needed to allow for this difference, but the original equivalents (as determined by Kater) were adopted without correction in the Weights and Measures Act of 1878.

A comparison of the corrected and adopted equivalents is given below :

	Metre.	Litre.	Kilogramme.
	Inches.*	Gallons.†	Pounds.
Corrected (Standards Com.), .	39·38203	·22018	2·20462
Adopted (Act of 1878), . .	39·37079	·2200967	2·20462

* At equal temperatures in ordinary air.
† At equal temperatures,—distilled water.

Table CXII.—Kilometres and Admiralty Knots.

Kilometres.	Admiralty Knots.	Kilometres.	Admiralty Knots.	Kilometres.	Admiralty Knots.	Kilometres.	Admiralty Knots.	Kilometres.	Admiralty Knots
7·5	4·047	17·5	9·443	27·5	14·839	37·5	20·235	47·5	25·632
8	4·317	18	9·713	28	15·109	38	20·505	48	25·901
8·5	4·587	18·5	9·983	28·5	15·379	38·5	20·775	48·5	26·171
9	4·857	19	10·253	29	15·649	39	21·045	49	26·441
9·5	5·126	19·5	10·523	29·5	15·919	39·5	21·315	49·5	26·711
10	5·396	20	10·792	30	16·188	40	21·585	50	26·981
10·5	5·666	20·5	11·062	30·5	16·458	40·5	21·854	50·5	27·251
11	5·936	21	11·332	31	16·728	41	22·124	51	27·520
11·5	6·206	21·5	11·602	31·5	16·998	41·5	22·394	51·5	27·790
12	6·475	22	11·872	32	17·268	42	22·664	52	28·060
12·5	6·745	22·5	12·141	32·5	17·538	42·5	22·934	52·5	28·330
13	7·015	23	12·411	33	17·807	43	23·203	53	28·600
13·5	7·285	23·5	12·681	33·5	18·077	43·5	23·473	53·5	28·869
14	7·555	24	12·951	34	18·347	44	23·743	54	29·139
14·5	7·824	24·5	13·221	34·5	18·617	44·5	24·013	54·5	29·409
15	8·094	25	13·490	35	18·887	45	24·283	55	29·679
15·5	8·364	25·5	13·760	35·5	19·156	45·5	24·552	55·5	29·949
16	8·634	26	14·030	36	19·426	46	24·822	56	30·218
16·5	8·904	26·5	14·300	36·5	19·696	46·5	25·092		
17	9·173	27	14·570	37	19·966	47	25·362		

$\frac{1}{10}$ Kilometre   = ·05396 of an Admiralty knot.
$\frac{1}{100}$,,   = ·005396   ,,   ,,
1 Admiralty knot = 6080 ft. per hour.
1 Statute mile   = 5280 ft.
1 Kilometre   = 3280·8992 ft.

TABLE CXIII.—MILLIMETRES AND INCHES.   343

Table CXIII.—Millimetres and Inches.

Milli-metres.	Inches.	Milli-metres.	Inches.	Milli-metres.	Inches.	Milli-metres.	Inches.	Milli-metres.	Inches.
1	·03937	41	1·6142	81	3·1890	121	4·7639	161	6·3387
2	·07874	42	1·6586	82	3·2284	122	4·8032	162	6·3781
3	·11811	43	1·6929	83	3·2678	123	4·8426	163	6·4174
4	·15748	44	1·7323	84	3·3071	124	4·8820	164	6·4568
5	·19685	45	1·7717	85	3·3465	125	4·9214	165	6·4962
6	·23622	46	1·8110	86	3·3859	126	4·9607	166	6·5356
7	·27560	47	1·8504	87	3·4252	127	5·0001	167	6·5749
8	·31497	48	1·8898	88	3·4646	128	5·0395	168	6·6143
9	·35434	49	1·9292	89	3·5040	129	5·0788	169	6·6537
10	·3937	50	1·9685	90	3·5434	130	5·1182	170	6·6930
11	·4331	51	2·0079	91	3·5827	131	5·1576	171	6·7324
12	·4724	52	2·0473	92	3·6221	132	5·1969	172	6·7718
13	·5118	53	2·0866	93	3·6614	133	5·2363	173	6·8111
14	·5512	54	2·1260	94	3·7008	134	5·2757	174	6·8505
15	·5906	55	2·1654	95	3·7402	135	5·3150	175	6·8899
16	·6299	56	2·2048	96	3·7796	136	5·3544	176	6·9293
17	·6693	57	2·2441	97	3·8190	137	5·3938	177	6·9686
18	·7087	58	2·2835	98	3·8583	138	5·4332	178	7·0080
19	·7480	59	2·3229	99	3·8977	139	5·4725	179	7·0474
20	·7874	60	2·3622	100	3·9371	140	5·5119	180	7·0867
21	·8268	61	2·4016	101	3·9764	141	5·5513	181	7·1261
22	·8662	62	2·4410	102	4·0158	142	5·5906	182	7·1655
23	·9055	63	2·4804	103	4·0552	143	5·6300	183	7·2048
24	·9449	64	2·5197	104	4·0946	144	5·6694	184	7·2442
25	·9843	65	2·5591	105	4·1339	145	5·7088	185	7·2836
26	1·0236	66	2·5985	106	4·1733	146	5·7481	186	7·3230
27	1·0630	67	2·6378	107	4·2127	147	5·7875	187	7·3623
28	1·1024	68	2·6772	108	4·2520	148	5·8269	188	7·4017
29	1·1418	69	2·7166	109	4·2914	149	5·8662	189	7·4411
30	1·1811	70	2·7560	110	4·3308	150	5·9056	190	7·4804
31	1·2205	71	2·7953	111	4·3702	151	5·9450	191	7·5198
32	1·2599	72	2·8347	112	4·4095	152	5·9844	192	7·5592
33	1·2992	73	2·8741	113	4·4489	153	6·0237	193	7·5986
34	1·3386	74	2·9134	114	4·4883	154	6·0631	194	7·6379
35	1·3780	75	2·9528	115	4·5276	155	6·1025	195	7·6773
36	1·4173	76	2·9922	116	4·5670	156	6·1418	196	7·7167
37	1·4567	77	3·0316	117	4·6064	157	6·1812	197	7·7560
38	1·4961	78	3·0709	118	4·6458	158	6·2206	198	7·7954
39	1·5355	79	3·1103	119	4·6851	159	6·2600	199	7·8348
40	1·5748	80	3·1497	120	4·7245	160	6·2993	200	7·8742

Table CXIII.—Millimetres and Inches—*continued.*

Milli-metres.	Inches.	Milli-metres.	Inches.	Milli-metres.	Inches.	Milli-metres.	Inches.	Milli-metres.	Inches.
201	7·9135	241	9·4884	281	11·0632	321	12·6380	361	14·2128
202	7·9529	242	9·5277	282	11·1026	322	12·6774	362	14·2522
203	7·9923	243	9·5671	283	11·1419	323	12·7168	363	14·2916
204	8·0316	244	9·6065	284	11·1813	324	12·7561	364	14·3310
205	8·0710	245	9·6458	285	11·2207	325	12·7955	365	14·3703
206	8·1104	246	9·6852	286	11·2600	326	12·8349	366	14·4097
207	8·1498	247	9·7246	287	11·2994	327	12·8742	367	14·4491
208	8·1891	248	9·7640	288	11·3388	328	12·9136	368	14·4884
209	8·2285	249	9·8033	289	11·3782	329	12·9530	369	14·5278
210	8·2679	250	9·8427	290	11·4175	330	12·9924	370	14·5672
211	8·3072	251	9·8821	291	11·4569	331	13·0317	371	14·6066
212	8·3466	252	9·9214	292	11·4963	332	13·0711	372	14·6459
213	8·3860	253	9·9608	293	11·5356	333	13·1105	373	14·6853
214	8·4254	254	10·0002	294	11·5750	334	13·1498	374	14·7247
215	8·4647	255	10·0396	295	11·6144	335	13·1892	375	14·7640
216	8·5041	256	10·0789	296	11·6538	336	13·2286	376	14·8034
217	8·5435	257	10·1183	297	11·6931	337	13·2680	377	14·8428
218	8·5828	258	10·1577	298	11·7325	338	13·3073	378	14·8822
219	8·6222	259	10·1970	299	11·7719	339	13·3467	379	14·9215
220	8·6616	260	10·2364	300	11·8112	340	13·3861	380	14·9609
221	8·7009	261	10·2758	301	11·8506	341	13·4254	381	15·0003
222	8·7403	262	10·3151	302	11·8900	342	13·4648	382	15·0396
223	8·7797	263	10·3545	303	11·9294	343	13·5042	383	15·0790
224	8·8190	264	10·3939	304	11·9687	344	13·5436	384	15·1184
225	8·8584	265	10·4333	305	12·0081	345	13·5829	385	15·1578
226	8·8978	266	10·4726	306	12·0475	346	13·6223	386	15·1971
227	8·9372	267	10·5120	307	12·0868	347	13·6617	387	15·2365
228	8·9765	268	10·5514	308	12·1262	348	13·7010	388	15·2759
229	9·0159	269	10·5907	309	12·1656	349	13·7404	389	15·3152
230	9·0553	270	10·6301	310	12·2049	350	13·7798	390	15·3546
231	9·0946	271	10·6695	311	12·2443	351	13·8192	391	15·3940
232	9·1340	272	10·7088	312	12·2837	352	13·8585	392	15·4334
233	9·1734	273	10·7482	313	12·3231	353	13·8979	393	15·4727
234	9·2128	274	10·7876	314	12·3624	354	13·9373	394	15·5121
235	9·2521	275	10·8270	315	12·4018	355	13·9766	395	15·5515
236	9·2915	276	10·8663	316	12·4412	356	14·0160	396	15·5908
237	9·3309	277	10·9057	317	12·4805	357	14·0554	397	15·6302
238	9·3702	278	10·9451	318	12·5199	358	14·0947	398	15·6696
239	9·4096	279	10·9844	319	12·5593	359	14·1341	399	15·7089
240	9·4490	280	11·0238	320	12·5986	360	14·1735	400	15·7483

TABLE CXIII.—MILLIMETRES AND INCHES. 345

Table CXIII.—Millimetres and Inches—*continued*.

Milli-metres	Inches	Milli-metres	Inches	Milli-metres	Inches	Milli-metres	Inches	Milli-metres	Inches
401	15·7877	441	17·3625	481	18·9374	521	20·5122	561	22·0870
402	15·8271	442	17·4019	482	18·9767	522	20·5516	562	22·1264
403	15·8664	443	17·4413	483	19·0161	523	20·5909	563	22·1658
404	15·9058	444	17·4806	484	19·0555	524	20·6303	564	22·2051
405	15·9452	445	17·5200	485	19·0948	525	20·6697	565	22·2445
406	15·9845	446	17·5594	486	19·1342	526	20·7090	566	22·2839
407	16·0239	447	17·5987	487	19·1736	527	20·7484	567	22·3232
408	16·0633	448	17·6381	488	19·2130	528	20·7878	568	22·3626
409	16·1026	449	17·6775	489	19·2523	529	20·8272	569	22·4020
410	16·1420	450	17·7169	490	19·2917	530	20·8665	570	22·4414
411	16·1814	451	17·7562	491	19·3311	531	20·9059	571	22·4807
412	16·2208	452	17·7956	492	19·3704	532	20·9453	572	22·5201
413	16·2601	453	17·8350	493	19·4098	533	20·9846	573	22·5595
414	16·2995	454	17·8743	494	19·4492	534	21·0240	574	22·5988
415	16·3389	455	17·9137	495	19·4885	535	21·0634	575	22·6382
416	16·3782	456	17·9531	496	19·5279	536	21·1027	576	22·6776
417	16·4176	457	17·9924	497	19·5673	537	21·1421	577	22·7170
418	16·4570	458	18·0318	498	19·6066	538	21·1815	578	22·7563
419	16·4964	459	18·0712	499	19·6460	539	21·2209	579	22·7957
420	16·5357	460	18·1106	500	19·6854	540	21·2602	580	22·8351
421	16·5751	461	18·1499	501	19·7248	541	21·2996	581	22·8744
422	16·6145	462	18·1893	502	19·7641	542	21·3390	582	22·9138
423	16·6538	463	18·2287	503	19·8035	543	21·3783	583	22·9532
424	16·6932	464	18·2680	504	19·8429	544	21·4177	584	22·9925
425	16·7326	465	18·3074	505	19·8822	545	21·4571	585	23·0319
426	16·7720	466	18·3468	506	19·9216	546	21·4964	586	23·0713
427	16·8113	467	18·3862	507	19·9610	547	21·5358	587	23·1106
428	16·8507	468	18·4255	508	20·0004	548	21·5752	588	23·1500
429	16·8901	469	18·4649	509	20·0397	549	21·6146	589	23·1894
430	16·9294	470	18·5043	510	20·0791	550	21·6539	590	23·2288
431	16·9688	471	18·5436	511	20·1185	551	21·6933	591	23·2681
432	17·0082	472	18·5830	512	20·1578	552	21·7327	592	23·3075
433	17·0476	473	18·6224	513	20·1972	553	21·7720	593	23·3469
434	17·0869	474	18·6617	514	20·2366	554	21·8114	594	23·3862
435	17·1263	475	18·7011	515	20·2760	555	21·8508	595	23·4256
436	17·1657	476	18·7405	516	20·3153	556	21·8902	596	23·4650
437	17·2050	477	18·7799	517	20·3547	557	21·9295	597	23·5044
438	17·2444	478	18·8192	518	20·3941	558	21·9689	598	23·5437
439	17·2838	479	18·8586	519	20·4334	559	22·0083	599	23·5831
440	17·3232	480	18·8980	520	20·4728	560	22·0476	600	23·6225

Table CXIII.—Millimetres and Inches—*continued*.

Milli-metres.	Inches.	Milli-metres.	Inches.	Milli-metres.	Inches.	Milli-metres.	Inches.	Milli-metres.	Inches.
601	23·6618	641	25·2367	681	26·8115	721	28·3863	761	29·9612
602	23·7012	642	25·2760	682	26·8509	722	28·4257	762	30·0005
603	23·7406	643	25·3154	683	26·8902	723	28·4651	763	30·0399
604	23·7800	644	25·3548	684	26·9296	724	28·5044	764	30·0793
605	23·8193	645	25·3942	685	26·9690	725	28·5438	765	30·1187
606	23·8587	646	25·4335	686	27·0084	726	28·5832	766	30·1580
607	23·8981	647	25·4729	687	27·0477	727	28·6226	767	30·1974
608	23·9374	648	25·5123	688	27·0871	728	28·6619	768	30·2368
609	23·9768	649	25·5516	689	27·1265	729	28·7013	769	30·2761
610	24·0162	650	25·5910	690	27·1658	730	28·7407	770	30·3155
611	24·0556	651	25·6304	691	27·2052	731	28·7800	771	30·3549
612	24·0949	652	25·6698	692	27·2446	732	28·8194	772	30·3942
613	24·1343	653	25·7091	693	27·2840	733	28·8588	773	30·4336
614	24·1737	654	25·7485	694	27·3233	734	28·8982	774	30·4730
615	24·2130	655	25·7879	695	27·3627	735	28·9375	775	30·5124
616	24·2524	656	25·8272	696	27·4021	736	28·9769	776	30·5517
617	24·2918	657	25·8666	697	27·4414	737	29·0163	777	30·5911
618	24·3312	658	25·9060	698	27·4808	738	29·0556	778	30·6305
619	24·3705	659	25·9454	699	27·5202	739	29·0950	779	30·6698
620	24·4099	660	25·9847	700	27·5596	740	29·1344	780	30·7092
621	24·4493	661	26·0241	701	27·5989	741	29·1738	781	30·7486
622	24·4886	662	26·0635	702	27·6383	742	29·2131	782	30·7880
623	24·5280	663	26·1028	703	27·6777	743	29·2525	783	30·8273
624	24·5674	664	26·1422	704	27·7170	744	29·2919	784	30·8667
625	24·6068	665	26·1816	705	27·7564	745	29·3312	785	30·9061
626	24·6461	666	26·2210	706	27·7958	746	29·3706	786	30·9454
627	24·6855	667	26·2603	707	27·8352	747	29·4100	787	30·9848
628	24·7249	668	26·2997	708	27·8745	748	29·4494	788	31·0242
629	24·7642	669	26·3391	709	27·9139	749	29·4887	789	31·0636
630	24·8036	670	26·3784	710	27·9533	750	29·5281	790	31·1029
631	24·8430	671	26·4178	711	27·9926	751	29·5675	791	31·1423
632	24·8823	672	26·4572	712	28·0320	752	29·6068	792	31·1817
633	24·9217	673	26·4965	713	28·0714	753	29·6462	793	31·2210
634	24·9611	674	26·5359	714	28·1108	754	29·6856	794	31·2604
635	25·0004	675	26·5753	715	28·1501	755	29·7250	795	31·2998
636	25·0398	676	26·6147	716	28·1895	756	29·7643	796	31·3392
637	25·0792	677	26·6540	717	28·2289	757	29·8037	797	31·3785
638	25·1186	678	26·6934	718	28·2682	758	29·8431	798	31·4179
639	25·1579	679	26·7328	719	28·3076	759	29·8824	799	31·4573
640	25·1973	680	26·7721	720	28·3470	760	29·9218	800	31·4966

TABLE CXIII.—MILLIMETRES AND INCHES. 347

Table CXIII.—Millimetres and Inches—*continued.*

Milli-metres.	Inches.	Milli-metres.	Inches.	Milli-metres.	Inches.	Milli-metres.	Inches.	Milli-metres.	Inches.
801	31·5360	841	33·1108	881	34·6857	921	36·2605	961	37·8258
802	31·5754	842	33·1502	882	34·7250	922	36·2999	962	37·8747
803	31·6148	843	33·1896	883	34·7644	923	36·3392	963	37·9141
804	31·6541	844	33·2290	884	34·8038	924	36·3786	964	37·9534
805	31·6935	845	33·2683	885	34·8432	925	36·4180	965	37·9928
806	31·7329	846	33·3077	886	34·8825	926	36·4574	966	38·0322
807	31·7722	847	33·3471	887	34·9219	927	36·4967	967	38·0716
808	31·8116	848	33·3864	888	34·9613	928	36·5361	968	38·1109
809	31·8510	849	33·4258	889	35·0006	929	36·5755	969	38·1503
810	31·8903	850	33·4652	890	35·0400	930	36·6148	970	38·1897
811	31·9297	851	33·5046	891	35·0794	931	36·6542	971	38·2290
812	31·9691	852	33·5439	892	35·1188	932	36·6936	972	38·2684
813	32·0085	853	33·5833	893	35·1581	933	36·7330	973	38·3078
814	32·0478	854	33·6227	894	35·1975	934	36·7723	974	38·3472
815	32·0872	855	33·6620	895	35·2369	935	36·8117	975	38·3865
816	32·1266	856	33·7014	896	35·2762	936	36·8511	976	38·4259
817	32·1659	857	33·7408	897	35·3156	937	36·8904	977	38·4653
818	32·2053	858	33·7801	898	35·3550	938	36·9298	978	38·5046
819	32·2447	859	33·8195	899	35·3943	939	36·9692	979	38·5440
820	32·2840	860	33·8589	900	35·4337	940	37·0086	980	38·5834
821	32·3234	861	33·8982	901	35·4731	941	37·0479	981	38·6228
822	32·3628	862	33·9376	902	35·5125	942	37·0873	982	38·6621
823	32·4022	863	33·9770	903	35·5518	943	37·1267	983	38·7015
824	32·4415	864	34·0164	904	35·5912	944	37·1660	984	38·7409
825	32·4809	865	34·0557	905	35·6306	945	37·2054	985	38·7802
826	32·5203	866	34·0951	906	35·6699	946	37·2448	986	38·8196
827	32·5596	867	34·1345	907	35·7093	947	37·2841	987	38·8590
828	32·5990	868	34·1738	908	35·7487	948	37·3235	988	38·8984
829	32·6384	869	34·2132	909	35·7880	949	37·3629	989	38·9377
830	32·6778	870	34·2526	910	35·8274	950	37·4023	990	38·9771
831	32·7171	871	34·2920	911	35·8668	951	37·4416	991	39·0165
832	32·7565	872	34·3313	912	35·9062	952	37·4810	992	39·0558
833	32·7959	873	34·3707	913	35·9455	953	37·5204	993	39·0952
834	32·8352	874	34·4101	914	35·9849	954	37·5597	994	39·1346
835	32·8746	875	34·4494	915	36·0243	955	37·5991	995	39·1739
836	32·9140	876	34·4888	916	36·0636	956	37·6385	996	39·2138
837	32·9534	877	34·5282	917	36·1030	957	37·6778	997	39·2527
838	32·9927	878	34·5676	918	36·1424	958	37·7172	998	39·2920
839	33·0321	879	34·6069	919	36·1818	959	37·7566	999	39·3314
840	33·0715	880	34·6463	920	36·2211	960	37·7960	1000	39·3708

Table CXIV.—Decimal equivalents of Fractions of an inch.

Fractions	Decimals	Fractions	Decimals	Fractions	Decimals	Fractions	Decimals
1/64	·015625	17/64	·265625	33/64	·515625	49/64	·765625
1/32	·03125	9/32	·28125	17/32	·53125	25/32	·78125
3/64	·046875	19/64	·296875	35/64	·546875	51/64	·796875
1/16	·0625	5/16	·3125	9/16	·5625	13/16	·8125
5/64	·078125	21/64	·328125	37/64	·578125	53/64	·828125
3/32	·09375	11/32	·34375	19/32	·59375	27/32	·84375
7/64	·109375	23/64	·359375	39/64	·609375	55/64	·859375
1/8	·125	3/8	·375	5/8	·625	7/8	·875
9/64	·140625	25/64	·390625	41/64	·640625	57/64	·890625
5/32	·15625	13/32	·40625	21/32	·65625	29/32	·90625
11/64	·171875	27/64	·421875	43/64	·671875	59/64	·921875
3/16	·1875	7/16	·4375	11/16	·6875	15/16	·9375
13/64	·208125	29/64	·453125	45/64	·703125	61/64	·953125
7/32	·21875	15/32	·46875	23/32	·71875	31/32	·96875
15/64	·234375	31/64	·484375	47/64	·734375	63/64	·984375
1/4	·25	1/2	·5	3/4	·75	1	1

Table CXV.—Metrical equivalents of Fractions of an inch, &c.

Fractions of an inch.	Milli-metres.	Fractions of an inch.	Milli-metres.	Inches.	Milli-metres.	Inches.	Milli-metres.
1/32	0·7987	17/32	13·4935	1	25·3995	17	431·7922
1/16	1·5875	9/16	14·2872	2	50·7991	18	457·1917
3/32	2·3812	19/32	15·0810	3	76·1986	19	482·5913
1/8	3·1749	5/8	15·8747	4	101·5982	20	507·9908
5/32	3·9687	21/32	16·6684	5	126·9977	21	533·3904
3/16	4·7624	11/16	17·4622	6	152·3972	22	558·7899
7/32	5·5561	23/32	18·2559	7	177·7968	23	584·1894
1/4	6·3499	3/4	19·0496	8	203·1963	24	609·5890
9/32	7·1436	25/32	19·8434	9	228·5959	25	634·9885
5/16	7·9874	13/16	20·6371	10	253·9954	26	660·3881
11/32	8·7311	27/32	21·4309	11	279·3950	27	685·7876
3/8	9·5248	7/8	22·2246	12	304·7945	28	711·1872
13/32	10·3186	29/32	23·0183	13	330·1940	29	736·5867
7/16	11·1123	15/16	23·8121	14	335·5936	30	761·9862
15/32	11·9060	31/32	24·6058	15	380·9981	31	787·3858
1/2	12·6998			16	406·3926	32	812·7853

TABLE CXVI.—SQUARE FEET AND SQUARE METRES.   349

Table CXVI.—Square Feet and Square Metres.

Square Feet.	Square Metres.	Square Feet.	Square Metres.	Square Feet.	Square Metres.	Square Feet.	Square Metres.
1	·0929	26	2·4154	51	4·7379	76	7·0604
2	·1858	27	2·5083	52	4·8308	77	7·1533
3	·2787	28	2·6012	53	4·9237	78	7·2462
4	·3716	29	2·6941	54	5·0166	79	7·3391
5	·4645	30	2·7870	55	5·1095	80	7·4320
6	·5574	31	2·8799	56	5·2024	81	7·5249
7	·6503	32	2·9728	57	5·2953	82	7·6178
8	·7432	33	3·0657	58	5·3882	83	7·7107
9	·8361	34	3·1586	59	5·4811	84	7·8036
10	·9290	35	3·2515	60	5·5740	85	7·8965
11	1·0219	36	3·3444	61	5·6669	86	7·9894
12	1·1148	37	3·4373	62	5·7598	87	8·0823
13	1·2077	38	3·5302	63	5·8527	88	8·1752
14	1·3006	39	3·6231	64	5·9456	89	8·2681
15	1·3935	40	3·7160	65	6·0385	90	8·3610
16	1·4864	41	3·8089	66	6·1314	91	8·4539
17	1·5793	42	3·9018	67	6·2243	92	8·5468
18	1·6722	43	3·9947	68	6·3172	93	8·6397
19	1·7651	44	4·0876	69	6·4101	94	8·7326
20	1·8580	45	4·1805	70	6·5030	95	8·8255
21	1·9509	46	4·2734	71	6·5969	96	8·9184
22	2·0438	47	4·3663	72	6·6888	97	9·0113
23	2·1367	48	4·4592	73	6·7817	98	9·1042
24	2·2296	49	4·5521	74	6·8746	99	9·1971
25	2·3225	50	4·6450	75	6·9675	100	9·2900

The above Table can, of course, be used for hundreds and thousands
f feet, or for hundredths and thousandths of feet, by altering the
osition of the decimal point: e.g.,—50 square feet = 4·645 square
netres, and 5000 square feet = 464·5 square metres; also ·5 square
oot = ·04645 square metre.

Table CXVII.—Square Metres and Square Feet.

Square Metres.	Square Feet.	Square Metres.	Square Feet.	Square Metres.	Square Feet.	Square Metres.	Square Feet.
1	10·764	26	279·872	51	548·979	76	818·087
2	21·529	27	290·636	52	559·744	77	828·851
3	32·293	28	301·400	53	570·508	78	839·615
4	43·057	29	312·165	54	581·272	79	850·380
5	53·822	30	322·929	55	592·036	80	861·144
6	64·586	31	333·693	56	602·801	81	871·908
7	75·350	32	344·458	57	613·565	82	882·673
8	86·114	33	355·222	58	624·329	83	893·437
9	96·879	34	365·986	59	635·094	84	904·201
10	107·643	35	376·750	60	645·858	85	914·966
11	118·407	36	387·515	61	656·622	86	925·730
12	129·172	37	398·279	62	667·387	87	936·494
13	139·936	38	409·043	63	678·151	88	947·258
14	150·700	39	419·808	64	688·915	89	958·023
15	161·464	40	430·572	65	699·680	90	968·787
16	172·229	41	441·336	66	710·444	91	979·551
17	182·993	42	452·101	67	721·208	92	990·316
18	193·757	43	462·865	68	731·972	93	1001·080
19	204·522	44	473·629	69	742·737	94	1011·844
20	215·286	45	484·394	70	753·501	95	1022·608
21	226·050	46	495·158	71	764·265	96	1033·373
22	236·815	47	505·922	72	775·030	97	1044·137
23	247·579	48	516·686	73	785·794	98	1054·901
24	258·343	49	527·451	74	796·558	99	1065·666
25	269·108	50	538·215	75	807·322	100	1076·430

The above Table can, of course, be used for hundreds and thousands of metres, or for hundredths and thousandths of metres, by altering the position of the decimal point: e.g.,—50 square metres = 538·215 square feet, and 5000 square metres = 53821·5 square feet; also ·5 square metre = 5·38215 square feet.

Table CXVIII.—English Weights and Metric Equivalents.

Lbs.	Kilogrammes.	Lbs.	Kilogrammes.	Lbs.	Kilogrammes.
1	·4536	42	19·0509	83	37·6482
2	·9072	43	19·5045	84	38·1018
3	1·3608	44	19·9581	85	38·5554
4	1·8144	45	20·4117	86	39·0089
5	2·2680	46	20·8653	87	39·4625
6	2·7216	47	21·3189	88	39·9161
7	3·1752	48	21·7724	89	40·3697
8	3·6287	49	22·2260	90	40·8233
9	4·0823	50	22·6796	91	41·2769
10	4·5359	51	23·1332	92	41·7305
11	4·9895	52	23·5868	93	42·1841
12	5·4431	53	24·0404	94	42·6877
13	5·8967	54	24·4940	95	43·0913
14	6·3503	55	24·9476	96	43·5449
15	6·8039	56	25·4012	97	43·9985
16	7·2575	57	25·8548	98	44·4521
17	7·7111	58	26·3084	99	44·9057
18	8·1647	59	26·7619	100	45·3593
19	8·6182	60	27·2155	101	45·8128
20	9·0718	61	27·6691	102	46·2664
21	9·5254	62	28·1227	103	46·7200
22	9·9790	63	28·5763	104	47·1736
23	10·4326	64	29·0299	105	47·6272
24	10·8862	65	29·4835	106	48·0808
25	11·3398	66	29·9371	107	48·5344
26	11·7934	67	30·3907	108	48·9880
27	12·2470	68	30·8443	109	49·4416
28	12·7006	69	31·2979	110	49·8952
29	13·1542	70	31·7515	111	50·3488
30	13·6078	71	32·3051	112	50·8024
31	14·0614	72	32·6587	200	90·7185
32	14·5149	73	33·1123	300	136·0778
33	14·9685	74	33·5658	400	181·4370
34	15·4221	75	34·0194	500	226·7963
35	15·8757	76	34·4730	600	272·1556
36	16·3292	77	34·9266	700	317·5148
37	16·7293	78	35·3802	800	362·8741
38	17·2365	79	35·8338	900	408·2384
39	17·6901	80	36·2874	1000	453·5926
40	18·1437	81	36·7410	2000	907·1853
41	18·5973	82	37·1946	2240	1016·0475

Table CXIX.—Metric Weights and English equivalents.

Kilo-grams.	Lbs.	Kilo-grams.	Lbs.	Kilo-grams.	Lbs.
1	2·2046	38	83·7756	75	165·3466
2	4·4092	39	85·9802	76	167·5512
3	6·6139	40	88·1848	77	169·7559
4	8·8185	41	90·3895	78	171·9605
5	11·0231	42	92·5941	79	174·1651
6	13·2277	43	94·7987	80	176·3697
7	15·4324	44	97·0034	81	178·5743
8	17·6370	45	99·2079	82	180·7789
9	19·8416	46	101·4126	83	182·9836
10	22·0462	47	103·6172	84	185·1882
11	24·2508	48	105·8218	85	187·3928
12	26·4554	49	108·0264	86	189·5974
13	28·6601	50	110·2311	87	191·8020
14	30·8647	51	112·4357	88	194·0067
15	33·0693	52	114·6403	89	196·2113
16	35·2739	53	116·8499	90	198·4159
17	37·4786	54	119·0495	91	200·6205
18	39·6832	55	121·2542	92	202·8251
19	41·8878	56	123·4588	93	205·0298
20	44·0924	57	125·6634	94	207·2344
21	46·2970	58	127·8680	95	209·4390
22	48·5017	59	130·6727	96	211·6431
23	50·7063	60	132·2773	97	213·8482
24	52·9109	61	134·4819	98	216·0529
25	55·1155	62	136·6865	99	218·2575
26	57·3202	63	138·8911	100	220·4621
27	59·5248	64	141·0958	200	440·9243
28	61·7294	65	143·3004	300	661·3864
29	63·9340	66	145·5050	400	881·8485
30	66·1386	67	147·7096	500	1102·3106
31	68·3433	68	149·9142	600	1322·7728
32	70·5479	69	152·1189	700	1543·2349
33	72·7525	70	154·3235	800	1763·6970
34	74·9571	71	156·5281	900	1984·1591
35	77·1617	72	158·7327	1000	2204·6213
36	79·3664	73	160·9374	1016	2239·8952
37	81·5709	74	163·1419		

Table CXX.—Pounds per square inch and Kilogrammes per square centimetre.

Lbs. per sq. inch.	Kilos. per sq. cm.	Lbs. per sq. inch.	Kilos. per sq. cm.	Lbs. per sq. inch.	Kilos. per sq. cm.	Lbs. per sq. inch.	Kilos. per sq. cm.	Lbs. per sq. inch.	Kilos. per sq. cm.
1	·0703	35	2·460	69	4·850	103	7·241	137	9·632
2	·1406	36	2·530	70	4·921	104	7·312	138	9·702
3	·2109	37	2·601	71	4·991	105	7·382	139	9·772
4	·2812	38	2·671	72	5·061	106	7·452	140	9·843
5	·3515	39	2·741	73	5·131	107	7·522	141	9·913
6	·4218	40	2·812	74	5·202	108	7·593	142	9·983
7	·4921	41	2·882	75	5·272	109	7·663	143	10·054
8	·5624	42	2·952	76	5·342	110	7·733	144	10·124
9	·6327	43	3·022	77	5·413	111	7·804	145	10·194
10	·7030	44	3·093	78	5·483	112	7·874	146	10·264
11	·7733	45	3·163	79	5·553	113	7·944	147	10·335
12	·8436	46	3·233	80	5·624	114	8·015	148	10·405
13	·9140	47	3·304	81	5·694	115	8·085	149	10·475
14	·9843	48	3·374	82	5·764	116	8·155	150	10·546
15	1·0546	49	3·444	83	5·834	117	8·226	155	10·897
16	1·1248	50	3·515	84	5·905	118	8·296	160	11·249
17	1·1952	51	3·585	85	5·975	119	8·366	165	11·600
18	1·265	52	3·655	86	6·045	120	8·436	170	11·952
19	1·335	53	3·725	87	6·116	121	8·507	175	12·303
20	1·406	54	3·796	88	6·186	122	8·577	180	12·655
21	1·476	55	3·866	89	6·256	123	8·647	185	13·006
22	1·546	56	3·936	90	6·327	124	8·718	190	13·358
23	1·616	57	4·007	91	6·397	125	8·788	195	13·710
24	1·687	58	4·077	92	6·467	126	8·858	200	14·061
25	1·757	59	4·147	93	6·537	127	8·929	210	14·76
26	1·827	60	4·218	94	6·608	128	8·999	220	15·46
27	1·898	61	4·288	95	6·678	129	9·069	230	16·16
28	1·968	62	4·358	96	6·748	130	9·140	240	16·87
29	2·038	63	4·428	97	6·819	131	9·210	250	17·57
30	2·109	64	4·499	98	6·889	132	9·280	260	18·27
31	2·179	65	4·569	99	6·959	133	9·350	270	18·98
32	2·249	66	4·639	100	7·030	134	9·421	280	19·68
33	2·319	67	4·710	101	7·101	135	9·491	290	20·38
34	2·390	68	4·780	102	7·171	136	9·561	300	21·09

Table CXXI.—Kilogrammes per square centimetre and pounds per square inch.

Kilos. per sq. cm.	Lbs. per square inch.	Kilos. per sq. cm.	Lbs. per square inch.	Kilos. per sq. cm.	Lbs. per square inch.	Kilos. per sq. cm.	Lbs. per square inch.
·1	1·422	3·1	44·091	6·1	86·761	9·1	129·431
·2	2·844	3·2	45·514	6·2	88·183	9·2	130·853
·3	4·266	3·3	46·936	6·3	89·606	9·3	132·275
·4	5·689	3·4	48·358	6·4	91·028	9·4	133·698
·5	7·111	3·5	49·781	6·5	92·450	9·5	135·120
·6	8·533	3·6	51·203	6·6	93·873	9·6	136·542
·7	9·956	3·7	52·625	6·7	95·295	9·7	137·965
·8	11·378	3·8	54·048	6·8	96·717	9·8	139·387
·9	12·800	3·9	55·470	6·9	98·140	9·9	140·809
1·0	14·223	4·0	56·892	7·0	99·562	10·0	142·232
1·1	15·645	4·1	58·315	7·1	100·984	10·5	149·343
1·2	17·067	4·2	59·737	7·2	102·407	11·0	156·455
1·3	18·490	4·3	61·159	7·3	103·829	11·5	163·566
1·4	19·912	4·4	62·582	7·4	105·251	12·0	170·678
1·5	21·334	4·5	64·004	7·5	106·674	12·5	177·790
1·6	22·757	4·6	65·426	7·6	108·096	13·0	184·901
1·7	24·179	4·7	66·849	7·7	109·518	13·5	192·013
1·8	25·601	4·8	68·271	7·8	110·940	14·0	199·124
1·9	27·024	4·9	69·693	7·9	112·363	14·5	206·236
2·0	28·446	5·0	71·116	8·0	113·785	15·0	213·348
2·1	29·868	5·1	72·538	8·1	115·207	15·5	220·459
2·2	31·291	5·2	73·960	8·2	116·630	16·0	227·571
2·3	32·713	5·3	75·382	8·3	118·052	16·5	234·682
2·4	34·135	5·4	76·805	8·4	119·474	17·0	241·794
2·5	35·558	5·5	78·227	8·5	120·897	17·5	248·906
2·6	36·980	5·6	79·649	8·6	122·319	18·0	256·017
2·7	38·402	5·7	81·072	8·7	123·741	18·5	263·129
2·8	39·824	5·8	82·494	8·8	125·164	19·0	270·240
2·9	41·247	5·9	83·916	8·9	126·586	19·5	277·352
3·0	42·669	6·0	85·339	9·0	128·008	20·0	284·464

TABLE CXXII.—AREAS OF SEGMENTS OF CIRCLES.   355

Table CXXII.—Areas of Segments of Circles.

To find the area of any segment of a circle,—Divide the versed sine or height of the segment (V) by the diameter of the circle of which it is a part (D), and multiply the square of the diameter by the value of x (see Table) corresponding to the value of $\frac{V}{D}$ obtained; that is,—

Area of Segment = Diameter $^2 \times x$.

$\frac{V}{D}$	x	$\frac{V}{D}$	x	$\frac{V}{D}$	x	$\frac{V}{D}$	x
·001	·000042	·038	·009763	·075	·026761	·112	·048262
·002	·000119	·039	·010148	·076	·027289	·113	·048894
·003	·000219	·040	·010537	·077	·027821	·114	·049528
·004	·000337	·041	·010931	·078	·028356	·115	·050165
·005	·000470	·042	·011330	·079	·028894	·116	·050804
·006	·000618	·043	·011734	·080	·029435	·117	·051446
·007	·000779	·044	·012142	·081	·029979	·118	·052090
·008	·000951	·045	·012554	·082	·030526	·119	·052736
·009	·001135	·046	·012971	·083	·031076	·120	·053385
·010	·001329	·047	·013392	·084	·031629	·121	·054036
·011	·001533	·048	·013818	·085	·032186	·122	·054689
·012	·001746	·049	·014247	·086	·032745	·123	·055345
·013	·001968	·050	·014681	·087	·033307	·124	·056003
·014	·002199	·051	·015119	·088	·033872	·125	·056663
·015	·002438	·052	·015561	·089	·034441	·126	·057326
·016	·002685	·053	·016007	·090	·035011	·127	·057991
·017	·002940	·054	·016457	·091	·035585	·128	·058658
·018	·003202	·055	·016911	·092	·036162	·129	·059327
·019	·003471	·056	·017369	·093	·036741	·130	·059999
·020	·003748	·057	·017831	·094	·037323	·131	·060672
·021	·004031	·058	·018296	·095	·037909	·132	·061348
·022	·004322	·059	·018766	·096	·038496	·133	·062026
·023	·004618	·060	·019239	·097	·039087	·134	·062707
·024	·004921	·061	·019716	·098	·039680	·135	·063389
·025	·005230	·062	·020196	·099	·040276	·136	·064074
·026	·005546	·063	·020680	·100	·040875	·137	·064760
·027	·005867	·064	·021168	·101	·041476	·138	·065449
·028	·006194	·065	·021659	·102	·042080	·139	·066140
·029	·006527	·066	·022154	·103	·042687	·140	·066833
·030	·006865	·067	·022652	·104	·043296	·141	·067528
·031	·007209	·068	·023154	·105	·043908	·142	·068225
·032	·007558	·069	·023659	·106	·044522	·143	·068924
·033	·007913	·070	·024168	·107	·045139	·144	·069625
·034	·008273	·071	·024680	·108	·045759	·145	·070328
·035	·008638	·072	·025195	·109	·046381	·146	·071033
·036	·009008	·073	·025714	·110	·047005	·147	·071741

Table CXXII.—Areas of Segments of Circles—*continued*.

$\frac{V}{D}$	z	$\frac{V}{D}$	z	$\frac{V}{D}$	z	$\frac{V}{D}$	z
·149	·078161	·193	·106261	·237	·142387	·281	·180918
·150	·073874	·194	·107051	·238	·143238	·282	·181817
·151	·074589	·195	·107842	·239	·144091	·283	·182718
·152	·075306	·196	·108636	·240	·144944	·284	·183619
·153	·076026	·197	·109430	·241	·145799	·285	·184521
·154	·076747	·198	·110226	·242	·146655	·286	·185425
·155	·077469	·199	·111024	·243	·147512	·287	·186329
·156	·078194	·200	·111823	·244	·148371	·288	·187234
·157	·078921	·201	·112624	·245	·149230	·289	·188140
·158	·079649	·202	·113426	·246	·150091	·290	·189047
·159	·080380	·203	·114230	·247	·150953	·291	·189955
·160	·081112	·204	·115035	·248	·151816	·292	·190864
·161	·081846	·205	·115842	·249	·152680	·293	·191775
·162	·082582	·206	·116650	·250	·153546	·294	·192684
·163	·083320	·207	·117460	·251	·154412	·295	·193596
·164	·084059	·208	·118271	·252	·155280	·296	·194509
·165	·084801	·209	·119083	·253	·156149	·297	·195422
·166	·085544	·210	·119897	·254	·157019	·298	·196337
·167	·086289	·211	·120712	·255	·157890	·299	·197252
·168	·087036	·212	·121529	·256	·158762	·300	·198168
·169	·087785	·213	·122347	·257	·159636	·301	·199085
·170	·088535	·214	·123167	·258	·160510	·302	·200003
·171	·089287	·215	·123988	·259	·161386	·303	·200922
·172	·090041	·216	·124810	·260	·162263	·304	·201841
·173	·090797	·217	·125634	·261	·163140	·305	·202761
·174	·091554	·218	·126459	·262	·164019	·306	·203683
·175	·092313	·219	·127285	·263	·164899	·307	·204605
·176	·093074	·220	·128113	·264	·165780	·308	·205527
·177	·093836	·221	·128942	·265	·166663	·309	·206451
·178	·094601	·222	·129773	·266	·167546	·310	·207376
·179	·095366	·223	·130605	·267	·168430	·311	·208301
·180	·096134	·224	·131438	·268	·169315	·312	·209227
·181	·096903	·225	·132272	·269	·170202	·313	·210154
·182	·097674	·226	·133108	·270	·171089	·314	·211082
·183	·098447	·227	·133945	·271	·171978	·315	·212011
·184	·099221	·228	·134784	·272	·172867	·316	·212940
185	·099997	·229	·135624	·273	·173758	·317	·213871
·186	100774	·230	·136465	·274	·174649	·318	·214802
·187	·101553	·231	·137307	·275	·175542	·319	·215733
·188	·102334	·232	·138150	·276	·176435	·320	·216666
·189	·103116	·233	·138995	·277	·177330	·321	·217599
·190	·103900	·234	·139841	·278	·178225	·322	·218533
·191	·104685	·235	·140688	·279	·179122	·323	·219468
192	·105472	·236	·141537	·280	·180019	·324	·220404

TABLE CXXII.—AREAS OF SEGMENTS OF CIRCLES. 357

Table CXXII.—Areas of Segments of Circles—*continued*.

$\dfrac{V}{D}$	x	$\dfrac{V}{D}$	x	$\dfrac{V}{D}$	x	$\dfrac{V}{D}$	x
·325	·221340	·369	·263213	·413	·306140	·457	·349752
·326	·222277	·370	·264178	·414	·307125	·458	·350748
·327	·223215	·371	·265144	·415	·308110	·459	·351745
·328	·224154	·372	·266111	·416	·309095	·460	·352742
·329	·225093	·373	·267078	·417	·310081	·461	·353739
·330	·226033	·374	·268045	·418	·311068	·462	·354736
·331	·226974	·375	·269018	·419	·312054	·463	·355732
·332	·227915	·376	·269982	·420	·313041	·464	·356730
·333	·228858	·377	·270951	·421	·314029	·465	·357727
·334	·229801	·378	·271920	·422	·315016	·466	·358725
·335	·230745	·379	·272890	·423	·316004	·467	·359723
·336	·231689	·380	·273861	·424	·316992	·468	·360721
·337	·232634	·381	·274832	·425	·317981	·469	·361719
·338	·233580	·382	·275803	·426	·318970	·470	·362717
·339	·234526	·383	·276775	·427	·319959	·471	·363715
·340	·235473	·384	·277748	·428	·320948	·472	·364713
·341	·236421	·385	·278721	·429	·321938	·473	·365712
·342	·237369	·386	·279694	·430	·322928	·474	·366710
·343	·238318	·387	·280668	·431	·323918	·475	·367709
·344	·239268	·388	·281642	·432	·324909	·476	·368708
·345	·240218	·389	·282617	·433	·325900	·477	·369707
·346	·241169	·390	·283592	·434	·326892	·478	·370706
·347	·242121	·391	·284568	·435	·327882	·479	·371705
·348	·243074	·392	·285544	·436	·328874	·480	·372704
·349	·244026	·393	·286521	·437	·329866	·481	·373703
·350	·244980	·394	·287498	·438	·330858	·482	·374702
·351	·245934	·395	·288476	·439	·331850	·483	·375702
·352	·246889	·396	·289453	·440	·332843	·484	·376702
·353	·247845	·397	·290482	·441	·333836	·485	·377701
·354	·248801	·398	·291411	·442	·334829	·486	·378701
·355	·249757	·399	·292390	·443	·335822	·487	·379700
·356	·250715	·400	·293369	·444	·336816	·488	·380700
·357	·251673	·401	·294349	·445	·337810	·489	·381699
·358	·252631	·402	·295330	·446	·338804	·490	·382699
·359	·253590	·403	·296311	·447	·339798	·491	·383699
·360	·254550	·404	·297292	·448	·340793	·492	·384699
·361	·255510	·405	·298273	·449	·341787	·493	·385699
·362	·256471	·406	·299255	·450	·342782	·494	·386699
·363	·257433	·407	·300238	·451	·343777	·495	·387699
·364	·258395	·408	·301220	·452	·344772	·496	·388699
·365	·259357	·409	·302203	·453	·345768	·497	·389699
·366	·260320	·410	·303187	·454	·346764	·498	·390699
·367	·261284	·411	·304171	·455	·347759	·499	·391699

Table CXXIII.—Areas of Circles.

Diameters	·0	·⅛	·¼	·⅜	·½	·⅝	·¾	·⅞	Diameters
0	·0	·01227	·0491	·1104	·1963	·3068	·4418	·6013	0
1	·78540	·99402	1·2272	1·4849	1·7671	2·0739	2·4053	2·7612	1
2	3·1416	3·5466	3·9761	4·4301	4·9087	5·4119	5·9396	6·4918	2
3	7·0686	7·6699	8·2958	8·9462	9·6211	10·3206	11·0447	11·7933	3
4	12·5664	13·3641	14·1863	15·0330	15·9043	16·8002	17·7206	18·6655	4
5	19·6350	20·6290	21·6476	22·6907	23·7583	24·8505	25·9673	27·1086	5
6	28·2744	29·4648	30·6797	31·9191	33·1831	34·4717	35·7848	37·1224	6
7	38·4846	39·8713	41·2826	42·7184	44·1787	45·6636	47·1731	48·7071	7
8	50·2656	51·8497	53·4563	55·0884	56·7451	58·4264	60·1322	61·8625	8
9	63·6174	65·3968	67·2008	69·0293	70·8823	72·7599	74·6621	76·5888	9
10	78·5400	80·5158	82·5161	84·5409	86·5903	88·6643	90·7628	92·8858	10
11	95·0334	97·2065	99·5022	101·6234	103·8691	106·1394	108·4343	110·7637	11
12	113·098	115·466	117·859	120·277	122·719	125·185	127·677	130·192	12
13	132·733	135·297	137·887	140·501	143·139	145·802	148·490	151·202	13
14	153·938	156·700	159·485	162·296	165·130	167·990	170·874	178·782	14
15	176·715	179·673	182·655	185·661	188·692	191·748	194·828	197·933	15
16	201·062	204·216	207·395	210·598	213·825	217·077	220·354	228·655	16
17	226·981	230·331	233·706	237·105	240·529	243·977	247·450	250·948	17
18	254·470	258·016	261·587	265·183	268·803	272·448	276·117	279·811	18
19	283·529	287·272	291·040	294·832	298·648	302·489	306·355	310·245	19
20	314·160	318·099	322·068	326·051	330·064	334·102	338·164	342·250	20
21	346·361	350·497	354·657	358·842	363·051	367·285	371·548	375·826	21
22	380·184	384·466	388·822	393·203	397·609	402·088	406·494	410·973	22

TABLE CXXIII.—AREAS OF CIRCLES. 359

Table CXXIII.—Areas of Circles—continued.

Areas.

Diameters	.0	.⅛	.¼	.⅜	.½	.⅝	.¾	.⅞	Diameters
23	415·477	420·004	424·558	429·135	433·737	438·364	443·015	447·690	23
24	452·390	457·115	461·864	466·638	471·436	476·259	481·107	485·979	24
25	490·875	495·796	500·742	505·712	510·706	515·726	520·769	525·838	25
26	530·930	536·048	541·190	546·356	551·547	556·763	562·003	567·267	26
27	572·557	577·870	583·209	588·571	593·959	599·371	604·807	610·268	27
28	615·754	621·264	626·798	632·357	637·941	643·549	649·182	654·840	28
29	660·521	666·228	671·959	677·714	683·494	689·299	695·128	700·982	29
30	706·860	712·763	718·690	724·642	730·618	736·619	742·645	748·695	30
31	754·769	760·869	766·992	773·140	779·313	785·510	791·732	797·979	31
32	804·250	810·545	816·865	823·210	829·579	835·972	842·391	848·833	32
33	855·301	861·792	868·309	874·850	881·415	888·005	894·620	901·259	33
34	907·922	914·611	921·323	928·061	934·822	941·609	948·420	955·255	34
35	962·115	969·000	975·909	982·842	989·800	996·783	1003·790	1010·822	35
36	1017·878	1024·960	1032·065	1039·195	1046·349	1053·528	1060·782	1067·960	36
37	1075·213	1082·490	1089·792	1097·118	1104·469	1111·844	1119·244	1126·669	37
38	1134·118	1141·591	1149·089	1156·612	1164·159	1171·781	1179·327	1186·948	38
39	1194·593	1202·263	1209·958	1217·677	1225·420	1233·188	1240·981	1248·798	39
40	1256·64	1264·51	1272·40	1280·31	1288·25	1296·22	1304·21	1312·22	40
41	1320·26	1328·32	1336·41	1344·52	1352·66	1360·82	1369·00	1377·21	41
42	1385·45	1393·70	1401·99	1410·30	1418·63	1426·99	1435·87	1443·77	42
43	1452·20	1460·66	1469·14	1477·64	1486·17	1494·73	1503·30	1511·91	43
44	1520·53	1529·19	1537·86	1546·56	1555·29	1564·04	1572·81	1581·61	44

Table CXXIII.—Areas of Circles—*continued*.

Diameters	·0	·⅛	·¼	·⅜	·½	·⅝	·¾	·⅞	Diameters
45	1590·43	1599·28	1608·16	1617·05	1625·97	1634·92	1643·89	1652·89	45
46	1661·91	1670·95	1680·02	1689·11	1698·23	1707·37	1716·54	1725·73	46
47	1734·95	1744·19	1753·45	1762·74	1772·06	1781·40	1790·76	1800·15	47
48	1809·56	1819·00	1828·46	1837·95	1847·46	1856·99	1866·55	1876·14	48
49	1885·75	1895·38	1905·04	1914·72	1924·43	1934·16	1943·91	1953·69	49
50	1963·50	1973·33	1983·18	1993·06	2002·97	2012·89	2022·85	2032·82	50
51	2042·83	2052·85	2062·90	2072·98	2083·08	2093·20	2103·35	2113·52	51
52	2123·72	2133·94	2144·19	2154·46	2164·76	2175·08	2185·42	2195·79	52
53	2206·19	2216·61	2227·05	2237·52	2248·01	2258·53	2269·07	2279·64	53
54	2290·23	2300·84	2311·48	2322·15	2332·83	2343·55	2354·29	2365·05	54
55	2375·83	2386·65	2397·48	2408·34	2419·23	2430·14	2441·07	2452·08	55
56	2463·01	2474·02	2485·05	2496·11	2507·19	2518·30	2529·43	2540·58	56
57	2551·76	2562·97	2574·20	2585·45	2596·73	2608·03	2619·36	2630·71	57
58	2642·09	2653·49	2664·91	2676·36	2687·84	2699·33	2710·86	2722·41	58
59	2733·98	2745·57	2757·20	2768·84	2780·51	2792·21	2803·93	2815·67	59
60	2827·44	2839·23	2851·05	2862·89	2874·76	2886·65	2898·57	2910·51	60
61	2922·47	2934·46	2946·48	2958·52	2970·58	2982·67	2994·78	3006·92	61
62	3019·08	3031·26	3043·47	3055·71	3067·97	3080·25	3092·56	3104·89	62
63	3117·25	3129·64	3142·04	3154·47	3166·93	3179·41	3191·91	3204·44	63
64	3217·00	3229·58	3242·18	3254·81	3267·46	3280·14	3292·84	3305·56	64
65	3318·31	3331·09	3343·89	3356·71	3369·56	3382·44	3395·33	3408·26	65
66	3421·20	3434·17	3447·17	3460·19	3473·24	3486·30	3499·40	3512·52	66

Areas.

TABLE CXXIII.—AREAS OF CIRCLES. 361

Table CXXIII.—Areas of Circles—*continued.*

Areas.

Diameters	0	⅛	¼	⅜	½	⅝	¾	⅞	Diameters
67	3525·66	3538·83	3552·02	3565·24	3578·48	3591·74	3605·04	3618·35	67
68	3631·69	3645·05	3658·44	3671·86	3685·29	3698·76	3712·24	3725·75	68
69	3739·29	3752·85	3766·43	3780·04	3793·68	3807·34	3821·02	3834·73	69
70	3848·46	3862·22	3876·00	3889·80	3903·63	3917·49	3931·37	3945·27	70
71	3959·20	3973·15	3987·13	4001·13	4015·16	4029·21	4043·29	4057·39	71
72	4071·51	4085·66	4099·84	4114·04	4128·26	4142·51	4156·78	4171·08	72
73	4185·40	4199·74	4214·11	4228·51	4242·93	4257·37	4271·84	4286·83	73
74	4300·85	4315·39	4329·96	4344·55	4359·17	4373·81	4388·47	4403·16	74
75	4417·87	4432·61	4447·38	4462·16	4476·98	4491·81	4506·67	4521·56	75
76	4536·47	4551·41	4566·36	4581·35	4596·36	4611·39	4626·45	4641·53	76
77	4656·64	4671·77	4686·92	4702·10	4717·31	4732·54	4747·79	4763·07	77
78	4778·37	4793·70	4809·05	4824·43	4839·83	4855·26	4870·71	4886·18	78
79	4901·68	4917·21	4932·75	4948·33	4963·92	4979·55	4995·19	5010·86	79
80	5026·56	5042·28	5058·03	5073·79	5089·59	5105·41	5121·25	5137·12	80
81	5153·01	5168·93	5184·87	5200·83	5216·82	5232·84	5248·88	5264·94	81
82	5281·03	5297·14	5313·28	5329·44	5345·63	5361·84	5378·08	5394·34	82
83	5410·62	5426·93	5443·26	5459·62	5476·01	5492·41	5508·84	5525·30	83
84	5541·78	5558·29	5574·82	5591·37	5607·95	5624·56	5641·18	5657·84	84
85	5674·51	5691·22	5707·94	5724·69	5741·47	5758·27	5775·10	5791·94	85
86	5808·82	5825·72	5842·64	5859·59	5876·56	5893·55	5910·58	5927·62	86
87	5944·69	5961·79	5978·91	5996·05	6013·22	6030·41	6047·63	6064·87	87
88	6082·14	6099·43	6116·74	6134·08	6151·45	6168·84	6186·25	6203·69	88

Table CXXIII.—Areas of Circles—*continued*.

Areas.

Diameters	0	·⅛	·¼	·⅜	·½	·⅝	·¾	·⅞	Diameters
89	6221·15	6238·64	6256·15	6273·69	6291·25	6308·84	6326·45	6344·08	89
90	6361·74	6379·42	6397·13	6414·86	6432·62	6450·40	6468·21	6486·04	90
91	6503·90	6521·78	6539·68	6557·61	6575·56	6593·54	6611·55	6629·57	91
92	6647·63	6665·70	6683·80	6701·93	6720·08	6738·25	6756·45	6774·68	92
93	6792·92	6811·20	6829·49	6847·82	6866·16	6884·53	6902·93	6921·35	93
94	6939·79	6958·26	6976·76	6995·28	7013·82	7032·39	7050·98	7069·59	94
95	7088·24	7106·90	7125·59	7144·31	7163·04	7181·81	7200·60	7219·41	95
96	7238·25	7257·11	7275·99	7294·91	7313·84	7332·80	7351·79	7370·79	96
97	7389·83	7408·89	7427·97	7447·08	7466·21	7485·37	7504·55	7523·75	97
98	7542·98	7562·24	7581·52	7600·82	7620·15	7639·50	7658·88	7678·28	98
99	7697·71	7717·16	7736·63	7756·18	7775·66	7795·21	7814·78	7834·38	99
100	7854·00	7873·64	7893·31	7913·00	7932·74	7952·47	7972·25	7992·03	100
101	8011·86	8031·69	8051·58	8071·46	8091·89	8111·32	8131·30	8151·28	101
102	8171·30	8191·33	8211·41	8231·49	8251·61	8271·73	8291·91	8312·09	102
103	8332·31	8352·53	8372·81	8393·08	8413·40	8433·73	8454·09	8474·47	103
104	8494·89	8515·31	8535·78	8556·25	8576·76	8597·28	8617·85	8638·42	104
105	8659·03	...	8700·32	...	8741·70	...	8783·18	...	105
106	8824·75	...	8866·43	...	8908·20	...	8950·07	...	106
107	8992·04	...	9034·11	...	9076·28	...	9118·54	...	107
108	9160·91	...	9208·87	...	9245·98	...	9288·58	...	108
109	9331·34	...	9374·19	...	9417·14	...	9460·19	...	109
110	9503·84	...	9546·59	...	9589·98	...	9638·87	...	110

TABLE CXXIII.—AREAS OF CIRCLES. 363

Table CXXIII.—Areas of Circles—*continued*.

Areas.

Diameters	.⅞	.¾	.⅝	.½	.⅜	.¼	.⅛	.0	Diameters
111	…	9808·12	…	9764·29	…	9720·55	…	9676·91	111
112	…	9984·45	…	9940·22	…	9896·09	…	9852·06	112
113	…	10162·34	…	10117·72	…	10073·20	…	10028·77	113
114	…	10341·80	…	10296·79	…	10251·88	…	10207·06	114
115	…	10522·84	…	10477·43	…	10432·12	…	10386·91	115
116	…	10705·44	…	10659·65	…	10613·94	…	10568·34	116
117	…	10889·62	…	10843·43	…	10797·84	…	10751·84	117
118	…	11075·87	…	11028·78	…	10982·30	…	10935·91	118
119	…	11262·69	…	11215·71	…	11168·83	…	11122·05	119
120	…	11461·57	…	11404·20	…	11356·93	…	11309·76	120
121	…	11642·03	…	11594·27	…	11546·61	…	11499·04	121
122	…	11834·06	…	11785·91	…	11737·85	…	11689·89	122
123	…	12027·66	…	11979·12	…	11980·67	…	11882·32	123
124	…	12222·84	…	12173·90	…	12125·05	…	12076·31	124
125	…	…	…	12370·25	…	…	…	12271·87	125
126	…	…	…	12568·17	…	…	…	12469·01	126
127	…	…	…	12767·66	…	…	…	12667·72	127
128	…	…	…	12968·72	…	…	…	12867·99	128
129	…	…	…	13171·85	…	…	…	13069·84	129
130	…	…	…	13375·56	…	…	…	13273·26	130
131	…	…	…	13581·33	…	…	…	13478·25	131
132	…	…	…	13788·68	…	…	…	13684·81	132

Table CXXIII.—Areas of Circles—*continued.*

Diameters.	Areas.								Diameters.
	·0	⅛	¼	⅜	½	⅝	¾	⅞	
133	13892·94	...	...	...	13997·60	...	...	...	133
134	14102·64	...	...	...	14208·08	...	...	...	134
135	14313·91	...	...	...	14420·14	...	...	...	135
136	14526·76	...	...	...	14633·77	...	...	...	136
137	14741·17	...	...	...	14848·97	...	...	...	137
138	14957·16	...	...	...	15065·74	...	...	...	138
139	15174·71	...	...	...	15284·08	...	...	...	139
140	15393·84	...	...	...	15503·99	...	...	...	140
141	15614·54	...	...	...	15725·48	...	...	...	141
142	15836·81	...	...	...	15948·53	...	...	...	142
143	16060·64	...	...	...	16173·15	...	...	...	143
144	16286·05	...	...	...	16399·35	...	...	...	144
145	16513·08	...	...	...	16627·11	...	...	...	145
146	16741·59	...	...	...	16856·45	...	...	...	146
147	16971·71	...	...	...	17087·36	...	...	...	147
148	17208·40	...	...	...	17319·84	...	...	...	148
149	17436·67	...	...	...	17553·89	...	...	...	149
150	17671·50	...	...	...	17789·51	...	...	...	150

Where the diameter contains an odd sixteenth, take one fourth of the area of twice the diameter ;

e.g., area of 2¹⁄₁₆ = $\dfrac{\text{Area of } 4⅛}{4}$, &c.　Areas of intermediate sizes may also be obtained by shifting the decimal point ; *e.g.*, area, 132 = 18684·81 ; area of 13·2 = 186·8481 ; area of 1·32 = 1·368481, &c.

TABLE CXXIV.—CIRCUMFERENCES OF CIRCLES. 365

Table CXXIV.—Circumferences of Circles.

Diameters	·0	·⅛	·¼	·⅜	·½	·⅝	·¾	·⅞	Diameters
0	·0	·3927	·7854	1·1781	1·5708	1·9635	2·3562	2·7489	0
1	3·1416	3·5343	3·9270	4·3197	4·7124	5·1051	5·4978	5·8905	1
2	6·2832	6·6759	7·0686	7·4613	7·8540	8·2467	8·6394	9·0321	2
3	9·4248	9·8175	10·2102	10·6029	10·9956	11·3883	11·7810	12·1737	3
4	12·5664	12·9591	13·3518	13·7445	14·1372	14·5299	14·9226	15·3153	4
5	15·7080	16·1007	16·4934	16·8861	17·2788	17·6715	18·0642	18·4569	5
6	18·8496	19·2423	19·6350	20·0277	20·4204	20·8131	21·2058	21·5985	6
7	21·9912	22·3839	22·7766	23·1698	23·5620	23·9547	24·3474	24·7401	7
8	25·1328	25·5255	25·9182	26·3109	26·7036	27·0963	27·4890	27·8817	8
9	28·2744	28·6671	29·0598	29·4525	29·8452	30·2379	30·6306	31·0233	9
10	31·4160	31·8087	32·2014	32·5941	32·9868	33·3795	33·7722	34·1649	10
11	34·5576	34·9503	35·3430	35·7357	36·1284	36·5211	36·9138	37·3065	11
12	37·6992	38·0919	38·4846	38·8773	39·2700	39·6627	40·0554	40·4481	12
13	40·8408	41·2335	41·6262	42·0189	42·4116	42·8043	43·1970	43·5897	13
14	43·9824	44·3751	44·7678	45·1605	45·5532	45·9459	46·3386	46·7313	14
15	47·1240	47·5167	47·9094	48·3021	48·6948	49·0875	49·4802	49·8729	15
16	50·2656	50·6583	51·0510	51·4437	51·8364	52·2291	52·6218	53·0145	16
17	53·4072	53·7999	54·1926	54·5853	54·9780	55·3707	55·7634	56·1561	17
18	56·5488	56·9415	57·3342	57·7269	58·1196	58·5123	58·9050	59·2977	18
19	59·6904	60·0831	60·4758	60·8685	61·2612	61·6539	62·0466	62·4393	19
20	62·8320	63·2247	63·6174	64·0101	64·4628	64·7955	65·1882	65·5809	20
21	65·9736	66·3663	66·7590	67·1517	67·5444	67·9371	68·3298	68·7225	21
22	69·1152	69·5079	69·9006	70·2983	70·6860	71·0787	71·4714	71·8641	22

Table CXXIV.—Circumferences of Circles—*continued.*

Diameters	.0	⅛	¼	⅜	½	⅝	¾	⅞	Diameters
				Circumferences.					
23	72·2568	72·6495	73·0422	73·4349	73·8276	74·2208	74·6130	75·0057	23
24	75·3984	75·7911	76·1838	76·5765	76·9692	77·3619	77·7546	78·1473	24
25	78·5400	78·9327	79·3254	79·7181	80·1108	80·5035	80·8962	81·2889	25
26	81·6816	82·0743	82·4670	82·8597	83·2524	83·6451	84·0378	84·4305	26
27	84·8232	85·2159	85·6086	86·0013	86·3940	86·7867	87·1794	87·5721	27
28	87·9648	88·3575	88·7502	89·1429	89·5356	89·9283	90·3210	90·7137	28
29	91·1064	91·4991	91·8918	92·2845	92·6772	93·0699	93·4626	93·8553	29
30	94·2480	94·6407	95·0334	95·4261	95·8188	96·2115	96·6042	96·9969	30
31	97·3896	97·7823	98·1750	98·5677	98·9604	99·3531	99·7458	100·1385	31
32	100·5312	100·9239	101·3166	101·7093	102·1020	102·4947	102·8874	103·2801	32
33	103·6728	104·0655	104·4582	104·8509	105·2436	105·6363	106·0290	106·4217	33
34	106·814	107·207	107·600	107·992	108·385	108·778	109·171	109·563	34
35	109·956	110·349	110·741	111·134	111·527	111·919	112·312	112·705	35
36	113·098	113·490	113·883	114·276	114·668	115·061	115·454	115·846	36
37	116·289	116·682	117·025	117·417	117·810	118·203	118·595	118·988	37
38	119·381	119·773	120·166	120·559	120·952	121·344	121·737	122·130	38
39	122·522	122·915	123·308	123·700	124·098	124·486	124·879	125·271	39
40	125·664	126·057	126·449	126·842	127·235	127·627	128·020	128·413	40
41	128·806	129·198	129·591	129·984	130·376	130·769	131·162	131·554	41
42	131·947	132·340	132·733	133·125	133·518	133·911	134·303	134·696	42
43	135·089	135·481	135·874	136·267	136·660	137·052	137·445	137·838	43
44	138·230	138·623	139·016	139·408	139·801	140·194	140·587	140·979	44

TABLE CXXIV.—CIRCUMFERENCES OF CIRCLES. 367

Table CXXIV.—Circumferences of Circles—*continued.*

Diameters	.0	.⅛	.¼	.⅜	.½	.⅝	.¾	.⅞	Diameters
				Circumferences.					
45	141·372	141·765	142·157	142·550	142·943	143·335	143·728	144·121	45
46	144·514	144·906	145·299	145·692	146·084	146·477	146·870	147·262	46
47	147·655	148·048	148·441	148·833	149·226	149·619	150·011	150·404	47
48	150·797	151·189	151·582	151·975	152·368	152·760	153·153	153·546	48
49	153·938	154·331	154·724	155·116	155·509	155·902	156·295	156·687	49
50	157·080	157·473	157·865	158·258	158·651	159·043	159·436	159·829	50
51	160·222	160·614	161·007	161·400	161·792	162·185	162·578	162·970	51
52	163·363	163·756	164·149	164·541	164·934	165·327	165·719	166·112	52
53	166·505	166·897	167·290	167·683	168·076	168·468	168·861	169·254	53
54	169·646	170·039	170·432	170·824	171·217	171·610	172·003	172·395	54
55	172·788	173·181	173·573	173·966	174·359	174·751	175·144	175·537	55
56	175·930	176·322	176·715	177·108	177·500	177·893	178·286	178·678	56
57	179·071	179·464	179·857	180·249	180·642	181·035	181·427	181·820	57
58	182·213	182·605	182·998	183·391	183·784	184·176	184·569	184·962	58
59	185·354	185·747	186·140	186·532	186·925	187·318	187·711	188·103	59
60	188·496	188·889	189·281	189·674	190·067	190·459	190·852	191·245	60
61	191·638	192·080	192·423	192·816	193·208	193·601	193·994	194·396	61
62	194·779	195·172	195·565	195·957	196·350	196·743	197·135	197·528	62
63	197·921	198·313	198·706	199·099	199·492	199·884	200·277	200·670	63
64	201·062	201·455	201·848	202·240	202·633	203·026	203·419	203·811	64
65	204·204	204·597	204·989	205·382	205·775	206·167	206·560	206·953	65
66	207·346	207·738	208·131	208·524	208·916	209·309	209·702	210·094	66

Table CXXIV.—Circumferences of Circles—continued.

Diameters	Circumferences							
	·0	·⅛	·¼	·⅜	·½	·⅝	·¾	·⅞
67	210·487	210·880	211·273	211·665	212·058	212·451	212·843	213·236
68	213·629	214·021	214·414	214·807	215·200	215·592	215·985	216·378
69	216·770	217·163	217·556	217·948	218·341	218·734	219·127	219·519
70	219·912	220·305	220·697	221·090	221·483	221·875	222·268	222·661
71	223·054	223·446	223·839	224·232	224·624	225·017	225·410	225·802
72	226·195	226·588	226·981	227·373	227·766	228·158	228·551	228·944
73	229·337	229·729	230·122	230·515	230·908	231·300	231·693	232·086
74	232·478	232·871	233·264	233·656	234·049	234·442	234·835	235·227
75	235·620	236·013	236·405	236·798	237·191	237·583	237·976	238·369
76	238·762	239·154	239·547	239·940	240·332	240·725	241·118	241·510
77	241·903	242·296	242·689	243·081	243·474	243·867	244·259	244·652
78	245·045	245·437	245·830	246·223	246·616	247·008	247·401	247·794
79	248·186	248·579	248·972	249·364	249·757	250·150	250·543	250·935
80	251·328	251·721	252·113	252·506	252·899	253·291	253·684	254·077
81	254·470	254·862	255·255	255·648	256·040	256·433	256·826	257·218
82	257·611	258·004	258·397	258·789	259·182	259·575	259·967	260·360
83	260·753	261·145	261·538	261·981	262·324	262·716	263·109	263·502
84	263·894	264·287	264·680	265·072	265·465	265·858	266·251	266·643
85	267·036	267·429	267·821	268·214	268·607	268·999	269·392	269·785
86	270·178	270·570	270·963	271·356	271·748	272·141	272·534	272·926
87	273·319	273·712	274·105	274·497	274·890	275·283	275·675	276·068
88	276·461	276·853	277·246	277·629	278·082	278·424	278·817	279·210

TABLE CXXIV.—CIRCUMFERENCES OF CIRCLES. 369

Table CXXIV.—Circumferences of Circles—*continued.*

Diameters		Circumferences								Diameters
	·⅞	·¾	·⅝	·½	·⅜	·¼	·⅛	·0		
89	282·351	281·969	281·566	281·173	280·780	280·388	279·995	279·602		89
90	285·493	285·100	284·707	284·315	283·922	283·529	283·137	282·744		90
91	288·634	288·242	287·849	287·456	287·064	286·671	286·278	285·886		91
92	291·776	291·383	290·991	290·598	290·205	289·813	289·420	289·027		92
93	294·918	294·525	294·132	293·740	293·347	292·954	292·562	292·169		93
94	298·059	297·667	297·274	296·881	296·488	296·096	295·703	295·310		94
95	301·201	300·808	300·415	300·023	299·630	299·237	298·845	298·452		95
96	304·342	303·950	303·557	303·164	302·772	302·379	301·986	301·594		96
97	307·484	307·091	306·699	306·306	305·913	305·521	305·128	304·735		97
98	310·626	310·233	309·840	309·448	309·055	308·662	308·270	307·877		98
99	313·767	313·375	312·982	312·589	312·196	311·804	311·411	311·018		99
100	316·909	316·516	316·123	315·731	315·338	314·945	314·553	314·160		100
101	320·050	319·658	319·265	318·872	318·480	318·087	317·694	317·302		101
102	323·192	322·799	322·407	322·014	321·621	321·229	320·836	320·443		102
103	326·334	325·941	325·548	325·156	324·763	324·370	323·977	323·585		103
104	329·475	329·083	328·690	328·297	327·904	327·512	327·119	326·726		104
105	...	332·224	...	331·439	...	330·653	...	329·868		105
106	...	335·366	...	334·580	...	333·795	...	333·010		106
107	...	338·507	...	337·722	...	336·937	...	336·151		107
108	...	341·649	...	340·864	...	340·078	...	339·298		108
109	...	344·791	...	344·005	...	343·220	...	342·434		109
110	...	347·932	...	347·147	...	346·361	...	345·576		110

24

Table CXXIV.—Circumferences of Circles—*continued.*

Diameters	0	⅛	¼	⅜	½	⅝	¾	⅞	Diameters
111	348·718	...	349·503	...	350·288	...	351·074	...	111
112	351·859	...	352·645	...	353·430	...	354·215	...	112
113	355·001	...	355·786	...	356·572	...	357·357	...	113
114	358·142	...	358·928	...	359·713	...	360·499	...	114
115	361·284	...	362·069	...	362·855	...	363·640	...	115
116	364·426	...	365·211	...	365·996	...	366·782	...	116
117	367·567	...	368·353	...	369·188	...	369·923	...	117
118	370·709	...	371·494	...	372·280	...	373·065	...	118
119	373·850	...	374·636	...	375·421	...	376·207	...	119
120	376·992	...	377·777	...	378·563	...	379·348	...	120
121	380·134	...	380·919	...	381·704	...	382·490	...	121
122	383·275	...	384·061	...	384·846	...	385·631	...	122
123	386·417	...	387·202	...	387·988	...	388·773	...	123
124	389·558	...	390·344	...	391·129	...	391·915	...	124
125	392·700	...	...	...	394·271	...	...	...	125
126	395·842	...	...	...	397·412	...	...	...	126
127	398·983	...	...	...	400·554	...	...	...	127
128	402·125	...	...	...	403·696	...	...	...	128
129	405·266	...	...	...	406·837	...	...	...	129
130	408·408	...	...	...	409·979	...	...	...	130
131	411·550	...	...	...	413·120	...	...	...	131
132	414·691	...	...	...	416·262	...	...	...	132

TABLE CXXIV.—CIRCUMFERENCES OF CIRCLES. 371

Table CXXIV.—Circumferences of Circles—*continued.*

Diameters	Circumferences							
	·0	·⅛	·¼	·⅜	·½	·⅝	·¾	·⅞
133	417·833				419·404			
134	420·974				422·545			
135	424·116				425·687			
136	427·258				428·828			
137	430·399				431·970			
138	433·541				435·112			
139	436·682				438·253			
140	439·824				441·395			
141	442·966				444·536			
142	446·107				447·678			
143	449·249				450·820			
144	452·390				453·961			
145	455·532				457·103			
146	458·674				460·244			
147	461·815				463·386			
148	464·957				466·528			
149	468·098				469·669			
150	471·240				472·811			

Table CXXV.—Areas and Circumferences of Small Circles.

Diameter.	Area.	Circum.	Diameter.	Area.	Circum.
1/32	·000767	·09817	17/32	·22166	1·6690
1/16	·003068	·19635	9/16	·24850	1·7671
3/32	·006903	·29452	19/32	·27688	1·8653
1/8	·012272	·39270	5/8	·30680	1·9635
5/32	·019175	·49087	21/32	·33824	2·0617
3/16	·027612	·58905	11/16	·37122	2·1598
7/32	·037583	·68722	23/32	·40574	2·2580
1/4	·049087	·78540	3/4	·44179	2·3562
9/32	·062126	·88357	25/32	·47937	2·4544
5/16	·076699	·98175	13/16	·51849	2·5525
11/32	092806	1·0799	27/32	·55914	2·6507
3/8	·11045	1·1781	7/8	·60132	2·7489
13/32	·12962	1·2763	29/32	·64504	2·8471
7/16	·15033	1·3744	15/16	·69029	2·9452
15/32	·17257	1·4726	31/32	·73708	3·0434
1/2	·19635	1·5708	1	·78540	3·1416

Table CXXVI.—Squares, Cubes, Square Roots, Cube Roots, and Reciprocals of all Integer Numbers from 1 to 2200.

No.	Square	Cube	Square Root	Cube Root	Reciprocal
1	1	1	1·0000000	1·0000000	1·000000000
2	4	8	1·4142136	1·2599210	·500000000
3	9	27	1·7320508	1·4422496	·333333333
4	16	64	2·0000000	1·5874011	·250000000
5	25	125	2·2360680	1·7099759	·200000000
6	36	216	2·4494897	1·8171206	·166666667
7	49	343	2·6457513	1·9129312	·142857143
8	64	512	2·8284271	2·0000000	·125000000
9	81	729	3·0000000	2·0800837	·111111111
10	100	1000	3·1622777	2·1544347	·100000000
11	121	1331	3·3166248	2·2239801	·090909091
12	144	1728	3·4641016	2·2894286	·083333333
13	169	2197	3·6055513	2·3513347	·076923077
14	196	2744	3·7416574	2·4101422	·071428571
15	225	3375	3·8729833	2·4662121	·066666667
16	256	4096	4·0000000	2·5198421	·062500000
17	289	4913	4·1231056	2·5712816	·058823529
18	324	5832	4·2426407	2·6207414	·055555556
19	361	6859	4·3588989	2·6684016	·052631579
20	400	8000	4·4721360	2·7144177	·050000000
21	441	9261	4·5825757	2·7589243	·047619048
22	484	10648	4·6904158	2·8020393	·045454545
23	529	12167	4·7958315	2·8438670	·043478261
24	576	13824	4·8989795	2·8844991	·041666667
25	625	15625	5·0000000	2·9240177	·040000000
26	676	17576	5·0990195	2·9624960	·038461538
27	729	19683	5·1961524	3·0000000	·037037037
28	784	21952	5·2915026	3·0365889	·035714286
29	841	24389	5·3851648	3·0723168	·034482759
30	900	27000	5·4772256	3·1072325	·033333333
31	961	29791	5·5677644	3·1413806	·032258065
32	1024	32768	5·6568542	3·1748021	·031250000
33	1089	35937	5·7445626	3·2075343	·030303030
34	1156	39304	5·8309519	3·2396118	·029411765
35	1225	42875	5·9160798	3·2710663	·028571429
36	1296	46656	6·0000000	3·3019272	·027777778
37	1369	50653	6·0827625	3·3322218	·027027027
38	1444	54872	6·1644140	3·3619754	·026315789
39	1521	59319	6·2449980	3·3912114	·025641026
40	1600	64000	6·3245553	3·4199519	·025000000
41	1681	68921	6·4031242	3·4482172	·024390244
42	1764	74088	6·4807407	3·4760266	·023809524
43	1849	79507	6·5574385	3·5033981	·023255814
44	1936	85184	6·6332496	3·5303483	·022727273
45	2025	91125	6·7082039	3·5568933	·022222222

No.	Square	Cube	Square Root	Cube Root	Reciprocal
46	2116	97336	6·7823300	3·5830479	·021739130
47	2209	103823	6·8556546	3·6088261	·021276600
48	2304	110592	6·9282032	3·6342411	·020833333
49	2401	117649	7·0000000	3·6593057	·020408163
50	2500	125000	7·0710678	3·6840314	·020000000
51	2601	132651	7·1414284	3·7084298	·019607843
52	2704	140608	7·2111026	3·7325111	·019230769
53	2809	148877	7·2801099	3·7562858	·018867925
54	2916	157464	7·3484692	3·7797631	·018518519
55	3025	166375	7·4161985	3·8029525	·018181818
56	3136	175616	7·4833148	3·8258624	·017857143
57	3249	185193	7·5498344	3·8485011	·017543860
58	3364	195112	7·6157731	3·8708766	·017241379
59	3481	205379	7·6811457	3·8929965	·016949153
60	3600	216000	7·7459667	3·9148676	·016666667
61	3721	226981	7·810249/	3·9364972	·016393443
62	3844	238328	7·8740079	3·9578915	·016129032
63	3969	250047	7·9372539	3·9790571	·015873016
64	4096	262144	8·0000000	4·0000000	·015625000
65	4225	274625	8·0622577	4·0207256	·015384615
66	4356	287496	8·1240884	4·0412401	·015151515
67	4489	300763	8·1853528	4·0615480	·014925373
68	4624	314432	8·2462113	4·0816551	·014705882
69	4761	328509	8·3066239	4·1015661	·014492754
70	4900	343000	8·3666003	4·1212853	·014285714
71	5041	357911	8·4261498	4·1408178	·014084507
72	5184	373248	8·4852814	4·1601676	·013888889
73	5329	389017	8·5440037	4·1793392	·013698630
74	5476	405224	8·6023253	4·1983364	·013513514
75	5625	421875	8·6602540	4·2171633	·013333333
76	5776	438976	8·7177979	4·2358236	·013157895
77	5929	456533	8·7749644	4·2543210	·012987013
78	6084	474552	8·8317609	4·2726586	·012820513
79	6241	493039	8·8881944	4·2908404	·012658228
80	6400	512000	8·9442719	4·3088695	·012500000
81	6561	531441	9·0000000	4·3267487	·012345679
82	6724	551368	9·0553851	4·3444815	·012195122
83	6889	571787	9·1104336	4·3620707	·012048193
84	7056	592704	9·1651514	4·3795191	·011904762
85	7225	614125	9·2195445	4·3968296	·011764706
86	7396	636056	9·2736185	4·4140049	·011627907
87	7569	658503	9·3273791	4·4310476	·011494253
88	7744	681472	9·3808315	4·4479602	·011363636
89	7921	704969	9·4339811	4·4647451	·011235955
90	8100	729000	9·4868330	4·4814047	·011111111
91	8281	753571	9·5393920	4·4979414	·010989011
92	8464	778688	9·5916630	4·5143574	·010869565
93	8649	804357	9·6436508	4·5306549	·010752688
94	8836	830584	9·6953597	4·5468359	·010638298

No.	Square	Cube	Square Root	Cube Root	Reciprocal
95	9025	857375	9·7467943	4·5629026	·010526316
96	9216	884736	9·7979590	4·5788570	·010416667
97	9409	912673	9·8488578	4·5947009	·010309278
98	9604	941192	9·8994949	4·6104363	·010204082
99	9801	970299	9·9498744	4·6260650	·010101010
100	10000	1000000	10·0000000	4·6415888	·010000000
101	10201	1030301	10·0498756	4·6570095	·009900990
102	10404	1061208	10·0995049	4·6723287	·009803922
103	10609	1092727	10·1488916	4·6875482	·009708738
104	10816	1124864	10·1980390	4·7026694	·009615385
105	11025	1157625	10·2469508	4·7176940	·009523810
106	11236	1191016	10·2956301	4·7326235	·009433962
107	11449	1225043	10·3440804	4·7474594	·009345794
108	11664	1259712	10·3923048	4·7622032	·009259259
109	11881	1295029	10·4403065	4·7768562	·009174312
110	12100	1331000	10·4880885	4·7914199	·009090909
111	12321	1367631	10·5356538	4·8058955	·009009009
112	12544	1404928	10·5830052	4·8202845	·008928571
113	12769	1442897	10·6301458	4·8345881	·008849558
114	12996	1481544	10·6770783	4·8488076	·008771930
115	13225	1520875	10·7238053	4·8629442	·008695652
116	13456	1560896	10·7703296	4·8769990	·008620690
117	13689	1601613	10·8166538	4·8909732	·008547009
118	13924	1643032	10·8627805	4·9048681	·008474576
119	14161	1685159	10·9087121	4·9186847	·008403361
120	14400	1728000	10·9544512	4·9324242	·008333333
121	14641	1771561	11·0000000	4·9460874	·008264463
122	14884	1815848	11·0453610	4·9596757	·008196721
123	15129	1860867	11·0905365	4·9731898	·008130081
124	15376	1906624	11·1355287	4·9866310	·008064516
125	15625	1953125	11·1803399	5·0000000	·008000000
126	15876	2000376	11·2249722	5·0132979	·007936508
127	16129	2048383	11·2694277	5·0265257	·007874016
128	16384	2097152	11·3137085	5·0396842	·007812500
129	16641	2146689	11·3578167	5·0527743	·007751938
130	16900	2197000	11·4017543	5·0657970	·007692308
131	17161	2248091	11·4455231	5·0787531	·007633588
132	17424	2299968	11·4891253	5·0916434	·007575759
133	17689	2352637	11·5325626	5·1044687	·007518797
134	17956	2406104	11·5758369	5·1172299	·007462687
135	18225	2460875	11·6189500	5·1299278	·007407407
136	18496	2515456	11·6619038	5·1425632	·007352941
137	18769	2571353	11·7046999	5·1551367	·007299270
138	19044	2628072	11·7473401	5·1676493	·007246377
139	19321	2685619	11·7898261	5·1801015	·007194245
140	19600	2744000	11·8321596	5·1924941	·007142857
141	19881	2803221	11·8743422	5·2048279	·007092199
142	20164	2863288	11·9163753	5·2171034	·007042254
143	20449	2924207	11·9582607	5·2293215	·006993007

No.	Square	Cube	Square Root	Cube Root	Reciprocal
144	20736	2985984	12·0000000	5·2414828	·006944444
145	21025	3048625	12·0415946	5·2535879	·006896552
146	21316	3112136	12·0830460	5·2656374	·006849315
147	21609	3176523	12·1243557	5·2776321	·006802721
148	21904	3241792	12·1655251	5·2895725	·006756757
149	22201	3307949	12·2065556	5·3014592	·006711409
150	22500	3375000	12·2474487	5·3132928	·006666667
151	22801	3442951	12·2882057	5·3250740	·006622517
152	23104	3511808	12·3288280	5·3368033	·006578947
153	23409	3581577	12·3693169	5·3484812	·006535948
154	23716	3652264	12·4096736	5·3601084	·006493506
155	24025	3723875	12·4498996	5·3716854	·006451613
156	24336	3796416	12·4899960	5·3832126	·006410256
157	24649	3869893	12·5299641	5·3946907	·006369427
158	24964	3944312	12·5698051	5·4061202	·006329114
159	25281	4019679	12·6095202	5·4175015	·006289308
160	25600	4096000	12·6491106	5·4288352	·006250000
161	25921	4173281	12·6885775	5·4401218	·006211180
162	26244	4251528	12·7279221	5·4513618	·006172840
163	26569	4330747	12·7671453	5·4625556	·006134969
164	26896	4410944	12·8062485	5·4737037	·006097561
165	27225	4492125	12·8452326	5·4848066	·006060606
166	27556	4574296	12·8840987	5·4958647	·006024096
167	27889	4657463	12·9228480	5·5068784	·005988024
168	28224	4741632	12·9614814	5·5178484	·005952381
169	28561	4826809	13·0000000	5·5287748	·005917160
170	28900	4913000	13·0384048	5·5396583	·005882353
171	29241	5000211	13·0766968	5·5504991	·005847953
172	29584	5088448	13·1148770	5·5612978	·005813953
173	29929	5177717	13·1529464	5·5720546	·005780347
174	30276	5268024	13·1909060	5·5827702	·005747126
175	30625	5359375	13·2287566	5·5934447	·005714286
176	30976	5451776	13·2664992	5·6040787	·005681818
177	31329	5545233	13·3041347	5·6146724	·005649718
178	31684	5639752	13·3416641	5·6252263	·005617978
179	32041	5735339	13·3790882	5·6357408	·005586592
180	32400	5832000	13·4164079	5·6462162	·005555556
181	32761	5929741	13·4536240	5·6566528	·005524862
182	33124	6028568	13·4907376	5·6670511	·005494505
183	33489	6128487	13·5277493	5·6774114	·005464481
184	33856	6229504	13·5646600	5·6877340	·005434783
185	34225	6331625	13·6014705	5·6980192	·005405405
186	34596	6434856	13·6381817	5·7082675	·005376344
187	34969	6539203	13·6747943	5·7184791	·005347594
188	35344	6644672	13·7113092	5·7286543	·005319149
189	35721	6751269	13·7477271	5·7387936	·005291005
190	36100	6859000	13·7840488	5·7488971	·005263158
191	36481	6967871	13·8202750	5·7589652	·005235602
192	36864	7077888	13·8564065	5·7689982	·005208333

No.	Square	Cube	Square Root	Cube Root	Reciprocal
193	37249	7189057	13·8924440	5·7789966	·005181347
194	37636	7301384	13·9283883	5·7889604	·005154639
195	38025	7414875	13·9642400	5·7988900	·005128205
196	38416	7529536	14·0000000	5·8087857	·005102041
197	38809	7645373	14·0356688	5·8186479	·005076142
198	39204	7762392	14·0712473	5·8284767	·005050505
199	39601	7880599	14·1067360	5·8382725	·005025126
200	40000	8000000	14·1421356	5·8480355	·005000000
201	40401	8120601	14·1774489	5·8577660	·004975124
202	40804	8242408	14·2126704	5·8674643	·004950495
203	41209	8365427	14·2478068	5·8771307	·004926108
204	41616	8489664	14·2828569	5·8867653	·004901961
205	42025	8615125	14·3178211	5·8963685	·004878049
206	42436	8741816	14·3527001	5·9059406	·004854369
207	42849	8869743	14·3874946	5·9154817	·004830918
208	43264	8998912	14·4222051	5·9249921	·004807692
209	43681	9129329	14·4568323	5·9344721	·004784689
210	44100	9261000	14·4913767	5·9439220	·004761905
211	44521	9393931	14·5258390	5·9533418	·004739336
212	44944	9528128	14·5602198	5·9627320	·004716981
213	45369	9663597	14·5945195	5·9720926	·004694836
214	45796	9800344	14·6287388	5·9814240	·004672897
215	46225	9938375	14·6628783	5·9907264	·004651163
216	46656	10077696	14·6969385	6·0000000	·004629630
217	47089	10218313	14·7309199	6·0092450	·004608295
218	47524	10360232	14·7648231	6·0184617	·004587156
219	47961	10503459	14·7986486	6·0276502	·004566210
220	48400	10648000	14·8323970	6·0368107	·004545455
221	48841	10793861	14·8660687	6·0459435	·004524887
222	49284	10941048	14·8996644	6·0550489	·004504505
223	49729	11089567	14·9331845	6·0641270	·004484305
224	50176	11239424	14·9666295	6·0731779	·004464286
225	50625	11390625	15·0000000	6·0822020	·004444444
226	51076	11543176	15·0332964	6·0911994	·004424779
227	51529	11697083	15·0665192	6·1001702	·004405286
228	51984	11852352	15·0996689	6·1091147	·004385965
229	52441	12008989	15·1327460	6·1180332	·004366812
230	52900	12167000	15·1657509	6·1269257	·004347826
231	53361	12326391	15·1986842	6·1357924	·004329004
232	53824	12487168	15·2315462	6·1446337	·004310345
233	54289	12649337	15·2643375	6·1534495	·004291845
234	54756	12812904	15·2970585	6·1622401	·004273504
235	55225	12977875	15·3297097	6·1710058	·004255319
236	55696	13144256	15·3622915	6·1797466	·004237288
237	56169	13312053	15·3948043	6·1884628	·004219409
238	56644	13481272	15·4272486	6·1971544	·004201681
239	57121	13651919	15·4596248	6·2058218	·004184100
240	57600	13824000	15·4919334	6·2144650	·004166667
241	58081	13997521	15·5241747	6·2230843	·004149378

No.	Square	Cube	Square Root	Cube Root	Reciprocal
242	58564	14172488	15·5563492	6·2316797	·004132231
243	59049	14348907	15·5884573	6·2402515	·004115226
244	59536	14526784	15·6204994	6·2487998	·004098361
245	60025	14706125	15·6524758	6·2573248	·004081633
246	60516	14886936	15·6843871	6·2658266	·004065041
247	61009	15069223	15·7162336	6·2743054	·004048583
248	61504	15252992	15·7480157	6·2827613	·004032258
249	62001	15438249	15·7797338	6·2911946	·004016064
250	62500	15625000	15·8113883	6·2996053	·004000000
251	63001	15813251	15·8429795	6·3079935	·003984064
252	63504	16003008	15·8745079	6·3163596	·003968254
253	64009	16194277	15·9059737	6·3247035	·003952569
254	64516	16387064	15·9373775	6·3330256	·003937008
255	65025	16581375	15·9687194	6·3413257	·003921569
256	65536	16777216	16·0000000	6·3496042	·003906250
257	66049	16974593	16·0312195	6·3578611	·003891051
258	66564	17173512	16·0623784	6·3660968	·003875969
259	67081	17373979	16·0934769	6·3743111	·003861004
260	67600	17576000	16·1245155	6·3825043	·003846154
261	68121	17779581	16·1554944	6·3906765	·003831418
262	68644	17984728	16·1864141	6·3988279	·003816794
263	69169	18191447	16·2172747	6·4069585	·003802281
264	69696	18399744	16·2480768	6·4150587	·003787879
265	70225	18609625	16·2788206	6·4231583	·003773585
266	70756	18821096	16·3095064	6·4312276	·003759398
267	71289	19034163	16·3401346	6·4392767	·003745318
268	71824	19248832	16·3707055	6·4473057	·003731343
269	72361	19465109	16·4012195	6·4553148	·003717472
270	72900	19683000	16·4316767	6·4633041	·003703704
271	73441	19902511	16·4620776	6·4712736	·003690037
272	73984	20123648	16·4924225	6·4792236	·003676471
273	74529	20346417	16·5227116	6·4871541	·003663004
274	75076	20570824	16·5529454	6·4950653	·003649635
275	75625	20796875	16·5831240	6·5029572	·003636364
276	76176	21024576	16·6132477	6·5108300	·003623188
277	76729	21253933	16·6433170	6·5186839	·003610108
278	77284	21484952	16·6733320	6·5265189	·003597122
279	77841	21717639	16·7032931	6·5343351	·003584229
280	78400	21952000	16·7332005	6·5421326	·003571429
281	78961	22188041	16·7630546	6·5499116	·003558719
282	79524	22425768	16·7928556	6·5576722	·003546099
283	80089	22665187	16·8226038	6·5654144	·003533569
284	80656	22906304	16·8522995	6·5731385	·003521127
285	81225	23149125	16·8819430	6·5808443	·003508772
286	81796	23393656	16·9115345	6·5885323	·003496503
287	82369	23639903	16·9410743	6·5962023	·003484321
288	82944	23887872	16·9705627	6·6038545	·003472222
289	83521	24137569	17·0000000	6·6114890	·003460208
290	84100	24389000	17·0293864	6·6191060	·003448276

No.	Square	Cube	Square Root	Cube Root	Reciprocal
291	84681	24642171	17·0587221	6·6267054	·003436426
292	85264	24897088	17·0880075	6·6342874	·003424658
293	85849	25153757	17·1172428	6·6418522	·003412969
294	86436	25412184	17·1464282	6·6493998	·003401361
295	87025	25672375	17·1755640	6·6569302	·003389831
296	87616	25934336	17·2046505	6·6644437	·003378378
297	88209	26198073	17·2336879	6·6719403	·003367003
298	88804	26463592	17·2626765	6·6794200	·003355705
299	89401	26730899	17·2916165	6·6868831	·003344482
300	90000	27000000	17·3205081	6·6943295	·003333333
301	90601	27270901	17·3493516	6·7017593	·003322259
302	91204	27543608	17·3781472	6·7091729	·003311258
303	91809	27818127	17·4068952	6·7165700	·003300330
304	92416	28094464	17·4355958	6·7239508	·003289474
305	93025	28372625	17·4642492	6·7313155	·003278689
306	93636	28652616	17·4928557	6·7386641	·003267974
307	94249	28934443	17·5214155	6·7459967	·003257329
308	94864	29218112	17·5499288	6·7533134	·003246753
309	95481	29503629	17·5783958	6·7606143	·003236246
310	96100	29791000	17·6068169	6·7678995	·003225806
311	96721	30080231	17·6351921	6·7751690	·003215434
312	97344	30371328	17·6635217	6·7824229	·003205128
313	97969	30664297	17·6918060	6·7896613	·003194888
314	98596	30959144	17·7200451	6·7968844	·003184713
315	99225	31255875	17·7482393	6·8040921	·003174603
316	99856	31554496	17·7763888	6·8112847	·003164557
317	100489	31855013	17·8044938	6·8184620	·003154574
318	101124	32157432	17·8325545	6·8256242	·003144654
319	101761	32461759	17·8605711	6·8327714	·003134796
320	102400	32768000	17·8885438	6·8399087	·003125000
321	103041	32076161	17·9164729	6·8470213	·003115265
322	103684	33386248	17·9443584	6·8541240	·003105590
323	104329	33698267	17·9722008	6·8612120	·003095975
324	104976	34012224	18·0000000	6·8682855	·003086420
325	105625	34328125	18·0277564	6·8753443	·003076923
326	106276	34645976	18·0554701	6·8823888	·003067485
327	106929	34965783	18·0831413	6·8894188	·003058104
328	107584	35287552	18·1107703	6·8964345	·003048780
329	108241	35611289	18·1383571	6·9034359	·003039514
330	108900	35937000	18·1659021	6·9104232	·003030303
331	109561	36264691	18·1934054	6·9173964	·003021148
332	110224	36594363	18·2208672	6·9243556	·003012048
333	110889	36926037	18·2482876	6·9313008	·003003003
334	111556	37259704	18·2756669	6·9382321	·002994012
335	112225	37595375	18·3030052	6·9451496	·002985075
336	112896	37933056	18·3303028	6·9520533	·002976190
337	113569	38272753	18·3575598	6·9589434	·002967359
338	114244	38614472	18·3847763	6·9658198	·002958580
339	114921	38958219	18·4119526	6·9726826	·002949853

No.	Square	Cube	Square Root	Cube Root	Reciprocal
340	115600	39304000	18·4390889	6·9795321	·002941176
341	116281	39651821	18·4661853	6·9863681	·002932551
342	116964	40001688	18·4932420	6·9931906	·002923977
343	117649	40353607	18·5202592	7·0000000	·002915452
344	118336	40707584	18·5472370	7·0067962	·002906977
345	119025	41063625	18·5741756	7·0135791	·002898551
346	119716	41421736	18·6010752	7·0203490	·002890173
347	120409	41781923	18·6279360	7·0271058	·002881844
348	121104	42144192	18·6547581	7·0338497	·002873563
349	121801	42508549	18·6815147	7·0405806	·002865330
350	122500	42875000	18·7082869	7·0472987	·002857143
351	123201	43243551	18·7349940	7·0540041	·002849003
352	123904	43614208	18·7616630	7·0606967	·002840909
353	124609	43986977	18·7882942	7·0673767	·002832861
354	125316	44361864	18·8148877	7·0740440	·002824859
355	126025	44738875	18·8414437	7·0806988	·002816901
356	126736	45118016	18·8679623	7·0873411	·002808989
357	127449	45499293	18·8944436	7·0939709	·002801120
358	128164	45882712	18·9208879	7·1005885	·002793296
359	128881	46268279	18·9472953	7·1071937	·002785515
360	129600	46656000	18·9736660	7·1137866	·002777778
361	130321	47045881	19·0000000	7·1203674	·002770083
362	131044	47437928	19·0262976	7·1269360	·002762431
363	131769	47832147	19·0525589	7·1334925	·002754821
364	132496	48228544	19·0787840	7·1400370	·002747253
365	133225	48627125	19·1049732	7·1465695	·002739726
366	133956	49027896	19·1311265	7·1530901	·002732240
367	134689	49430863	19·1572441	7·1595988	·002724796
368	135424	49836032	19·1833261	7·1660957	·002717391
369	136161	50243409	19·2093727	7·1725809	·002710027
370	136900	50653000	19·2353841	7·1790544	·002702703
371	137641	51064811	19·2613603	7·1855162	·002695418
372	138384	51478848	19·2873015	7·1919663	·002688172
373	139129	51895117	19·3132079	7·1984050	·002680965
374	139876	52313624	19·3390796	7·2048322	·002673797
375	140625	52734375	19·3649167	7·2112479	·002666667
376	141376	53157376	19·3907194	7·2176522	·002659574
377	142129	53582633	19·4164878	7·2240450	·002652520
378	142884	54010152	19·4422221	7·2304268	·002645503
379	143641	54439939	19·4679223	7·2367972	·002638522
380	144400	54872000	19·4935887	7·2431565	·002631579
381	145161	55306341	19·5192213	7·2495045	·002624672
382	145924	55742968	19·5448203	7·2558415	·002617801
383	146689	56181887	19·5703858	7·2621675	·002610966
384	147456	56623104	19·5959179	7·2684824	·C02604167
385	148225	57066625	19·6214169	7·2747864	·002597403
386	148996	57512456	19·6468827	7·2810794	·002590674
387	149769	57960603	19·6723156	7·2873617	·002583979
388	150544	58411072	19·6977156	7·2936330	·002577320

No.	Square	Cube	Square Root	Cube Root	Reciprocal
389	151321	58863869	19·7230829	7·2998936	·002570694
390	152100	59319000	19·7484177	7·3061436	·002564103
391	152881	59776471	19·7737199	7·3123828	·002557545
392	153664	60236288	19·7989899	7·3186114	·002551020
393	154449	60698457	19·8242276	7·3248295	·002544529
394	155236	61162984	19·8494332	7·3310369	·002538071
395	156025	61629875	19·8746069	7·3372339	·002531646
396	156816	62099136	19·8997487	7·3434205	·002525253
397	157609	62570773	19·9248588	7·3495966	·002518892
398	158404	63044792	19·9499373	7·3557624	·002512563
399	159201	63521199	19·9749844	7·3619178	·002506266
400	160000	64000000	20·0000000	7·3680630	·002500000
401	160801	64481201	20·0249844	7·3741979	·002493766
402	161604	64964808	20·0499377	7·3803227	·002487562
403	162409	65450827	20·0748599	7·3864373	·002481390
404	163216	65939264	20·0997512	7·3925418	·002475248
405	164025	66430125	20·1246118	7·3986363	·002469136
406	164836	66923416	20·1494417	7·4047206	·002463054
407	165649	67419143	20·1742410	7·4107950	·002457002
408	166464	67917312	20·1990099	7·4168595	·002450980
409	167281	68417929	20·2237484	7·4229142	·002444988
410	168100	68921000	20·2484567	7·4289589	·002439024
411	168921	69426531	20·2731349	7·4349938	·002433090
412	169744	69934528	20·2977831	7·4410189	·002427184
413	170569	70444997	20·3224014	7·4470342	·002421308
414	171396	70957944	20·3469899	7·4530399	·002415459
415	172225	71473375	20·3715488	7·4590359	·002409639
416	173056	71991296	20·3960781	7·4650223	·002403846
417	173889	72511713	20·4205779	7·4709991	·002398082
418	174724	73034632	20·4450483	7·4769664	·002392344
419	175561	73560059	20·4694895	7·4829242	·002386635
420	176400	74088000	20·4939015	7·4888724	·002380952
421	177241	74618461	20·5182845	7·4948113	·002375297
422	178084	75151448	20·5426386	7·5007406	·002369668
423	178929	75686967	20·5669638	7·5066607	·002364066
424	179776	76225024	20·5912608	7·5125715	·002358491
425	180625	76765625	20·6155281	7·5184730	·002352941
426	181476	77308776	20·6397674	7·5243652	·002347418
427	182329	77854483	20·6639783	7·5302482	·002341920
428	183184	78402752	20·6881609	7·5361221	·002336449
429	184041	78952589	20·7123152	7·5419867	·002331002
430	184900	79507000	20·7364414	7·5478423	·002325581
431	185761	80062991	20·7605395	7·5536888	·002320186
432	186624	80621568	20·7846097	7·5595263	·002314815
433	187489	81182737	20·8086520	7·5653548	·002309469
434	188356	81746504	20·8326667	7·5711743	·002304147
435	189225	82312875	20·8566536	7·5769849	·002298851
436	190096	82881856	20·8806130	7·5827865	·002293578
437	190969	83453453	20·9045450	7·5885793	·002288330

No.	Square	Cube	Square Root	Cube Root	Reciprocal
438	191844	84027672	20·9284495	7·5943633	·002283105
439	192721	84604519	20·9523268	7·6001385	·002277004
440	193600	85184000	20·9761770	7·6059049	·002272727
441	194481	85766121	21·0000000	7·6116626	·002267574
442	195364	86350888	21·0237960	7·6174116	·002262443
443	196249	86938307	21·0475652	7·6231519	·002257336
444	197136	87528384	21·0713075	7·6288837	·002252252
445	198025	88121125	21·0950231	7·6346067	·002247191
446	198916	88716536	21·1187121	7·6403213	·002242152
447	199809	89314623	21·1423745	7·6460272	·002237136
448	200704	89915392	21·1660105	7·6517247	·002232143
449	201601	90518849	21·1896201	7·6574138	·002227171
450	202500	91125000	21·2132034	7·6630943	·002222222
451	203401	91733851	21·2367606	7·6687665	·002217295
452	204304	92345408	21·2602916	7·6744303	·002212389
453	205209	92959677	21·2837967	7·6800857	·002207506
454	206116	93576664	21·3072758	7·6857329	·002202643
455	207025	94196375	21·3307290	7·6913717	·002197802
456	207936	94818816	21·3541565	7·6970023	·002192982
457	208849	95443993	21·3775583	7·7026246	·002188184
458	209764	96071912	21·4009346	7·7082388	·002183406
459	210681	96702579	21·4242853	7·7138448	·002178649
460	211600	97336000	21·4476106	7·7194426	·002173913
461	212521	97972181	21·4709106	7·7250325	·002169197
462	213444	98611128	21·4941853	7·7306141	·002164502
463	214369	99252847	21·5174348	7·7361877	·002159827
464	215296	99897344	21·5406592	7·7417532	·002155172
465	216225	100544625	21·5638587	7·7473109	·002150538
466	217156	101194696	21·5870331	7·7528606	·002145923
467	218089	101847563	21·6101828	7·7584023	·002141328
468	219024	102503232	21·6333077	7·7639361	·002136752
469	219961	103161709	21·6564078	7·7694620	·002132196
470	220900	103823000	21·6794834	7·7749801	·002127660
471	221841	104487111	21·7025344	7·7804904	·002123142
472	222784	105154048	21·7255610	7·7859928	·002118644
473	223729	105823817	21·7485632	7·7914875	·002114165
474	224676	106496424	21·7715411	7·7969745	·002109705
475	225625	107171875	21·7944947	7·8024538	·002105263
476	226576	107850176	21·8174242	7·8079254	·002100840
477	227529	108531333	·21·8403297	7·8133892	·002096436
478	228484	109215352	21·8632111	7·8188456	·002092050
479	229441	109902239	21·8860686	7·8242942	·002087683
480	230400	110592000	21·9089023	7·8297353	·002083333
481	231361	111284641	21·9317122	7·8351688	·002079002
482	232324	111980168	21·9544984	7·8405949	·002074689
483	233289	112678587	21·9772610	7·8460134	·002070393
484	234256	113379904	22·0000000	7·8514244	·002066116
485	235225	114084125	22·0227155	7·8568281	·002061856
486	236196	114791256	22·0454077	7·8622242	·002057613

No.	Square	Cube	Square Root	Cube Root	Reciprocal
487	237169	115501303	22·0680765	7·8676130	·002053388
488	238144	116214272	22·0907220	7·8729944	·002049180
489	239121	116930169	22·1133444	7·8783684	·002044990
490	240100	117649000	22·1359436	7·8837352	·002040816
491	241081	118370771	22·1585198	7·8890946	·002036660
492	242064	119095488	22·1810730	7·8944468	·002032520
493	243049	119823157	22·2036033	7·8997917	·002028398
494	244036	120553784	22·2261108	7·9051294	·002024291
495	245025	121287375	22·2485955	7·9104599	·002020202
496	246016	122023936	22·2710575	7·9157832	·002016129
497	247009	122763473	22·2934968	7·9210994	·002012072
498	248004	123505992	22·3159136	7·9264085	·002008082
499	249001	124251499	22·3383079	7·9317104	·002004008
500	250000	125000000	22·3606798	7·9370053	·002000000
501	251001	125751501	22·3830293	7·9422931	·001996008
502	252004	126506008	22·4053565	7·9475739	·001992032
503	253009	127263527	22·4276615	7·9528477	·001988072
504	254016	128024064	22·4499443	7·9581144	·001984127
505	255025	128787625	22·4722051	7·9633743	·001980198
506	256036	129554216	22·4944438	7·9686271	·001976285
507	257049	130323843	22·5166605	7·9738731	·001972387
508	258064	131096512	22·5388553	7·9791122	·001968504
509	259081	131872229	22·5610283	7·9843444	·001964637
510	260100	132651000	22·5831796	7·9895697	·001960784
511	261121	133432831	22·6053091	7·9947883	·001956947
512	262144	134217728	22·6274170	8·0000000	·001953125
513	263169	135005697	22·6495033	8·0052049	·001949318
514	264196	135796744	22·6715681	8·0104032	·001945525
515	265225	136590875	22·6936114	8·0155946	·001941748
516	266256	137388096	22·7156334	8·0207794	·001937984
517	267289	138188413	22·7376340	8·0259574	·001934236
518	268324	138991832	22·7596134	8·0311287	·001930502
519	269361	139798359	22·7815715	8·0362935	·001926782
520	270400	140608000	22·8035085	8·0414515	·001923077
521	271441	141420761	22·8254244	8·0466030	·001919386
522	272484	142236648	22·8473193	8·0517479	·001915709
523	273529	143055667	22·8691933	8·0568862	·001912046
524	274576	143877824	22·8910463	8·0620180	·001908397
525	275625	144703125	22·9128785	8·0671432	·001904762
526	276676	145531576	22·9346899	8·0722620	·001901141
527	277729	146363183	22·9564806	8·0773743	·001897533
528	278784	147197952	22·9782506	8·0824800	·001893939
529	279841	148035889	23·0000000	8·0875794	·001890359
530	280900	148877000	23·0217289	8·0926723	·001886792
531	281961	149721291	23·0434372	8·0977589	·001883239
532	283024	150568768	23·0651252	8·1028390	·001879699
533	284089	151419437	23·0867928	8·1079128	·001876173
534	285156	152273304	23·1084400	8·1129803	·001872659
535	286225	153130375	23·1300670	8·1180414	·001869159

No.	Square	Cube	Square Root	Cube Root	Reciprocal
536	287296	153990656	23·1516738	8·1230962	·001865672
537	288369	154854153	23·1732605	8·1281447	·001862197
538	289444	155720872	23·1948270	8·1331870	·001858736
539	290521	156590819	23·2163735	8·1382230	·001855288
540	291600	157464000	23·2379001	8·1432529	·001851852
541	292681	158340421	23·2594067	8·1482765	·001848429
542	293764	159220088	23·2808985	8·1532939	·001845018
543	294849	160103007	23·3023604	8·1583051	·001841621
544	295936	160989184	23·3238076	8·1633102	·001838235
545	297025	161878625	23·3452351	8·1683092	·001834862
546	298116	162771336	23·3666429	8·1733020	·001831502
547	299209	163667323	23·3880311	8·1782888	·001828154
548	300304	164566592	23·4093998	8·1832695	·001824818
549	301401	165469149	23·4307490	8·1882441	·001821494
550	302500	166375000	23·4520788	8·1932127	·001818182
551	303601	167284151	23·4733892	8·1981753	·001814882
552	304704	168196608	23·4946802	8·2031319	·001811594
553	305809	169112377	23·5159520	8·2080825	·001808318
554	306916	170031464	23·5372046	8·2130271	·001805054
555	308025	170953875	23·5584380	8·2179657	·001801802
556	309136	171879616	23·5796522	8·2228985	·001798561
557	310249	172808693	23·6008474	8·2278254	·001795332
558	311364	173741112	23·6220236	8·2327463	·001792115
559	312481	174676879	23·6431808	8·2376614	·001788909
560	313600	175616000	23·6643191	8·2425706	·001785714
561	314721	176558481	23·6854386	8·2474740	·001782531
562	315844	177504328	23·7065392	8·2523715	·001779359
563	316969	178453547	23·7276210	8·2572633	·001776199
564	318096	179406144	23·7486842	8·2621492	·001773050
565	319225	180362125	23·7697286	8·2670294	·001769912
566	320356	181321496	23·7907545	8·2719039	·001766784
567	321489	182284263	23·8117618	8·2767726	·001763668
568	322624	183250432	23·8327506	8·2816355	·001760563
569	323761	184220009	23·8537209	8·2864928	·001757469
570	324900	185193000	23·8746728	8·2913444	·001754386
571	326041	186169411	23·8956063	8·2961903	·001751313
572	327184	187149248	23·9165215	8·3010304	·001748252
573	328329	188132517	23·9374184	8·3058651	·001745201
574	329476	189119224	23·9582971	8·3106941	·001742160
575	330625	190109875	23·9791576	8·3155175	·001739130
576	331776	191102976	24·0000000	8·3203353	·001736111
577	332929	192100033	24·0208243	8·3251475	·001733102
578	334084	193100552	24·0416306	8·3299542	·001730104
579	335241	194104539	24·0624188	8·3347553	·001727116
580	336400	195112000	24·0831891	8·3395509	·001724138
581	337561	196122941	24·1039416	8·3443410	·001721170
582	338724	197137368	24·1246762	8·3491256	·001718213
583	339889	198155287	24·1453929	8·3539047	
584	341056	199176704			

No.	Square	Cube	Square Root	Cube Root	Reciprocal
977	954529	932574833	31·2569992	9·9227879	·001023541
978	956484	935441352	31·2729915	9·9261222	·001022495
979	958441	938313739	31·2889757	9·9295042	·001021450
980	960400	941192000	31·3049517	9·9328839	·001020408
981	962361	944076141	31·3209195	9·9362613	·001019368
982	964324	946966168	31·3368792	9·9396363	·001018330
983	966289	949862087	31·3528308	9·9430092	·001017294
984	968256	952763904	31·3687743	9·9463797	·001016260
985	970225	955671625	31·3847097	9·9497479	·001015228
986	972196	958585256	31·4006369	9·9531138	·001014199
987	974169	961504803	31·4165561	9·9564775	·001013171
988	976144	964430272	31·4324673	9·9598389	·001012146
989	978121	967361669	31·4483704	9·9631981	·001011122
990	980100	970299000	31·4642654	9·9665549	·001010101
991	982081	973242271	31·4801525	9·9699095	·001009082
992	984064	976191488	31·4960315	9·9732619	·001008065
993	986049	979146657	31·5119025	9·9766120	·001007049
994	988036	982107784	31·5277655	9·9799599	·001006036
995	990025	985074875	31·5436206	9·9833055	·001005025
996	992016	988047936	31·5594677	9·9866488	·001004016
997	994009	991026973	31·5753068	9·9899900	·001003009
998	996004	994011992	31·5911380	9·9933289	·001002004
999	998001	997002999	31·6069613	9·9966656	·001001001
1000	1000000	1000000000	31·6227766	10·0000000	·0010000000
1001	1002001	1003003001	31·6385840	10·0033322	·0009990010
1002	1004004	1006012008	31·6543836	10·0066622	·0009980040
1003	1006009	1009027027	31·6701752	10·0099899	·0009970090
1004	1008016	1012048064	31·6859590	10·0133155	·0009960159
1005	1010025	1015075125	31·7017349	10·0166389	·0009950249
1006	1012036	1018108216	31·7175030	10·0199601	·0009940358
1007	1014049	1021147343	31·7332633	10·0232791	·0009930487
1008	1016064	1024192512	31·7490157	10·0265958	·0009920635
1009	1018081	1027243729	31·7647603	10·0299104	·0009910803
1010	1020100	1030301000	31·7804972	10·0332228	·0009900990
1011	1022121	1033364331	31·7962262	10·0365330	·0009891197
1012	1024144	1036433728	31·8119474	10·0398410	·0009881423
1013	1026169	1039509197	31·8276609	10·0431469	·0009871668
1014	1028196	1042590744	31·8433666	10·0464506	·0009861933
1015	1030225	1045678375	31·8590646	10·0497521	·0009852217
1016	1032256	1048772096	31·8747549	10·0530514	·0009842520
1017	1034289	1051871913	31·8904374	10·0563485	·0009832842
1018	1036324	1054977832	31·9061123	10·0596435	·0009823183
1019	1038361	1058089859	31·9217794	10·0629364	·0009813543
1020	1040400	1061208000	31·9374388	10·0662271	·0009803922
1021	1042441	1064332261	31·9530906	10·0695156	·0009794319
1022	1044484	1067462648	31·9687347	10·0728020	·0009784736
1023	1046529	1070599167	31·9843712	10·0760863	·0009775171
1024	1048576	1073741824	32·0000000	10·0793684	·0009765625
1025	1050625	1076890625	32·0156212	10·0826484	·0009756098

No.	Square	Cube	Square Root	Cube Root	Reciprocal
634	401956	254840104	25·1793566	8·5907238	·001577287
635	403225	256047875	25·1992063	8·5952380	·001574803
636	404496	257259456	25·2190404	8·5997476	·001572327
637	405769	258474853	25·2388589	8·6042525	·001569859
638	407044	259694072	25·2586619	8·6087526	·001567398
639	408321	260917119	25·2784493	8·6132480	·001564945
640	409600	262144000	25·2982213	8·6177388	·001562500
641	410881	263374721	25·3179778	8·6222248	·001560062
642	412164	264609288	25·3377189	8·6267063	·001557632
643	413449	265847707	25·3574447	8·6311830	·001555210
644	414736	267089984	25·3771551	8·6356551	·001552795
645	416025	268336125	25·3968502	8·6401226	·001550388
646	417316	269586136	25·4165301	8·6445855	·001547988
647	418609	270840023	25·4361947	8·6490437	·001545595
648	419904	272097792	25·4558441	8·6534974	·001543210
649	421201	273359449	25·4754784	8·6579465	·001540832
650	422500	274625000	25·4950976	8·6623911	·001538462
651	423801	275894451	25·5147016	8·6668310	·001536098
652	425104	277167808	25·5342907	8·6712665	·001533742
653	426409	278445077	25·5538647	8·6756974	·001531394
654	427716	279726264	25·5734237	8·6801237	·001529052
655	429025	281011375	25·5929678	8·6845456	·001526718
656	430336	282300416	25·6124969	8·6889630	·001524390
657	431649	283593393	25·6320112	8·6933759	·001522070
658	432964	284890312	25·6515107	8·6977843	·001519757
659	434281	286191179	25·6709953	8·7021882	·001517451
660	435600	287496000	25·6904652	8·7065877	·001515152
661	436921	288804781	25·7099203	8·7109827	·001512859
662	438244	290117528	25·7293607	8·7153734	·001510574
663	439569	291434247	25·7487864	8·7197596	·001508296
664	440896	292754944	25·7681975	8·7241414	·001506024
665	442225	294079625	25·7875939	8·7285187	·001503759
666	443556	295408296	25·8069758	8·7328918	·001501502
667	444889	296740963	25·8263431	8·7372604	·001499250
668	446224	298077632	25·8456960	8·7416246	·001497006
669	447561	299418309	25·8650343	8·7459846	·001494768
670	448900	300763000	25·8843582	8·7503401	·001492537
671	450241	302111711	25·9036677	8·7546913	·001490313
672	451584	303464448	25·9229628	8·7590383	·001488095
673	452929	304821217	25·9422435	8·7633809	·001485884
674	454276	306182024	25·9615100	8·7677192	·001483680
675	455625	307546875	25·9807621	8·7720532	·001481481
676	456976	308915776	26·0000000	8·7763830	·001479290
677	458329	310288733	26·0192237	8·7807084	·001477105
678	459684	311665752	26·0384331	8·7850296	·001474926
679	461041	313046839	26·0576284	8·7893466	·001472754
680	462400	314432000	26·0768096	8·7936593	·001470588
681	463761	315821241	26·0959767	8·7979679	·001468429
682	465124	317214568	26·1151297	8·8022721	·001466276

No.	Square	Cube	Square Root	Cube Root	Reciprocal
683	466489	318611987	26·1342687	8·8065722	·001464129
684	467856	320013504	26·1533937	8·8108681	·001461988
685	469225	321419125	26·1725047	8·8151598	·001459854
686	470596	322828856	26·1916017	8·8194474	·001457726
687	471969	324242703	26·2106848	8·8237307	·001455604
688	473344	325660672	26·2297541	8·8280099	·001453488
689	474721	327082769	26·2488095	8·8322850	·001451379
690	476100	328509000	26·2678511	8·8365559	·001449275
691	477481	329939371	26·2868789	8·8408227	·001447178
692	478864	331373888	26·3058929	8·8450854	·001445087
693	480249	332812557	26·3248932	8·8493440	·001443001
694	481636	334255384	26·3438797	8·8535985	·001440922
695	483025	335702375	26·3628527	8·8578489	·001438849
696	484416	337153536	26·3818119	8·8620952	·001436782
697	485809	338608873	26·4007576	8·8663375	·001434720
698	487204	340068392	26·4196896	8·8705757	·001432665
699	488601	341532099	26·4386081	8·8748099	·001430615
700	490000	343000000	26·4575131	8·8790400	·001428571
701	491401	344472101	26·4764046	8·8832661	·001426534
702	492804	345948408	26·4952826	8·8874882	·001424501
703	494209	347428927	26·5141472	8·8917063	·001422475
704	495616	348913664	26·5329983	8·8959204	·001420455
705	497025	350402625	26·5518361	8·9001804	·001418440
706	498436	351895816	26·5706605	8·9043366	·001416431
707	499849	353393243	26·5894716	8·9085387	·001414427
708	501264	354894912	26·6082694	8·9127369	·001412429
709	502681	356400829	26·6270539	8·9169311	·001410437
710	504100	357911000	26·6458252	8·9211214	·001408451
711	505521	359425431	26·6645833	8·9253078	·001406470
712	506944	360944128	26·6833281	8·9294902	·001404494
713	508369	362467097	26·7020598	8·9336687	·001402525
714	509796	363994344	26·7207784	8·9378433	·001400560
715	511225	365525875	26·7394839	8·9420140	·001398601
716	512656	367061696	26·7581763	8·9461809	·001396648
717	514089	368601813	26·7768557	8·9503438	·001394700
718	515524	370146232	26·7955220	8·9545029	·001392758
719	516961	371694959	26·8141754	8·9586581	·001390821
720	518400	373248000	26·8328157	8·9628095	·001388889
721	519841	374805361	26·8514432	8·9669570	·001386963
722	521284	376367048	26·8700577	8·9711007	·001385042
723	522729	377933067	26·8886593	8·9752406	·001383126
724	524176	379503424	26·9072481	8·9793766	·001381215
725	525625	381078125	26·9258240	8·9835089	·001379310
726	527076	382657176	26·9443872	8·9876373	·001377410
727	528529	384240583	26·9629375	8·9917620	·001375516
728	529984	385828352	26·9814751	8·9958829	·001373626
729	531441	387420489	27·0000000	9·0000000	·001371742
730	532900	389017000	27·0185122	9·0041134	·001369863
731	534361	390617891	27·0370117	9·0082229	·001367989

No.	Square	Cube	Square Root	Cube Root	Reciprocal
732	535824	392223168	27·0554985	9·0123238	·001366120
733	537289	393832837	27·0739727	9·0164309	·001364256
734	538756	395446904	27·0924344	9·0205293	·001362398
735	540225	397065375	27·1108834	9·0246239	·001360544
736	541696	398688256	27·1293199	9·0287149	·001358696
737	543169	400315553	27·1477439	9·0328021	·001356852
738	544644	401947272	27·1661554	9·0368857	·001355014
739	546121	403583419	27·1845544	9·0409655	·001353180
740	547600	405224000	27·2029410	9·0450417	·001351351
741	549081	406869021	27·2213152	9·0491142	·001349528
742	550564	408518488	27·2396769	9·0531831	·001347709
743	552049	410172407	27·2580263	9·0572482	·001345895
744	553536	411830784	27·2763634	9·0613098	·001344086
745	555025	413493625	27·2946881	9·0653677	·001342282
746	556516	415160936	27·3130006	9·0694220	·001340483
747	558009	416832723	27·3313007	9·0734726	·001338688
748	559504	418508992	27·3495887	9·0775197	·001336898
749	561001	420189749	27·3678644	9·0815631	·001335113
750	562500	421875000	27·3861279	9·0856030	·001333333
751	564001	423564751	27·4043792	9·0896392	·001331558
752	565504	425259008	27·4226184	9·0936719	·001329787
753	567009	426957777	27·4408455	9·0977010	·001328021
754	568516	428661064	27·4590604	9·1017265	·001326260
755	570025	430368875	27·4772633	9·1057485	·001324503
756	571536	432081216	27·4954542	9·1097669	·001322751
757	573049	433798093	27·5136330	9·1137818	·001321004
758	574564	435519512	27·5317998	9·1177931	·001319261
759	576081	437245479	27·5499546	9·1218010	·001317523
760	577600	438976000	27·5680975	9·1258053	·001315789
761	579121	440711081	27·5862284	9·1298061	·001314060
762	580644	442450728	27·6043475	9·1338034	·001312336
763	582169	444194947	27·6224546	9·1377971	·001310616
764	583696	445943744	27·6405499	9·1417874	·001308901
765	585225	447697125	27·6586334	9·1457742	·001307190
766	586756	449455096	27·6767050	9·1497576	·001305483
767	588289	451217663	27·6947648	9·1537375	·001303781
768	589824	452984832	27·7128129	9·1577139	·001302083
769	591361	454756609	27·7308492	9·1616869	·001300390
770	592900	456533000	27·7488739	9·1656565	·001298701
771	594441	458314011	27·7668868	9·1696225	·001297017
772	595984	460099648	27·7848880	9·1735852	·001295337
773	597529	461889917	27·8028775	9·1775445	·001293661
774	599076	463684824	27·8208555	9·1815003	·001291990
775	600625	465484375	27·8388218	9·1854527	·001290323
776	602176	467288576	27·8567766	9·1894018	·001288660
777	603729	469097433	27·8747197	9·1933474	·001287001
778	605284	470910952	27·8926514	9·1972897	·001285347
779	606841	472729139	27·9105715	9·2012286	·001283697
780	608400	474552000	27·9284801	9·2051641	·001282051

No.	Square	Cube	Square Root	Cube Root	Reciprocal
781	609961	476379541	27·9463772	9·2090962	·001280410
782	611524	478211768	27·9642629	9·2130250	·001278772
783	613089	480048687	27·9821872	9·2169505	·001277139
784	614656	481890304	28·0000000	9·2208726	·001275510
785	616225	483736625	28·0178515	9·2247914	·001273885
786	617796	485587656	28·0356915	9·2287068	·001272265
787	619369	487443403	28·0535203	9·2326189	·001270648
788	620944	489303872	28·0713377	9·2365277	·001269036
789	622521	491169069	28·0891438	9·2404333	·001267427
790	624100	493039000	28·1069386	9·2443355	·001265823
791	625681	494913671	28·1247222	9·2482344	·001264223
792	627264	496793088	28·1424946	9·2521300	·001262626
793	628849	498677257	28·1602557	9·2560224	·001261084
794	630436	500566184	28·1780056	9·2599114	·001259446
795	632025	502459875	28·1957444	9·2637973	·001257862
796	633616	504358336	28·2134720	9·2676798	·001256281
797	635209	506261573	28·2311884	9·2715592	·001254705
798	636804	508169592	28·2488938	9·2754352	·001253133
799	638401	510082399	28·2665881	9·2793081	·001251564
800	640000	512000000	28·2842712	9·2831777	·001250000
801	641601	513922401	28·3019434	9·2870440	·001248439
802	643204	515849608	28·3196045	9·2909072	·001246883
803	644809	517781627	28·3372546	9·2947671	·001245330
804	646416	519718464	28·3548938	9·2986239	·001243781
805	648025	521660125	28·3725219	9·3024775	·001242236
806	649636	523606616	28·3901391	9·3063278	·001240695
807	651249	525557943	28·4077454	9·3101750	·001239157
808	652864	527514112	28·4253408	9·3140190	·001237624
809	654481	529475129	28·4429253	9·3178599	·001236094
810	656100	531441000	28·4604989	9·3216975	·001234568
811	657721	533411731	28·4780617	9·3255320	·001233046
812	659344	535387328	28·4956137	9·3293634	·001231527
813	660969	537367797	28·5131549	9·3331916	·001230012
814	662596	539353144	28·5306852	9·3370167	·001228501
815	664225	541343375	28·5482048	9·3408386	·001226994
816	665856	543338496	28·5657137	9·3446575	·001225490
817	667489	545338513	28·5832119	9·3484731	·001223990
818	669124	547343432	28·6006993	9·3522857	·001222494
819	670761	549353259	28·6181760	9·3560952	·001221001
820	672400	551368000	28·6356421	9·3599016	·001219512
821	674041	553387661	28·6530976	9·3637049	·001218027
822	675684	555412248	28·6705424	9·3675051	·001216545
823	677329	557441767	28·6879766	9·3713022	·001215067
824	678976	559476224	28·7054002	9·3750963	·001213592
825	680625	561515625	28·7228132	9·3788873	·001212121
826	682276	563559976	28·7402157	9·3826752	·001210654
827	683929	565609283	28·7576077	9·3864600	·001209190
828	685584	567663552	28·7749891	9·3902419	·001207729
829	687241	569722789	28·7923601	9·3940206	·001206272

No.	Square	Cube	Square Root	Cube Root	Reciprocal
830	688900	571787000	28·8097206	9·3977964	·001204819
831	690561	573856191	28·8270706	9·4015691	·001203369
832	692224	575930368	28·8444102	9·4053387	·001201923
833	693889	578009537	28·8617394	9·4091054	·001200480
834	695556	580093704	28·8790582	9·4128690	·001199041
835	697225	582182875	28·8963666	9·4166297	·001197605
836	698896	584277056	28·9136646	9·4203873	·001196172
837	700569	586376253	28·9309523	9·4241420	·001194743
838	702244	588480472	28·9482297	9·4278936	·001193317
839	703921	590589719	28·9654967	9·4316423	·001191895
840	705600	592704000	28·9827535	9·4353880	·001190476
841	707281	594823321	29·0000000	9·4391307	·001189061
842	708964	596947688	29·0172363	9·4428704	·001187648
843	710649	599077107	29·0344623	9·4466072	·001186240
844	712336	601211584	29·0516781	9·4503410	·001184834
845	714025	603351125	29·0688837	9·4540719	·001183432
846	715716	605495736	29·0860791	9·4577999	·001182033
847	717409	607645423	29·1032644	9·4615249	·001180638
848	719104	609800192	29·1204396	9·4652470	·001179245
849	720801	611960049	29·1376046	9·4689661	·001177856
850	722500	614125000	29·1547595	9·4726824	·001176471
851	724201	616295051	29·1719043	9·4763957	·001175088
852	725904	618470208	29·1890390	9·4801061	·001173709
853	727609	620650477	29·2061637	9·4838136	·001172333
854	729316	622835864	29·2232784	9·4875182	·001170960
855	731025	625026375	29·2403830	9·4912200	·001169591
856	732736	627222016	29·2574777	9·4949188	·001168224
857	734449	629422793	29·2745623	9·4986147	·001166861
858	736164	631628712	29·2916370	9·5023078	·001165501
859	737881	633839779	29·3087018	9·5059980	·001164144
860	739600	636056000	29·3257566	9·5096854	·001162791
861	741321	638277381	29·3428015	9·5133699	·001161440
862	743044	640503928	29·3598365	9·5170515	·001160093
863	744769	642735647	29·3768616	9·5207303	·001158749
864	746496	644972544	29·3938769	9·5244063	·001157407
865	748225	647214625	29·4108823	9·5280794	·001156069
866	749956	649461896	29·4278779	9·5317497	·001154734
867	751689	651714363	29·4448637	9·5354172	·001153403
868	753424	653972032	29·4618397	9·5390818	·001152074
869	755161	656234909	29·4788059	9·5427437	·001150748
870	756900	658503000	29·4957624	9·5464027	·001149425
871	758641	660776311	29·5127091	9·5500589	·001148106
872	760384	663054848	29·5296461	9·5537123	·001146789
873	762129	665338617	29·5465734	9·5573630	·001145475
874	763876	667627624	29·5634910	9·5610108	·001144165
875	765625	669921875	29·5803989	9·5646559	·001142857
876	767376	672221376	29·5972972	9·5682982	·001141553
877	769129	674526133	29·6141858	9·5719377	·001140251
878	770884	676836152	29·6310648	9·5755745	·001138952

No.	Square	Cube	Square Root	Cube Root	Reciprocal
879	772641	679151439	29·6479342	9·5792085	·001137656
880	774400	681472000	29·6647939	9·5828397	·001136364
881	776161	683797841	29·6816442	9·5864682	·001135074
882	777924	686128968	29·6984848	9·5900939	·001133787
883	779689	688465387	29·7153159	9·5937169	·001132503
884	781456	690807104	29·7321375	9·5973373	·001131222
885	783225	693154125	29·7489496	9·6009548	·001129944
886	784996	695506456	29·7657521	9·6045696	·001128668
887	786769	697864103	29·7825452	9·6081817	·001127396
888	788544	700227072	29·7993289	9·6117911	·001126126
889	790321	702595369	29·8161030	9·6153977	·001124859
890	792100	704969000	29·8328678	9·6190017	·001123596
891	793881	707347971	29·8496231	9·6226030	·001122334
892	795664	709732288	29·8663690	9·6262016	·001121076
893	797449	712121957	29·8831056	9·6297975	·001119821
894	799236	714516984	29·8998328	9·6333907	·001118568
895	801025	716917375	29·9165506	9·6369812	·001117318
896	802816	719323136	29·9332591	9·6405690	·001116071
897	804609	721734273	29·9499583	9·6441542	·001114827
898	806404	724150792	29·9666481	9·6477367	·001113586
899	808201	726572699	29·9833287	9·6513166	·001112347
900	810000	729000000	30·0000000	9·6548938	·001111111
901	811801	731432701	30·0166620	9·6584684	·001109878
902	813604	733870808	30·0333148	9·6620403	·001108647
903	815409	736314327	30·0499584	9·6656096	·001107420
904	817216	738763264	30·0665928	9·6691762	·001106195
905	819025	741217625	30·0832179	9·6727403	·001104972
906	820836	743677416	30·0998339	9·6763017	·001103753
907	822649	746142643	30·1164407	9·6798604	·001102536
908	824464	748613312	30·1330383	9·6834166	·001101322
909	826281	751089429	30·1496269	9·6869701	·001100110
910	828100	753571000	30·1662063	9·6905211	·001098901
911	829921	756058031	30·1827765	9·6940694	·001097695
912	831744	758550528	30·1993377	9·6976151	·001096491
913	833569	761048497	30·2158899	9·7011583	·001095290
914	835396	763551944	30·2324329	9·7046989	·001094092
915	837225	766060875	30·2489669	9·7082369	·001092896
916	839056	768575296	30·2654919	9·7117723	·001091703
917	840889	771095213	30·2820079	9·7153051	·001090513
918	842724	773620632	30·2985148	9·7188354	·001089325
919	844561	776151559	30·3150128	9·7223631	·001088139
920	846400	778688000	30·3315018	9·7258883	·001086957
921	848241	781229961	30·3479818	9·7294109	·001085776
922	850084	783777448	30·3644529	9·7329309	·001084599
923	851929	786330467	30·3809151	9·7364484	·001083424
924	853776	788889024	30·3973683	9·7399634	·001082251
925	855625	791453125	30·4138127	9·7434758	·001081081
926	857476	794022776	30·4302481	9·7469857	·001079914
927	859329	796597983	30·4466747	9·7504930	·001078749

No.	Square	Cube	Square Root	Cube Root	Reciprocal
928	861184	799178752	30·4630924	9·7539979	·001077586
929	863041	801765089	30·4795013	9·7575002	·001076426
930	864900	804357000	30·4959014	9·7610001	·001075269
931	866761	806954491	30·5122926	9·7644974	·001074114
932	868624	809557568	30·5286750	9·7679922	·001072961
933	870489	812166237	30·5450487	9·7714845	·001071811
934	872356	814780504	30·5614136	9·7749743	·001070664
935	874225	817400375	30·5777697	9·7784616	·001069519
936	876096	820025856	30·5941171	9·7819466	·001068376
937	877969	822656953	30·6104557	9·7854288	·001067236
938	879844	825293672	30·6267857	9·7889087	·001066098
939	881721	827936019	30·6431069	9·7923861	·001064963
940	883600	830584000	30·6594194	9·7958611	·001063830
941	885481	833237621	30·6757233	9·7993336	·001062699
942	887364	835896888	30·6920185	9·8028036	·001061571
943	889249	888561807	30·7083051	9·8062711	·001060445
944	891136	841232384	30·7245830	9·8097362	·001059322
945	893025	843908625	30·7408523	9·8131989	·001058201
946	894916	846590536	30·7571130	9·8166591	·001057082
947	896809	849278123	30·7733651	9·8201169	·001055966
948	898704	851971392	30·7896086	9·8235723	·001054852
949	900601	854670349	30·8058436	9·8270252	·001053741
950	902500	857375000	30·8220700	9·8304757	·001052632
951	904401	860085351	30·8382879	9·8339238	·001051525
952	906304	862801408	30·8544972	9·8373695	·001050420
953	908209	865523177	30·8706981	9·8408127	·001049318
954	910116	868250664	30·8868904	9·8442536	·001048218
955	912025	870983875	30·9030743	9·8476920	·001047120
956	913936	873722816	30·9192497	9·8511280	·001046025
957	915849	876467493	30·9354166	9·8545617	·001044932
958	917764	879217912	30·9515751	9·8579929	·001043841
959	919681	881974079	30·9677251	9·8614218	·001042753
960	921600	884736000	30·9838668	9·8648483	·001041667
961	923521	887503681	31·0000000	9·8682724	·001040583
962	925444	890277128	31·0161248	9·8716941	·001039501
963	927369	893056347	31·0322413	9·8751135	·001038422
964	929296	895841344	31·0483494	9·8785305	·001037344
965	931225	898632125	31·0644491	9·8819451	·001036269
966	933156	901428696	31·0805405	9·8853574	·001035197
967	935089	904231063	31·0966236	9·8887673	·001034126
968	937024	907039232	31·1126984	9·8921749	·001033058
969	938961	909853209	31·1287648	9·8955801	·001031992
970	940900	912673000	31·1448230	9·8989830	·001030928
971	942841	915498611	31·1608729	9·9023835	·001029866
972	944784	918330048	31·1769145	9·9057817	·001028807
973	946729	921167317	31·1929479	9·9091776	·001027749
974	948676	924010424	31·2089731	9·9125712	·001026694
975	950625	926859375	31·2249900	9·9159624	·001025641
976	952576	929714176	31·2409987	9·9193513	·001024590

No.	Square	Cube	Square Root	Cube Root	Reciprocal
977	954529	932574833	31·2569992	9·9227879	·001023541
978	956484	935441352	31·2729915	9·9261222	·001022495
979	958441	938313739	31·2889757	9·9295042	·001021450
980	960400	941192000	31·3049517	9·9328839	·001020408
981	962361	944076141	31·3209195	9·9362613	·001019368
982	964324	946966168	31·3368792	9·9396363	·001018330
983	966289	949862087	31·3528308	9·9430092	·001017294
984	968256	952763904	31·3687743	9·9463797	·001016260
985	970225	955671625	31·3847097	9·9497479	·001015228
986	972196	958585256	31·4006369	9·9531138	·001014199
987	974169	961504803	31·4165561	9·9564775	·001013171
988	976144	964430272	31·4324673	9·9598389	·001012146
989	978121	967361669	31·4483704	9·9631981	·001011122
990	980100	970299000	31·4642654	9·9665549	·001010101
991	982081	973242271	31·4801525	9·9699095	·001009082
992	984064	976191488	31·4960315	9·9732619	·001008065
993	986049	979146657	31·5119025	9·9766120	·001007049
994	988036	982107784	31·5277655	9·9799599	·001006036
995	990025	985074875	31·5436206	9·9833055	·001005025
996	992016	988047936	31·5594677	9·9866488	·001004016
997	994009	991026973	31·5753068	9·9899900	·001003009
998	996004	994011992	31·5911380	9·9933289	·001002004
999	998001	997002999	31·6069613	9·9966656	·001001001
1000	1000000	1000000000	31·6227766	10·0000000	·0010000000
1001	1002001	1003003001	31·6385840	10·0033322	·0009990010
1002	1004004	1006012008	31·6543836	10·0066622	·0009980040
1003	1006009	1009027027	31·6701752	10·0099899	·0009970090
1004	1008016	1012048064	31·6859590	10·0133155	·0009960159
1005	1010025	1015075125	31·7017349	10·0166389	·0009950249
1006	1012036	1018108216	31·7175030	10·0199601	·0009940358
1007	1014049	1021127343	31·7332633	10·0232791	·0009930487
1008	1016064	1024192512	31·7490157	10·0265958	·0009920635
1009	1018081	1027243729	31·7647603	10·0299104	·0009910803
1010	1020100	1030301000	31·7804972	10·0332228	·0009900990
1011	1022121	1033365331	31·7962262	10·0365330	·0009891197
1012	1024144	1036433728	31·8119474	10·0398410	·0009881423
1013	1026169	1039509197	31·8276609	10·0431469	·0009871668
1014	1028196	1042590744	31·8433666	10·0464506	·0009861933
1015	1030225	1045678375	31·8590646	10·0497521	·0009852217
1016	1032256	1048772096	31·8747549	10·0530514	·0009842520
1017	1034289	1051871913	31·8904374	10·0563485	·0009832842
1018	1036324	1054977832	31·9061123	10·0596435	·0009823183
1019	1038361	1058089859	31·9217794	10·0629364	·0009813543
1020	1040400	1061208000	31·9374388	10·0662271	·0009803922
1021	1042441	1064332261	31·9530906	10·0695156	·0009794319
1022	1044484	1067462648	31·9687347	10·0728020	·0009784736
1023	1046529	1070599167	31·9843712	10·0760863	·0009775171
1024	1048576	1073741824	32·0000000	10·0793684	·0009765625
1025	1050625	1076890625	32·0156212	10·0826484	·0009756098

No.	Square	Cube	Square Root	Cube Root	Reciprocal
1026	1052676	1080045576	32·0312348	10·0859262	·0009746589
1027	1054729	1083206683	32·0468407	10·0892019	·0009737098
1028	1056784	1086373952	32·0624391	10·0924755	·0009727626
1029	1058841	1089547389	32·0780298	10·0957469	·0009718173
1030	1060900	1092727000	32·0936131	10·0990163	·0009708738
1031	1062961	1095912791	32·1091887	10·1022835	·0009699321
1032	1065024	1099104768	32·1247568	10·1055487	·0009689922
1033	1067089	1102302937	32·1403173	10·1088117	·0009680542
1034	1069156	1105507304	32·1558704	10·1120726	·0009671180
1035	1071225	1108717875	32·1714159	10·1153314	·0009661836
1036	1073296	1111934656	32·1869539	10·1185882	·0009652510
1037	1075369	1115157653	32·2024844	10·1218428	·0009643202
1038	1077444	1118386872	32·2180074	10·1250953	·0009633911
1039	1079521	1121622319	32·2335229	10·1283457	·0009624639
1040	1081600	1124864000	32·2490310	10·1315941	·0009615385
1041	1083681	1128111921	32·2645316	10·1348403	·0009606148
1042	1085764	1131366088	32·2800248	10·1380845	·0009596929
1043	1087849	1134626507	32·2955105	10·1413266	·0009587728
1044	1089936	1137893184	32·3109888	10·1445667	·0009578544
1045	1092025	1141166125	32·3264598	10·1478047	·0009569378
1046	1094116	1144445336	32·3419238	10·1510406	·0009560229
1047	1096209	1147730823	32·3573794	10·1542744	·0009551098
1048	1098304	1151022592	32·3728281	10·1575062	·0009541985
1049	1100401	1154320649	32·3882695	10·1607359	·0009532888
1050	1102500	1157625000	32·4087035	10·1639636	·0009523810
1051	1104601	1160935651	32·4191301	10·1671893	·0009514748
1052	1106704	1164252608	32·4345495	10·1704129	·0009505703
1053	1108809	1167575877	32·4499615	10·1736344	·0009496676
1054	1110916	1170905464	32·4653662	10·1768539	·0009487666
1055	1113025	1174241375	32·4807635	10·1800714	·0009478673
1056	1115136	1177583616	32·4961536	10·1832868	·0009469697
1057	1117249	1180932193	32·5115364	10·1865002	·0009460738
1058	1119364	1184287112	32·5269119	10·1897116	·0009451796
1059	1121481	1187648379	32·5422802	10·1929209	·0009442871
1060	1123600	1191016000	32·5576412	10·1961283	·0009433962
1061	1125721	1194389981	32·5729949	10·1993336	·0009425071
1062	1127844	1197770328	32·5883415	10·2025369	·0009416196
1063	1129969	1201157047	32·6036807	10·2057382	·0009407338
1064	1132096	1204550144	32·6190129	10·2089375	·0009398496
1065	1134225	1207949625	32·6343377	10·2121347	·0009389671
1066	1136356	1211355496	32·6496554	10·2153300	·0009380863
1067	1138489	1214767763	32·6649659	10·2185233	·0009372071
1068	1140624	1218186432	32·6802693	10·2217146	·0009363296
1069	1142761	1221611509	32·6955654	10·2249039	·0009354537
1070	1144900	1225043000	32·7108544	10·2280912	·0009345794
1071	1147041	1228480911	32·7261363	10·2312766	·0009337063
1072	1149184	1231925248	32·7414111	10·2344599	·0000328358
1073	1151329	1235376017	32·7566787	10·2376413	·0009319664
1074	1153476	1238833224	32·7719392	10·2408207	·0009310987

No.	Square	Cube	Square Root	Cube Root	Reciprocal
1075	1155625	1242296875	32·7871926	10·2439981	·0009302326
1076	1157776	1245766976	32·8024389	10·2471735	·0009293680
1077	1159929	1249243533	32·8176782	10·2503470	·0009285051
1078	1162084	1252726552	32·8329103	10·2535186	·0009276438
1079	1164241	1256216039	32·8481354	10·2566881	·0009267841
1080	1166400	1259712000	32·8633535	10·2598557	·0009259259
1081	1168561	1263214441	32·8785644	10·2630213	·0009250694
1082	1170724	1266723368	32·8937684	10·2661850	·0009242144
1083	1172889	1270238787	32·9089653	10·2693467	·0009233610
1084	1175056	1273760704	32·9241553	10·2725065	·0009225092
1085	1177225	1277289125	32·9393382	10·2756644	·0009216590
1086	1179396	1280824056	32·9545141	10·2788203	·0009208103
1087	1181569	1284365503	32·9696830	10·2819743	·0009199632
1088	1183744	1287913472	32·9848450	10·2851264	·0009191176
1089	1185921	1291467969	33·0000000	10·2882765	·0009182736
1090	1188100	1295029000	33·0151480	10·2914247	·0009174312
1091	1190281	1298596571	33·0302891	10·2945709	·0009165903
1092	1192464	1302170688	33·0454233	10·2977158	·0009157509
1093	1194649	1305751357	33·0605505	10·3008577	·0009149131
1094	1196836	1309338584	33·0756708	10·3039982	·0009140768
1095	1199025	1312932375	33·0907842	10·3071368	·0009132420
1096	1201216	1316532736	33·1058907	10·3102735	·0009124088
1097	1203409	1320139673	33·1209903	10·3134088	·0009115770
1098	1205604	1323753192	33·1360830	10·3165411	·0009107468
1099	1207801	1327373299	33·1511689	10·3196721	·0009099181
1100	1210000	1331000000	33·1662479	10·3228012	·0009090909
1101	1212201	1334633301	33·1813200	10·3259284	·0009082652
1102	1214404	1338273208	33·1963853	10·3290537	·0009074410
1103	1216609	1341919727	33·2114438	10·3321770	·0009066183
1104	1218816	1345572864	33·2264955	10·3352985	·0009057971
1105	1221025	1349232625	33·2415403	10·3384181	·0009049774
1106	1223236	1352899016	33·2565783	10·3415358	·0009041591
1107	1225449	1356572043	33·2716095	10·3446517	·0009033424
1108	1227664	1360251712	33·2866339	10·3477657	·0009025271
1109	1229881	1363938029	33·3016516	10·3508778	·0009017133
1110	1232100	1367631000	33·3166625	10·3539880	·0009009009
1111	1234321	1371330631	33·3316666	10·3570964	·0009000900
1112	1236544	1375036928	33·3466640	10·3602029	·0008992806
1113	1238769	1378749897	33·3616546	10·3633076	·0008984726
1114	1240996	1382469544	33·3766385	10·3664103	·0008976661
1115	1243225	1386195875	33·3916157	10·3695113	·0008968610
1116	1245456	1389928896	33·4065862	10·3726103	·0008960573
1117	1247689	1393668613	33·4215499	10·3757076	·0008952551
1118	1249924	1397415032	33·4365070	10·3788030	·0008944544
1119	1252161	1401168159	33·4514573	10·3818965	·0008936550
1120	1254400	1404928000	33·4664011	10·3849882	·0008928571
1121	1256641	1408694561	33·4813381	10·3880781	·0008920607
1122	1258884	1412467848	33·4962684	10·3911661	·0008912656
1123	1261129	1416247867	33·5111921	10·3942523	·0008904720

No.	Square	Cube	Square Root	Cube Root	Reciprocal
1124	1263876	1420034624	33·5261092	10·3973366	·0008896797
1125	1265625	1423828125	33·5410196	10·4004192	·0008888889
1126	1267876	1427628376	33·5559234	10·4034999	·0008880995
1127	1270129	1431435383	33·5708206	10·4065787	·0008873114
1128	1272384	1435249152	33·5857112	10·4096557	·0008865248
1129	1274641	1439069689	33·6005952	10·4127310	·0008857396
1130	1276900	1442897000	33·6154726	10·4158044	·0008849558
1131	1279161	1446731091	33·6303434	10·4188760	·0008841733
1132	1281424	1450571968	33·6452077	10·4219458	·0008833922
1133	1283689	1454419637	33·6600653	10·4250138	·0008826125
1134	1285956	1458274104	33·6749165	10·4280800	·0008818342
1135	1288225	1462135375	33·6897610	10·4311443	·0008810573
1136	1290496	1466003456	33·7045991	10·4342069	·0008802817
1137	1292769	1469878353	33·7194306	10·4372677	·0008795075
1138	1295044	1473760072	33·7342556	10·4403267	·0008787346
1139	1297321	1477648619	33·7490741	10·4433839	·0008779631
1140	1299600	1481544000	33·7638860	10·4464393	·0008771930
1141	1301881	1485446221	33·7786915	10·4494929	·0008764242
1142	1304164	1489355288	33·7934905	10·4525448	·0008756567
1143	1306449	1493271207	33·8082830	10·4555948	·0008748906
1144	1308736	1497193984	33·8230691	10·4586431	·0008741259
1145	1311025	1501123625	33·8378486	10·4616896	·0008733624
1146	1313316	1505080136	33·8526218	10·4647343	·0008726003
1147	1315609	1509003523	33·8673884	10·4677773	·0008718396
1148	1317904	1512953792	33·8821487	10·4708185	·0008710801
1149	1320201	1516910949	33·8969025	10·4738579	·0008703220
1150	1322500	1520875000	33·9116499	10·4768955	·0008695652
1151	1324801	1524845951	33·9263909	10·4799314	·0008688097
1152	1327104	1528823808	33·9411255	10·4829656	·0008680556
1153	1329409	1532808577	33·9558537	10·4859980	·0008673027
1154	1331716	1536800264	33·9705755	10·4890286	·0008665511
1155	1334025	1540798875	33·9852910	10·4920575	·0008658009
1156	1336336	1544804416	34·0000000	10·4950847	·0008650519
1157	1338649	1548816893	34·0147027	10·4981101	·0008643042
1158	1340964	1552836312	34·0293990	10·5011337	·0008635579
1159	1343281	1556862679	34·0440890	10·5041556	·0008628128
1160	1345600	1560896000	34·0587727	10·5071757	·0008620690
1161	1347921	1564936281	34·0734501	10·5101942	·0008613264
1162	1350244	1568983528	34·0881211	10·5132109	·0008605852
1163	1352569	1573037747	34·1027858	10·5162259	·0008598452
1164	1354896	1577098944	34·1174442	10·5192391	·0008591065
1165	1357225	1581167125	34·1320963	10·5222506	·0008583691
1166	1359556	1585242296	34·1467422	10·5252604	·0008576329
1167	1361889	1589324463	34·1613817	10·5282685	·0008568980
1168	1364224	1593413632	34·1760150	10·5312749	·0008561644
1169	1366561	1597509809	34·1906420	10·5342795	·0008554320
1170	1368900	1601613000	34·2052627	10·5372825	·0008547009
1171	1371241	1605723211	34·2198773	10·5402837	·0008539710
1172	1373584	1609840448	34·2344855	10·5432832	·0008532423

No.	Square	Cube	Square Root	Cube Root	Reciprocal
1173	1375929	1613964717	34·2490875	10·5462810	·0008525149
1174	1378276	1618096024	34·2636834	10·5492771	·0008517888
1175	1380625	1622234375	34·2782730	10·5522715	·0008510638
1176	1382976	1626379776	34·2928564	10·5552642	·0008503401
1177	1385329	1630532233	34·3074336	10·5582552	·0008496177
1178	1387684	1634691752	34·3220046	10·5612445	·0008488964
1179	1390041	1638858339	34·3365694	10·5642322	·0008481764
1180	1392400	1643032000	34·3511281	10·5672181	·0008474576
1181	1394761	1647212741	34·3656805	10·5702024	·0008467401
1182	1397124	1651400568	34·3802268	10·5731849	·0008460237
1183	1399489	1655595487	34·3947670	10·5761658	·0008453085
1184	1401856	1659797504	34·4093011	10·5791449	·0008445946
1185	1404225	1664006625	34·4238289	10·5821225	·0008438819
1186	1406596	1668222856	34·4383507	10·5850983	·0008431703
1187	1408969	1672446203	34·4528663	10·5880725	·0008424600
1188	1411344	1676676672	34·4673759	10·5910450	·0008417508
1189	1413721	1680914269	34·4818793	10·5940158	·0008410429
1190	1416100	1685159000	34·4963766	10·5969850	·0008403361
1191	1418481	1689410871	34·5108678	10·5999525	·0008396306
1192	1420864	1693669888	34·5253530	10·6029184	·0008389262
1193	1423249	1697936057	34·5398321	10·6058826	·0008382230
1194	1425636	1702209384	34·5543051	10·6088451	·0008375209
1195	1428025	1706489875	34·5687720	10·6118060	·0008368201
1196	1430416	1710777536	34·5832329	10·6147652	·0008361204
1197	1432809	1715072373	34·5976879	10·6177228	·0008354219
1198	1435204	1719374392	34·6121366	10·6206788	·0008347245
1199	1437601	1723683599	34·6265794	10·6236331	·0008340284
1200	1440000	1728000000	34·6410162	10·6265857	·0008333333
1201	1442401	1732323601	34·6554469	10·6295367	·0008326395
1202	1444804	1736654408	34·6698716	10·6324860	·0008319468
1203	1447209	1740992427	34·6842904	10·6354338	·0008312552
1204	1449616	1745337664	34·6987031	10·6383799	·0008305648
205	1452025	1749690125	34·7131099	10·6413244	·0008298755
206	1454436	1754049816	34·7275107	10·6442672	·0008291874
207	1456849	1758416743	34·7419055	10·6472085	·0008285004
208	1459264	1762790912	34·7562944	10·6501480	·0008278146
209	1461681	1767172329	34·7706773	10·6530860	·0008271299
210	1464100	1771561000	34·7850543	10·6560223	·0008264463
211	1466521	1775956931	34·7994253	10·6589570	·0008257638
212	1468944	1780360128	34·8137904	10·6618902	·0008250825
213	1471369	1784770597	34·8281495	10·6648217	·0008244023
214	1473796	1789188344	34·8425028	10·6677516	·0008237232
215	1476225	1793613375	34·8568501	10·6706799	·0008230453
216	1478656	1798045696	34·8711915	10·6736066	·0008223684
217	1481089	1802485313	34·8855271	10·6765317	·0008216927
218	1483524	1806932232	34·8998567	10·6794552	·0008210181
219	1485961	1811386459	34·9141805	10·6823771	·0008203445
220	1488400	1815848000	34·9284984	10·6852973	·0008196721
221	1490841	1820316861	34·9428104	10·6882160	·0008190008

No.	Square	Cube	Square Root	Cube Root	Reciprocal
1222	1493284	1824793048	34·9571166	10·6911331	·0008183306
1223	1495729	1829276567	34·9714169	10·6940486	·0008176615
1224	1498176	1833767424	34·9857114	10·6969625	·0008169935
1225	1500625	1838265625	35·0000000	10·6998748	·0008163265
1226	1503076	1842771176	35·0142828	10·7027855	·0008156607
1227	1505529	1847234083	35·0285598	10·7056947	·0008149959
1228	1507984	1851804352	35·0428309	10·7086023	·0008143322
1229	1510441	1856331989	35·0570963	10·7115083	·0008136696
1230	1512900	1860867000	35·0713558	10·7144127	·0008130081
1231	1515361	1865409391	35·0856096	10·7173155	·0008123477
1232	1517824	1869959168	35·0998575	10·7202168	·0008116883
1233	1520289	1874516337	35·1140997	10·7231165	·0008110300
1234	1522756	1879080904	35·1283361	10·7260146	·0008103728
1235	1525225	1883652875	35·1425668	10·7289112	·0008097166
1236	1527696	1888232256	35·1567917	10·7318062	·0008090615
1237	1530169	1892819053	35·1710108	10·7346997	·0008084074
1238	1532644	1897413272	35·1852242	10·7375916	·0008077544
1239	1535121	1902014919	35·1994318	10·7404819	·0008071025
1240	1537600	1906624000	35·2136337	10·7433707	·0008064516
1241	1540081	1911240521	35·2278299	10·7462579	·0008058018
1242	1542564	1915864488	35·2420204	10·7491436	·0008051530
1243	1545049	1920495907	35·2562051	10·7520277	·0008045052
1244	1547536	1925134784	35·2703842	10·8549103	·0008038585
1245	1550025	1929781125	35·2845575	10·7577913	·0008032129
1246	1552516	1934434936	35·2987252	10·7606708	·0008025682
1247	1555009	1939096223	35·3128872	10·7635488	·0008019246
1248	1557504	1943764992	35·3270435	10·7664252	·0008012821
1249	1560001	1948441249	35·3411941	10·7693001	·0008006405
1250	1562500	1953125000	35·3553391	10·7721735	·0008000000
1251	1565001	1957816251	35·3694784	10·7750453	·0007993605
1252	1567504	1962515008	35·3836120	10·7779156	·0007987220
1253	1570009	1967221277	35·3977400	10·7807843	·0007980846
1254	1572516	1971935064	35·4118624	10·7836516	·0007974482
1255	1575025	1976656375	35·4259792	10·7865173	·0007968127
1256	1577536	1981385216	35·4400903	10·7893815	·0007961783
1257	1580049	1986121593	35·4541958	10·7922441	·0007955449
1258	1582564	1990865512	35·4682957	10·7951053	·0007949126
1259	1585081	1995616979	35·4823900	10·7979649	·0007942812
1260	1587600	2000376000	35·4964787	10·8008230	·0007936508
1261	1590121	2005142581	35·5105618	10·8036797	·0007930214
1262	1592644	2009916728	35·5246393	10·8065348	·0007923930
1263	1595169	2014698447	35·5387113	10·8093884	·0007917656
1264	1597696	2019487744	35·5527777	10·8122404	·0007911392
1265	1600225	2024284625	35·5668385	10·8150909	·0007905138
1266	1602756	2029089096	35·5808937	10·8179400	·0007898894
1267	1605289	2033901163	35·5949434	10·8207876	·0007892660
1268	1607824	2038720832	35·6089876	10·8236336	·0007886435
1269	1610361	2043548109	35·6230262	10·8264782	·0007880221
1270	1612900	2048383000	35·6370593	10·8293213	·0007874016

No.	Square	Cube	Square Root	Cube Root	Reciprocal
1271	1615441	2053225511	35·6510869	10·8321629	·0007867821
1272	1617984	2058075648	35·6651090	10·8350030	·0007861635
1273	1620529	2062933417	35·6791255	10·8378416	·0007855460
1274	1623076	2067798824	35·6931366	10·8406788	·0007849294
1275	1625625	2072671875	35·7071421	10·8435144	·0007843137
1276	1628176	2077552576	35·7211422	10·8463485	·0007836991
1277	1630729	2082440933	35·7351367	10·8491812	·0007830854
1278	1633284	2087336952	35·7491258	10·8520125	·0007824726
1279	1635841	2092240639	35·7631095	10·8548422	·0007818608
1280	1638400	2097152000	35·7770876	10·8576704	·0007812500
1281	1640961	2102071041	35·7910603	10·8604972	·0007806401
1282	1643524	2106997768	35·8050276	10·8633225	·0007800312
1283	1646089	2111932187	35·8189894	10·8661464	·0007794232
1284	1648656	2116874304	35·8329457	10·8689687	·0007788162
1285	1651225	2121824125	35·8468966	10·8717897	·0007782101
1286	1653796	2126781656	35·8608421	10·8746091	·0007776050
1287	1656369	2131746903	35·8747822	10·8774271	·0007770008
1288	1658944	2136719872	35·8887169	10·8802436	·0007763975
1289	1661521	2141700569	35·9026461	10·8830587	·0007757952
1290	1664100	2146689000	35·9165699	10·8858723	·0007751938
1291	1666681	2151685171	35·9304884	10·8886845	·0007745933
1292	1669264	2156689088	35·9444015	10·8914952	·0007739938
1293	1671849	2161700757	35·9583092	10·8943044	·0007733952
1294	1674436	2166720184	33·9722115	10·8971123	·0007727975
1295	1677025	2171747375	35·9861084	10·8999186	·0007722008
1296	1679616	2176782336	36·0000000	10·9027235	·0007716049
1297	1682209	2181825073	36·0138862	10·9055269	·0007710100
1298	1684804	2186875592	36·0277671	10·9083290	·0007704160
1299	1687401	2191933899	36·0416426	10·9111296	·0007698229
1300	1690000	2197000000	36·0555128	10·9139287	·0007692308
1301	1692601	2202073901	36·0693776	10·9167265	·0007686395
1302	1695204	2207155608	36·0832371	10·9195228	·0007680492
1303	1697809	2212245127	36·0970913	10·9223177	·0007674597
1304	1700416	2217342464	36·1109402	10·9251111	·0007668712
1305	1703025	2222447625	36·1247837	10·9279031	·0007662835
1306	1705636	2227560616	36·1386220	10·9306937	·0007656968
1307	1708249	2232681443	36·1524550	10·9334829	·0007651109
1308	1710864	2237810112	36·1662826	10·9362706	·0007645260
1309	1713481	2242946629	36·1801050	10·9390569	·0007639419
1310	1716100	2248091000	36·1939221	10·9418418	·0007633588
1311	1718721	2253243231	36·2077340	10·9446253	·0007627765
1312	1721344	2258403328	36·2215406	10·9474074	·0007621951
1313	1723969	2263571297	36·2353419	10·9501880	·0007616146
1314	1726596	2268747144	36·2491379	10·9529673	·0007610350
1315	1729225	2273930875	36·2629287	10·9557451	·0007604563
1316	1731856	2279122496	36·2767143	10·9585215	·0007598784
1317	1734489	2284322013	36·2904946	10·9612965	·0007593014
1318	1737124	2289529432	36·3042697	10·9640701	·0007587253
1319	1739761	2294744759	36·3180396	10·9668423	·0007581501

No.	Square	Cube	Square Root	Cube Root	Reciprocal
1320	1742400	2299968000	36·3318042	10·9696131	·0007575758
1321	1745041	2305199161	36·3455637	10·9723825	·0007570023
1322	1747684	2310438248	36·3593179	10·9751505	·0007564297
1323	1750329	2315685267	36·3730670	10·9779171	·0007558579
1324	1752976	2320940224	36·3868108	10·9806823	·0007552870
1325	1755625	2326203125	36·4005494	10·9834462	·0007547170
1326	1758276	2331473976	36·4142829	10·9862086	·0007541478
1327	1760929	2336752783	36·4280112	10·9889696	·0007535795
1328	1763584	2342039552	36·4417343	10·9917293	·0007530120
1329	1766241	2347334289	36·4554523	10·9944876	·0007524454
1330	1768900	2352637000	36·4691650	10·9972445	·0007518797
1331	1771561	2357947691	36·4828727	11·0000000	·0007513148
1332	1774224	2363266368	36·4965752	11·0027541	·0007507508
1333	1776889	2368593037	36·5102725	11·0055069	·0007501875
1334	1779556	2373927704	36·5239647	11·0082583	·0007496252
1335	1782225	2379270375	36·5376518	11·0110082	·0007490637
1336	1784896	2384621056	36·5513338	11·0137569	·0007485030
1337	1787569	2389979753	36·5650106	11·0165041	·0007479432
1338	1790244	2395346472	36·5786823	11·0192500	·0007473842
1339	1792921	2400721219	36·5923489	11·0219945	·0007468260
1340	1795600	2406104000	36·6060104	11·0247377	·0007462687
1341	1798281	2411494821	36·6196668	11·0274795	·0007457122
1342	1800964	2416893688	36·6333181	11·0302199	·0007451565
1343	1803649	2422300607	36·6469644	11·0329590	·0007446016
1344	1806336	2427715584	36·6606056	11·0356967	·0007440476
1345	1809025	2433138625	36·6742416	11·0384330	·0007434944
1346	1811716	2438569736	36·6878726	11·0411680	·0007429421
1347	1814409	2444008923	36·7014986	11·0439017	·0007423905
1348	1817104	2449456192	36·7151195	11·0466339	·0007418398
1349	1819801	2454911549	36·7287353	11·0493649	·0007412898
1350	1822500	2460375000	36·7423461	11·0520945	·0007407407
1351	1825201	2465846551	36·7559519	11·0548227	·0007401924
1352	1827904	2471326208	36·7695526	11·0575497	·0007396450
1353	1830609	2476813977	36·7831483	11·0602752	·0007390983
1354	1833316	2482309864	36·7967390	11·0629994	·0007385524
1355	1836025	2487813875	36·8103246	11·0657222	·0007380074
1356	1838736	2493326016	36·8239053	11·0684437	·0007374631
1357	1841449	2498846293	36·8374809	11·0711639	·0007369197
1358	1844164	2504374712	36·8510515	11·0738828	·0007363770
1359	1846881	2509911279	36·8646172	11·0766003	·0007358352
·1360	1849600	2515456000	36·8781778	11·0793165	·0007352941
1361	1852321	2521008881	36·8917335	11·0820314	·0007347539
1362	1855044	2526569928	36·9052842	11·0847449	·0007342144
1363	1857769	2532139147	36·9188299	11·0874571	·0007336757
1364	1860496	2537716544	36·9323706	11·0901679	·0007331378
1365	1863225	2543302125	36·9459064	11·0928775	·0007326007
1366	1865956	2548895896	36·9594372	11·0955857	·0007320644
1367	1868689	2554497863	36·9729631	11·0982926	·0007315289
1368	1871424	2560108032	36·9864840	11·1009982	·0007309942

No.	Square	Cube	Square Root	Cube Root	Reciprocal
1369	1874161	2565726409	37·0000000	11·1037025	·0007304602
1370	1876900	2571353000	37·0135110	11·1064054	·0007299270
1371	1879641	2576987811	37·0270172	11·1091070	·0007293946
1372	1882384	2582630848	37·0405184	11·1118073	·0007288630
1373	1885129	2588282117	37·0540146	11·1145064	·0007283321
1374	1887876	2593941624	37·0675060	11·1172041	·0007278020
1375	1890625	2599609375	37·0809924	11·1199004	·0007272727
1376	1893376	2605285376	37·0944740	11·1225955	·0007267442
1377	1896129	2610969633	37·1079506	11·1252893	·0007262164
1378	1898884	2616662152	37·1214224	11·1279817	·0007256894
1379	1901641	2622362939	37·1348893	11·1306729	·0007251632
1380	1904400	2628072000	37·1483512	11·1333628	·0007246377
1381	1907161	2633789341	37·1618084	11·1360514	·0007241130
1382	1909924	2639514968	37·1752606	11·1387386	·0007235890
1383	1912689	2645248887	37·1887079	11·1414246	·0007230658
1384	1915456	2650991104	37·2021505	11·1441093	·0007225434
1385	1918225	2656741625	37·2155881	11·1467926	·0007220217
1386	1920996	2662500456	37·2290209	11·1494747	·0007215007
1387	1923769	2668267603	37·2424489	11·1521555	·0007209805
1388	1926544	2674043072	37·2558720	11·1548350	·0007204611
1389	1929321	2679826869	37·2692903	11·1575133	·0007199424
1390	1932100	2685619000	37·2827037	11·1601903	·0007194245
1391	1934881	2691419471	37·2961124	11·1628659	·0007189073
1392	1937664	2697228288	37·3095162	11·1655403	·0007183908
1393	1940449	2703045457	37·3229152	11·1682134	·0007178751
1394	1943236	2708870984	37·3363094	11·1708852	·0007173601
1395	1946025	2714704875	37·3496988	11·1735558	·0007168459
1396	1948816	2720547136	37·3630834	11·1762250	·0007163324
1397	1951609	2726397773	37·3764632	11·1788930	·0007158196
1398	1954404	2732256792	37·3898382	11·1815598	·0007153076
1399	1957201	2738124199	37·4032084	11·1842252	·0007147963
1400	1960000	2744000000	37·4165738	11·1868894	·0007142857
1401	1962801	2749884201	37·4299345	11·1895523	·0007137759
1402	1965604	2755776808	37·4432904	11·1922139	·0007132668
1403	1968409	2761677827	37·4566414	11·1948743	·0007127584
1404	1971216	2767587264	37·4699880	11·1975334	·0007122507
1405	1974025	2773505125	37·4833296	11·2001913	·0007117438
1406	1976836	2779431416	37·4966665	11·2028479	·0007112376
1407	1979649	2785366143	37·5099987	11·2055032	·0007107321
1408	1982464	2791309312	37·5233261	11·2081573	·0007102273
1409	1985281	2797260929	37·5366487	11·2108101	·0007097232
1410	1988100	2803221000	37·5499667	11·2134617	·0007092199
1411	1990921	2809189531	37·5632799	11·2161120	·0007087172
1412	1993744	2815166528	37·5765885	11·2187611	·0007082153
1413	1996569	2821151997	37·5898922	11·2214089	·0007077141
1414	1999396	2827145944	37·6031913	11·2240554	·0007072136
1415	2002225	2833148375	37·6164857	11·2267007	·0007067138
1416	2005056	2839159296	37·6297754	11·2293448	·0007062147
1417	2007889	2845178713	37·6430604	11·2319876	·0007057163

26

No.	Square	Cube	Square Root	Cube Root	Reciprocal
1418	2010724	2851206632	37·6563407	11·2346292	·0007052186
1419	2013561	2857243059	37·6696164	11·2372696	·0007047216
1420	2016400	2863288000	37·6828874	11·2399087	·0007042254
1421	2019241	2869341461	37·6961596	11·2425465	·0007037298
1422	2022084	2875403448	37·7094153	11·2451831	·0007032349
1423	2024929	2881473967	37·7226722	11·2478185	·0007027407
1424	2027776	2887553024	37·7359245	11·2504527	·0007022472
1425	2030625	2893640625	37·7491722	11·2530856	·0007017544
1426	2033476	2899736776	37·7624152	11·2557173	·0007012623
1427	2036329	2905841483	37·7756535	11·2583478	·0007007708
1428	2039184	2911954752	37·7888873	11·2609770	·0007002801
1429	2042041	2918076589	37·8021163	11·2636050	·0006997901
1430	2044900	2924207000	37·8153408	11·2662318	·0006993007
1431	2047761	2930345991	37·8285606	11·2688573	·0006988120
1432	2050624	2936493568	37·8417759	11·2714816	·0006983240
1433	2053489	2942649737	37·8549864	11·2741047	·0006978367
1434	2056356	2948814504	37·8681924	11·2767266	·0006973501
1435	2059225	2954987875	37·8813938	11·2793472	·0006968641
1436	2062096	2961169856	37·8945906	11·2819666	·0006963788
1437	2064969	2967360453	37·9077828	11·2845849	·0006958942
1438	2067844	2973559672	37·9209704	11·2872019	·0006954103
1439	2070721	2979767519	37·9341535	11·2898177	·0006949270
1440	2073600	2985984000	37·9473319	11·2924323	·0006944444
1441	2076481	2992209121	37·9605058	11·2950457	·0006939625
1442	2079364	2998442888	37·9736751	11·2976579	·0006934813
1443	2082249	3004685207	37·9868398	11·3002688	·0006930007
1444	2085136	3010936384	38·0000000	11·3028786	·0006925208
1445	2088025	3017196125	38·0131556	11·3054871	·0006920415
1446	2090916	3023464536	38·0263067	11·3080945	·0006915629
1447	2093809	3029741623	38·0394532	11·3107006	·0006910850
1448	2096704	3036027392	38·0525952	11·3133056	·0006906078
1449	2099601	3042321849	38·0657326	11·3159094	·0006901312
1450	2102500	3048625000	38·0788655	11·3185119	·0006896552
1451	2105401	3054936851	38·0919939	11·3211132	·0006891799
1452	2108304	3061257408	38·1051178	11·3237134	·0006887052
1453	2111209	3067586677	38·1182371	11·3263124	·0006882312
1454	2114116	3073924664	38·1313519	11·3289102	·0006877579
1455	2117025	3080271375	38·1444622	11·3315067	·0006872852
1456	2119936	3086626816	38·1575681	11·3341022	·0006868132
1457	2122849	3092990993	38·1706693	11·3366964	·0006863418
1458	2125764	3099363912	38·1837662	11·3392894	·0006858711
1459	2128681	3105745579	38·1968585	11·3418813	·0006854010
1460	2131600	3112136000	38·2099463	11·3444719	·0006849315
1461	2134521	3118535181	38·2230297	11·3470614	·0006844627
1462	2137444	3124943128	38·2361085	11·3496497	·0006839945
1463	2140369	3131359847	38·2491829	11·3522368	·0006835270
1464	2143296	3137785344	38·2622529	11·3548227	·0006830601
1465	2146225	3144219625	38·2753184	11·3574075	·0006825939
1466	2149156	3150662696	38·2883794	11·3599911	·0006821282

No.	Square	Cube	Square Root	Cube Root	Reciprocal
1467	2152089	3157114563	38·3014360	11·3625785	·0006816633
1468	2155024	3163575232	38·3144881	11·3651547	·0006811989
1469	2157961	3170044709	38·3275358	11·3677347	·0006807352
1470	2160900	3176523000	38·3405790	11·3703136	·0006802721
1471	2163841	3183010111	38·3536178	11·3728914	·0006798097
1472	2166784	3189506048	38·3666522	11·3754679	·0006793478
1473	2169729	3196010817	38·3796821	11·3780433	·0006788866
1474	2172676	3202524424	38·3927076	11·3806175	·0006784261
1475	2175625	3209046875	38·4057287	11·3831906	·0006779661
1476	2178576	3215578176	38·4187454	11·3857625	·0006775068
1477	2181529	3222118333	38·4317577	11·3883332	·0006770481
1478	2184484	3228667352	38·4447656	11·3909028	·0006765900
1479	2187441	3235225239	38·4577691	11·3934712	·0006761325
1480	2190400	3241792000	38·4707681	11·3960384	·0006756757
1481	2193361	3248367641	38·4837627	11·3986045	·0006752194
1482	2196324	3254952168	38·4967530	11·4011695	·0006747638
1483	2199289	3261545587	38·5097390	11·4037332	·0006743088
1484	2202256	3268147904	38·5227206	11·4062959	·0006738544
1485	2205225	3274759125	38·5356977	11·4088574	·0006734007
1486	2208196	3281379256	38·5486705	11·4114177	·0006729475
1487	2211169	3288008303	38·5616389	11·4139769	·0006724950
1488	2214144	3294646272	38·5746030	11·4165349	·0006720430
1489	2217121	3301293169	38·5875627	11·4190918	·0006715917
1490	2220100	3307949000	38·6005181	11·4216476	·0006711409
1491	2223081	3314613771	38·6134691	11·4242022	·0006706908
1492	2226064	3321287488	38·6264158	11·4267556	·0006702413
1493	2229049	3327970157	38·6393582	11·4293079	·0006697924
1494	2232036	3334661784	38·6522962	11·4318591	·0006693440
1495	2235025	3341362375	38·6652299	11·4344092	·0006688963
1496	2238016	3348071936	38·6781593	11·4369581	·0006684492
1497	2241009	3354790473	38·6910843	11·4395059	·0006680027
1498	2244004	3361517992	38·7040050	11·4420525	·0006675567
1499	2247001	3368254499	38·7169214	11·4445980	·0006671114
1500	2250000	3375000000	38·7298335	11·4471424	·0006666667
1501	2253001	3381754501	38·7427412	11·4496857	·0006662225
1502	2256004	3388518008	38·7556447	11·4522278	·0006657790
1503	2259009	3395290527	38·7685439	11·4547688	·0006643360
1504	2262016	3402072064	38·7814389	11·4573087	·0006648936
1505	2265294	3408862625	38·7943294	11·4598474	·0006644518
1506	2268036	3415662216	38·8072158	11·4623850	·0006640106
1507	2271049	3422470843	38·8200978	11·4649215	·0006635700
1508	2274064	3429288512	38·8329757	11·4674568	·0006631300
1509	2277081	3436115229	38·8458491	11·4699911	·0006626905
1510	2280100	3442951000	38·8587184	11·4725242	·0006622517
1511	2283121	3449795831	38·8715834	11·4750562	·0006618134
1512	2286144	3456649728	38·8844442	11·4775871	·0006613757
1513	2289169	3463512697	38·8973006	11·4801169	·0006609385
1514	2292196	3470384744	38·9101529	11·4826455	·0006605020
1515	2295225	3477265875	38·9230009	11·4851731	·0006600660

No.	Square	Cube	Square Root	Cube Root	Reciprocal
1516	2298256	3484156096	38·9358447	11·4876995	·0006596306
1517	2301289	3491055413	38·9486841	11·4902249	·0006591958
1518	2304324	3497963832	38·9615194	11·4927491	·0006587615
1519	2307361	3504881359	38·9743505	11·4952722	·0006583278
1520	2310400	3511808000	38·9871774	11·4977942	·0006578947
1521	2313441	3518743761	39·0000000	11·5003151	·0006574622
1522	2316484	3525688648	39·0128184	11·5028348	·0006570302
1523	2319529	3532642667	39·0256326	11·5053535	·0006565988
1524	2322576	3539605824	39·0384426	11·5078711	·0006561680
1525	2325625	3546578125	39·0512483	11·5103876	·0006557377
1526	2328676	3553559576	39·0640499	11·5129030	·0006553080
1527	2331729	3560550183	39·0768473	11·5154173	·0006548788
1528	2334784	3567549952	39·0896406	11·5179305	·0006544503
1529	2337841	3574558889	39·1024296	11·5204425	·0006540222
1530	2340900	3581577000	39·1152144	11·5229535	·0006535948
1531	2343961	3588604291	39·1279951	11·5254634	·0006531679
1532	2347024	3595640768	39·1407716	11·5279722	·0006527415
1533	2350089	3602686437	39·1535439	11·5304799	·0006523157
1534	2353156	3609741304	39·1663120	11·5329865	·0006518905
1535	2356225	3616805375	39·1790760	11·5354920	·0006514658
1536	2359296	3623878656	39·1918359	11·5379965	·0006510417
1537	2362369	3630961153	39·2045915	11·5404998	·0006506181
1538	2365444	3638052872	39·2173431	11·5430021	·0006501951
1539	2368521	3645153819	39·2300905	11·5455033	·0006497726
1540	2371600	3652264000	39·2428337	11·5480034	·0006493506
1541	2374681	3659383421	39·2555728	11·5505025	·0006489293
1542	2377764	3666512088	39·2683078	11·5530004	·0006485084
1543	2380849	3673650007	39·2810387	11·5554973	·0006480881
1544	2383936	3680797184	39·2937654	11·5579931	·0006476684
1545	2387025	3687953625	39·3064880	11·5604878	·0006472492
1546	2390116	3695119336	39·3192065	11·5629815	·0006468305
1547	2393209	3702294323	39·3319208	11·5654740	·0006464124
1548	2396304	3709478592	39·3446311	11·5679655	·0006459948
1549	2399401	3716672149	39·3573373	11·5704559	·0006455778
1550	2402500	3723875000	39·3700394	11·5729453	·0006451613
1551	2405601	3731087151	39·3827378	11·5754336	·0006447453
1552	2408704	3738308608	39·3954312	11·5779208	·0006443299
1553	2411809	3745539377	39·4081210	11·5804069	·0006439150
1554	2414916	3752779464	39·4208067	11·5828919	·0006435006
1555	2418025	3760028875	39·4334883	11·5853759	·0006430868
1556	2421136	3767287616	39·4461658	11·5878588	·0006426735
1557	2424249	3774555693	39·4588393	11·5903407	·0006422608
1558	2427364	3781833112	39·4715087	11·5928215	·0006418485
1559	2430481	3789119879	39·4841740	11·5953013	·0006414368
1560	2433600	3796416000	39·4968353	11·5977799	·0006410256
1561	2436721	3803721481	39·5094925	11·6002576	·0006406150
1562	2439844	3811036328	39·5221457	11·6027342	·0006402049
1563	2442969	3818360547	39·5347948	11·6052097	·0006397953
1564	2446096	3825694144	39·5474399	11·6076341	·0006393862

No.	Square	Cube	Square Root	Cube Root	Reciprocal
1565	2449225	3833037125	39·5600809	11·6101575	·0006389776
1566	2452356	3840389496	39·5727179	11·6126299	·0006385696
1567	2455489	3847751263	39·5853508	11·6151012	·0006381621
1568	2458624	3855122432	39·5979797	11·6175715	·0006377551
1569	2461761	3862503009	39·6106046	11·6200407	·0006373486
1570	2464900	3869893000	39·6232255	11·6225088	·0006369427
1571	2468041	3877292411	39·6358424	11·6249759	·0006365872
1572	2471184	3884701248	39·6484552	11·6274420	·0006361823
1573	2474329	3892119517	39·6610640	11·6299070	·0006357279
1574	2477476	3899547224	39·6736688	11·6323710	·0006353240
1575	2480625	3906984375	39·6862696	11·6348339	·0006349026
1576	2483776	3914430976	39·6988665	11·6372957	·0006345178
1577	2486929	3921887033	39·7114593	11·6397566	·0006341154
1578	2490084	3929352552	39·7240481	11·6422164	·0006337136
1579	2493241	3936827589	39·7366329	11·6446751	·0006333122
1580	2496400	3944312000	39·7492138	11·6471329	·0006329114
1581	2499561	3951805941	39·7617907	11·6495895	·0006325111
1582	2502724	3959309368	39·7743636	11·6520452	·0006321113
1583	2505889	3966822287	39·7869325	11·6544998	·0006317119
1584	2509056	3974344704	39·7994975	11·6569534	·0006343131
1585	2512225	3981876625	39·8120585	11·6594059	·0006309148
1586	2515396	3989418056	39·8246155	11·6618574	·0006305170
1587	2518569	3996969003	39·8371686	11·6643079	·0006301197
1588	2521744	4004529472	39·8497177	11·6667574	·0006297229
1589	2524921	4012099469	39·8622628	11·6692058	·0006293266
1590	2528100	4019679000	39·8748040	11·6716532	·0006289308
1591	2531281	4027268071	39·8873413	11·6740996	·0006285355
1592	2534464	4034866688	39·8998747	11·6765449	·0006281407
1593	2537649	4042474857	39·9124041	11·6789892	·0006277464
1594	2540836	4050092584	39·9249295	11·6814325	·0006273526
1595	2544025	4057719875	39·9374511	11·6838748	·0006269592
1596	2547216	4065356736	39·9499687	11·6863161	·0006265664
1597	2550409	4073003173	39·9624824	11·6887563	·0006261741
1598	2553604	4080659192	39·9749922	11·6911955	·0006257822
1599	2556801	4088324799	39·9874980	11·6936337	·0006253909
1600	2560000	4096000000	40·0000000	11·6960709	·0006250000
1601	2563201	4103684801	40·0124980	11·6985071	·0006246096
1602	2566404	4111379208	40·0249922	11·7009422	·0006242197
1603	2569609	4119083227	40·0374824	11·7033764	·0006238303
1604	2572816	4126796864	40·0499688	11·7058095	·0006234414
1605	2576025	4134520125	40·0624512	11·7082417	·0006230530
1606	2579236	4142253016	40·0749298	11·7106728	·0006226650
1607	2582449	4149995543	40·0874045	11·7131029	·0006222775
1608	2585664	4157747712	40·0998753	11·7155320	·0006218905
1609	2588881	4165509529	40·1123423	11·7179601	·0006215040
1610	2592100	4173281000	40·1248053	11·7203872	·0006211180
1611	2595321	4181062131	40·1372645	11·7228133	·0006207325
1612	2598544	4188852928	40·1497198	11·7252384	·0006203474
1613	2601769	4196653397	40·1621713	11·7276625	·0006199628

No.	Square	Cube	Square Root	Cube Root	Reciprocal
1614	2604996	4204463544	40·1746188	11·7300855	·0006195787
1615	2608225	4212283375	40·1870626	11·7325076	·0006191950
1616	2611456	4220112896	40·1995025	11·7349286	·0006188119
1617	2614689	4227952113	40·2119385	11·7373487	·0006184292
1618	2617924	4235801032	40·2243707	11·7397677	·0006180470
1619	2621161	4243659659	40·2367990	11·7421858	·0006176652
1620	2624400	4251528000	40·2492236	11·7446029	·0006172840
1621	2627641	4259406061	40·2616443	11·7470190	·0006169031
1622	2630884	4267293848	40·2740611	11·7494341	·0006165228
1623	2634129	4275191367	40·2864742	11·7518482	·0006161429
1624	2637376	4283098624	40·2988834	11·7542613	·0006157635
1625	2640625	4291015625	40·3112888	11·7566734	·0006153846
1626	2643876	4298942376	40·3236903	11·7590846	·0006150062
1627	2647129	4306878883	40·3360881	11·7614947	·0006146282
1628	2650384	4314825152	40·3484820	11·7639039	·0006142506
1629	2653641	4322781189	40·3608721	11·7663121	·0006138735
1630	2656900	4330747000	40·3732585	11·7687193	·0006134969
1631	2660161	4338722591	40·3856410	11·7711255	·0006131208
1632	2663424	4346707968	40·3980198	11·7735306	·0006127451
1633	2666689	4354703137	40·4103947	11·7759349	·0006123699
1634	2669956	4362708104	40·4227658	11·7783381	·0006119951
1635	2673225	4370722875	40·4351332	11·7807404	·0006116208
1636	2676496	4378747456	40·4474968	11·7831417	·0006112469
1637	2679769	4386781853	40·4598566	11·7855420	·0006108735
1638	2683044	4394826072	40·4722127	11·7879414	·0006105006
1639	2686321	4402880119	40·4845649	11·7903397	·0006101281
1640	2689600	4410944000	40·4969135	11·7927371	·0006097561
1641	2692881	4419017721	40·5092582	11·7951335	·0006093845
1642	2696164	4427101288	40·5215992	11·7975289	·0006090134
1643	2699449	4435194707	40·5339364	11·7999234	·0006086427
1644	2702736	4443297984	40·5462699	11·8023169	·0006082725
1645	2706025	4451411125	40·5585996	11·8047094	·0006079027
1646	2709316	4459534136	40·5709255	11·8071010	·0006075334
1647	2712609	4467667023	40·5832477	11·8094916	·0006071645
1648	2715904	4475809792	40·5955663	11·8118812	·0006067961
1649	2719201	4483962449	40·6078810	11·8142698	·0006064281
1650	2722500	4492125000	40·6201920	11·8166576	·0006060606
1651	2725801	4500297451	40·6324993	11·8190443	·0006056935
1652	2729104	4508479808	40·6448029	11·8214301	·0006053269
1653	2732409	4516672077	40·6571027	11·8238149	·0006049607
1654	2735716	4524874264	40·6693988	11·8261987	·0006045949
1655	2739025	4533086375	40·6816912	11·8285816	·0006042296
1656	2742336	4541308416	40·6939799	11·8309634	·0006038647
1657	2745649	4549540393	40·7062648	11·8333444	·0006035003
1658	2748964	4557782312	40·7185461	11·8357244	·0006031363
1659	2752281	4566034179	40·7308237	11·8381034	·0006027728
1660	2755600	4574296000	40·7430976	11·8404815	·0006024096
1661	2758921	4582567781	40·7553677	11·8428586	·0006020470
1662	2762244	4590849528	40·7676342	11·8452348	·0006016847

No.	Square	Cube	Square Root	Cube Root	Reciprocal
1663	2765569	4599141247	40·7798970	11·8476100	·0006013229
1664	2768896	4607442994	40·7921561	11·8499843	·0006009615
1665	2772225	4615754625	40·8044115	11·8523576	·0006006006
1666	2775556	4624076296	40·8166633	11·8547299	·0006002401
1667	2778889	4632407963	40·8289113	11·8571014	·0005998800
1668	2782224	4640749632	40·8411557	11·8594719	·0005995204
1669	2785561	4649101309	40·8533964	11·8618414	·0005991612
1670	2788900	4657463000	40·8656335	11·8642100	·0005988024
1671	2792241	4665834711	40·8778669	11·8665776	·0005984440
1672	2795584	4674216448	40·8900966	11·8689443	·0005980861
1673	2798929	4682608217	40·9023227	11·8713100	·0005977286
1674	2802276	4691010024	40·9145451	11·8736748	·0005973716
1675	2805625	4699421875	40·9267638	11·8760387	·0005970149
1676	2808976	4707843776	40·9389790	11·8784016	·0005966587
1677	2812329	4716275733	40·9511905	11·8807636	·0005963029
1678	2815684	4724717752	40·9683983	11·8831246	·0005959476
1679	2819041	4733169839	40·9756025	11·8854847	·0005955926
1680	2822400	4741632000	40·9878081	11·8878439	·0005952381
1681	2825761	4750104241	41·0000000	11·8902022	·0005948840
1682	2829124	4758586568	41·0121983	11·8925595	·0005945303
1683	2832489	4767078987	41·0243830	11·8949159	·0005941771
1684	2835856	4775581504	41·0365691	11·8972713	·0005938242
1685	2839225	4784094125	41·0487515	11·8996258	·0005934718
1686	2842596	4792616856	41·0609303	11·9019793	·0005931198
1687	2845969	4801149703	41·0731055	11·9043319	·0005927682
1688	2849344	4809692672	41·0852772	11·9066836	·0005924171
1689	2852721	4818245769	41·0974452	11·9090344	·0005920663
1690	2856100	4826809000	41·1096096	11·9113843	·0005917160
1691	2859481	4835382371	41·1217704	11·9137332	·0005913661
1692	2862864	4843965888	41·1339276	11·9160812	·0005910165
1693	2866249	4852559557	41·1460812	11·9184283	·0005906675
1694	2869636	4861168384	41·1582813	11·9207744	·0005903188
1695	2873025	4869777375	41·1703777	11·9231196	·0005899705
1696	2876416	4878401536	41·1825206	11·9254639	·0005896226
1697	2879809	4887035873	41·1946599	11·9278073	·0005892752
1698	2883204	4895680392	41·2067956	11·9301497	·0005889282
1699	2886601	4904335099	41·2189277	11·9324913	·0005885815
1700	2890000	4913000000	41·2310563	11·9348319	·0005882353
1701	2893401	4921675101	41·2431812	11·9371716	·0005878895
1702	2896804	4930360408	41·2553027	11·9395104	·0005875441
1703	2900209	4939055927	41·2674205	11·9418482	·0005871991
1704	2903616	4947761664	41·2795349	11·9441852	·0005868545
1705	2907025	4956477625	41·2916456	11·9465213	·0005865108
1706	2910436	4965203816	41·3037529	11·9488564	·0005861665
1707	2913849	4973940243	41·3158565	11·9511906	·0005858231
1708	2917264	4982686912	41·3279566	11·9535289	·0005854801
1709	2920681	4991443829	41·3400532	11·9558563	·0005851375
1710	2924100	5000211000	41·3521463	11·9581878	·0005847953
1711	2927521	5008988431	41·3642358	11·9605184	·0005844535

No.	Square	Cube	Square Root	Cube Root	Reciprocal
1712	2930944	5017776128	41·3763217	11·9628481	·0005841121
1713	2934369	5026574097	41·3884042	11·9651768	·0005837712
1714	2937796	5035382344	41·4004831	11·9675047	·0005834306
1715	2941225	5044200875	41·4125585	11·9698317	·0005830904
1716	2944656	5053029696	41·4246304	11·9721577	·0005827506
1717	2948089	5061868813	41·4366987	11·9744829	·0005824112
1718	2951524	5070718232	41·4487636	11·9768071	·0005820722
1719	2954961	5079577959	41·4608249	11·9791304	·0005817336
1720	2958400	5088448000	41·4728827	11·9814528	·0005813953
1721	2961841	5097328361	41·4849370	11·9837744	·0005810575
1722	2965284	5106219048	41·4969878	11·9860950	·0005807201
1723	2968729	5115120067	41·5090351	11·9884148	·0005803831
1724	2972176	5124031424	41·5210790	11·9907336	·0005800464
1725	2975625	5132953125	41·5331193	11·9930516	·0005797101
1726	2979076	5141885176	41·5451561	11·9953686	·0005793743
1727	2982529	5150827583	41·5571895	11·9976848	·0005790388
1728	2985984	5159780352	41·5692194	12·0000000	·0005787037
1729	2989441	5168743489	41·5812457	12·0023144	·0005783690
1730	2992900	5177717000	41·5932686	12·0046278	·0005780347
1731	2996361	5186700891	41·6052881	12·0069404	·0005777008
1732	2999824	5195695168	41·6173041	12·0092521	·0005773672
1733	3003289	5204699837	41·6293166	12·0115629	·0005770340
1734	3006756	5213714904	41·6413256	12·0138728	·0005767013
1735	3010225	5222740375	41·6533312	12·0161818	·0005763689
1736	3013696	5231776256	41·6653333	12·0184900	·0005760369
1737	3017169	5240822553	41·6773319	12·0207973	·0005757052
1738	3020644	5249979272	41·6893271	12·0231037	·0005753740
1739	3024121	5258946419	41·7013189	12·0254092	·0005750431
1740	3027600	5268024000	41·7133072	12·0277138	·0005747126
1741	3031081	5277112021	41·7252921	12·0300175	·0005743825
1742	3034564	5286210488	41·7372735	12·0323204	·0005740528
1743	3038049	5295319407	41·7492515	12·0346223	·0005737235
1744	3041536	5304438784	41·7612260	12·0369233	·0005733945
1745	3045025	5313568625	41·7731971	12·0392235	·0005730659
1746	3048516	5322708936	41·7851648	12·0415229	·0005727377
1747	3052009	5331859723	41·7971291	12·0438213	·0005724098
1748	3055504	5341020992	41·8090899	12·0461189	·0005720824
1749	3059001	5350192749	41·8210473	12·0484156	·0005717553
1750	3062500	5359375000	41·8330013	12·0507114	·0005714286
1751	3066001	5368567751	41·8449519	12·0530063	·0005711022
1752	3069504	5377771008	41·8568991	12·0553003	·0005707763
1753	3073009	5386984777	41·8688428	12·0575935	·0005704507
1754	3076516	5396209064	41·8807832	12·0598859	·0005701254
1755	3080025	5405443875	41·8927201	12·0621773	·0005698006
1756	3083536	5414689216	41·9046537	12·0644679	·0005694761
1757	3087049	5423945093	41·9165838	12·0667576	·0005691520
1758	3090564	5433211512	41·9285106	12·0690464	·0005688282
1759	3094081	5442488479	41·9404339	12·0713344	·0005685048
1760	3097600	5451776000	41·9523539	12·0736215	·0005681818

No.	Square	Cube	Square Root	Cube Root	Reciprocal
1761	3101121	5461074081	41·9642705	12·0759077	·0005678592
1762	3104644	5470382728	41·9761837	12·0781930	·0005675369
1763	3108169	5479701947	41·9880935	12·0804775	·0005672150
1764	3111696	5489031744	42·0000000	12·0827612	·0005668934
1765	3115225	5498372125	42·0119031	12·0850439	·0005665722
1766	3118756	5507723096	42·0238028	12·0873258	·0005662514
1767	3122289	5517084663	42·0356991	12·0896069	·0005659310
1768	3125824	5526456832	42·0475921	12·0918870	·0005656109
1769	3129361	5535839609	42·0594817	12·0941664	·0005652911
1770	3132900	5545233000	42·0713679	12·0964449	·0005649718
1771	3136441	5554637011	42·0832508	12·0987226	·0005646527
1772	3139984	5564051648	42·0951304	12·1009993	·0005643341
1773	3143529	5573476917	42·1070065	12·1032753	·0005640158
1774	3147076	5582912824	42·1188794	12·1055503	·0005636979
1775	3150625	5592359875	42·1307488	12·1078245	·0005633803
1776	3154176	5601816576	42·1426150	12·1100979	·0005630631
1777	3157729	5611284433	42·1544778	12·1123704	·0005627462
1778	3161284	5620762952	42·1663373	12·1146420	·0005624297
1779	3164841	5630252139	42·1781934	12·1169128	·0005621135
1780	3168400	5639752000	42·1900462	12·1191827	·0005617978
1781	3171961	5649262541	42·2018957	12·1214518	·0005614823
1782	3175524	5658783768	42·2137418	12·1237200	·0005611672
1783	3179089	5668315687	42·2255846	12·1259874	·0005608525
1784	3182656	5677858304	42·2374242	12·1282539	·0005605381
1785	3186225	5687411625	42·2492603	12·1305197	·0005602241
1786	3189796	5696975656	42·2610932	12·1327845	·0005599104
1787	3193369	5706550403	42·2729227	12·1350485	·0005595971
1788	3196944	5716135872	42·2847490	12·1373117	·0005592841
1789	3200521	5725732069	42·2965719	12·1395740	·0005589715
1790	3204100	5735339000	42·3083916	12·1418355	·0005586592
1791	3207681	5744956671	42·3202079	12·1440961	·0005583473
1792	3211264	5754585088	42·3320210	12·1463559	·0005580357
1793	3214849	5764224257	42·3438307	12·1486148	·0005577245
1794	3218436	5773874184	42·3556371	12·1508729	·0005574136
1795	3222025	5783534875	42·3674403	12·1531302	·0005571031
1796	3225616	5793206336	42·3792402	12·1553866	·0005567929
1797	3229209	5802888573	42·3910368	12·1576422	·0005564830
1798	3232804	5812581592	42·4028301	12·1598970	·0005561735
1799	3236401	5822285399	42·4146201	12·1621509	·0005558644
1800	3240000	5832000000	42·4264069	12·1644040	·0005555556
1801	3243601	5841725401	42·4381903	12·1666562	·0005552471
1802	3247204	5851461608	42·4499705	12·1689076	·0005549390
1803	3250809	5861208627	42·4617475	12·1711582	·0005546312
1804	3254416	5870966464	42·4735212	12·1734079	·0005543237
1805	3258025	5880735125	42·4852916	12·1756569	·0005540166
1806	3261636	5890514616	42·4970587	12·1779050	·0005537099
1807	3265249	5900304943	42·5088226	12·1801522	·0005534034
1808	3268864	5910106112	42·5205833	12·1823987	·0005530973
1809	3272481	5919918129	42·5323406	12·1846443	·0005527916

No.	Square	Cube	Square Root	Cube Root	Reciprocal
1810	3276100	5929741000	42·5440948	12·1868891	·0005524862
1811	3279721	5939574731	42·5558456	12·1891831	·0005521811
1812	3283344	5949419328	42·5675933	12·1913762	·0005518764
1813	3286969	5959274797	42·5793377	12·1936185	·0005515720
1814	3290596	5969141144	42·5910789	12·1958599	·0005512679
1815	3294225	5979018375	42·6028168	12·1981006	·0005509642
1816	3297856	5988906496	42·6145515	12·2003404	·0005506608
1817	3301489	5998805513	42·6262829	12·2025794	·0005503577
1818	3305124	6008715432	42·6380112	12·2048176	·0005500550
1819	3308761	6018636259	42·6497362	12·2070549	·0005497525
1820	3312400	6028568000	42·6614580	12·2092915	·0005494505
1821	3316041	6038510661	42·6731766	12·2115272	·0005491488
1822	3319684	6048464248	42·6848919	12·2137621	·0005488474
1823	3323329	6058428767	42·6966040	12·2159962	·0005485464
1824	3326976	6068404224	42·7083130	12·2182295	·0005482456
1825	3330625	6078390625	42·7200187	12·2204620	·0005479452
1826	3334276	6088387976	42·7317212	12·2226936	·0005476451
1827	3337929	6098396283	42·7434206	12·2249244	·0005473454
1828	3341584	6108415552	42·7551167	12·2271544	·0005470460
1829	3345241	6118445789	42·7668095	12·2293836	·0005467469
1830	3348900	6128487000	42·7784992	12·2316120	·0005464481
1831	3352561	6138539191	42·7901858	12·2338396	·0005461496
1832	3356224	6148602368	42·8018691	12·2360663	·0005458515
1833	3359889	6158676537	42·8135492	12·2382923	·0005455537
1834	3363556	6168761704	42·8252262	12·2405174	·0005452563
1835	3367225	6178857875	42·8368999	12·2427418	·0005449591
1836	3370896	6188965056	42·8485706	12·2449653	·0005446623
1837	3374569	6199083253	42·8602380	12·2471880	·0005443658
1838	3378244	6209212472	42·8719022	12·2494099	·0005440696
1839	3381921	6219352719	42·8835633	12·2516310	·0005437738
1840	3385600	6229504000	42·8952212	12·2538513	·0005434783
1841	3389281	6239666321	42·9068759	12·2560708	·0005431831
1842	3392964	6249839688	42·9185275	12·2582895	·0005428882
1843	3396649	6260024107	42·9301759	12·2605074	·0005425936
1844	3400336	6270219584	42·9418211	12·2627245	·0005422993
1845	3404025	6280426125	42·9534632	12·2649408	·0005420054
1846	3407716	6290643736	42·9651021	12·2671563	·0005417118
1847	3411409	6300872423	42·9767379	12·2693710	·0005414185
1848	3415104	6311112192	42·9883705	12·2715849	·0005411255
1849	3418801	6321363049	43·0000000	12·2737980	·0005408329
1850	3422500	6331625000	43·0116263	12·2760103	·0005405405
1851	3426201	6341898051	43·0232495	12·2782218	·0005402485
1852	3429904	6352182208	43·0348696	12·2804325	·0005399568
1853	3433609	6362477477	43·0464865	12·2826424	·0005396654
1854	3437316	6372783864	43·0581003	12·2848515	·0005393743
1855	3441025	6383101375	43·0697109	12·2870598	·0005390836
1856	3444736	6393430016	43·0813185	12·2892673	·0005387931
1857	3448449	6403769793	43·0929228	12·2914740	·0005385030
1858	3452164	6414120712	43·1045241	12·2936800	·0005382131

No.	Square	Cube	Square Root	Cube Root	Reciprocal
1859	3455881	6424482779	43·1161223	12·2958851	·0005379236
1860	3459600	6434856000	43·1277173	12·2980895	·0005376344
1861	3463321	6445240381	43·1393092	12·3002980	·0005373455
1862	3467044	6455635928	43·1508980	12·3024958	·0005370569
1863	3470769	6466042647	43·1624837	12·3046978	·0005367687
1864	3474496	6476460544	43·1740663	12·3068990	·0005364807
1865	3478225	6486889625	43·1856458	12·3090994	·0005361930
1866	3481956	6497329896	43·1972221	12·3112991	·0005359057
1867	3485689	6507781363	43·2087954	12·3134979	·0005356186
1868	3489424	6518244032	43·2203656	12·3156959	·0005353319
1869	3493161	6528717909	43·2319326	12·3178932	·0005350455
1870	3496900	6539203000	43·2434966	12·3200897	·0005347594
1871	3500641	6549699311	43·2550575	12·3222854	·0005344735
1872	3504384	6560206848	43·2666153	12·3244803	·0005341880
1873	3508129	6570725617	43·2781700	12·3266744	·0005339028
1874	3511876	6581255624	43·2897216	12·3288678	·0005336179
1875	3515625	6591796875	43·3012702	12·3310604	·0005333333
1876	3519376	6602349376	43·3128157	12·3332522	·0005330490
1877	3523129	6612913133	43·3243580	12·3354432	·0005327651
1878	3526884	6623488152	43·3358973	12·3376334	·0005324814
1879	3530641	6634074439	43·3474336	12·3398229	·0005321980
1880	3534400	6644672000	43·3589668	12·3420116	·0005319149
1881	3538161	6655280841	43·3704969	12·3441995	·0005316321
1882	3541924	6665900968	43·3820239	12·3463866	·0005313496
1883	3545689	6676582387	43·3935479	12·3485730	·0005310674
1884	3549456	6687175104	43·4050688	12·3507586	·0005307856
1885	3553225	6697829125	43·4165867	12·3529434	·0005305040
1886	3556996	6708494456	43·4281015	12·3551274	·0005302227
1887	3560769	6719171103	43·4396132	12·3573107	·0005299417
1888	3564544	6729859072	43·4511220	12·3594932	·0005296610
1889	3568321	6740558369	43·4626276	12·3616749	·0005293806
1890	3572100	6751269000	43·4741302	12·3638559	·0005291005
1891	3575881	6761990971	43·4856298	12·3660361	·0005288207
1892	3579664	6772724288	43·4971263	12·3682155	·0005285412
1893	3583449	6783468957	43·5086198	12·3703941	·0005282620
1894	3587236	6794224984	43·5201103	12·3725721	·0005279831
1895	3591025	6804992375	43·5315977	12·3747492	·0005277045
1896	3594816	6815771136	43·5430821	12·3769255	·0005274262
1897	3598609	6826561273	43·5545635	12·3791011	·0005271481
1898	3602404	6837362792	43·5660418	12·3812759	·0005268704
1899	3606201	6848175699	43·5775171	12·3884500	·0005265929
1900	3610000	6859000000	43·5889894	12·3856233	·0005263158
1901	3613801	6869835701	43·6004587	12·3877959	·0005260389
1902	3617604	6880682808	43·6119249	12·3899676	·0005257624
1903	3621409	6891541327	43·6233882	12·3921386	·0005254861
1904	3625216	6902411264	43·6348485	12·3943089	·0005252101
1905	3629025	6913292625	43·6463057	12·3964784	·0005249344
1906	3632836	6924185416	43·6577599	12·3986471	·0005246590
1907	3636649	6935089643	43·6692111	12·4008151	·0005243888

No.	Square	Cube	Square Root	Cube Root	Reciprocal
1908	3640464	6946005312	43·6806593	12·4029823	·0005241090
1909	3644281	6956932429	43·6921045	12·4051488	·0005238845
1910	3648100	6967871000	43·7035467	12·4073145	·0005235602
1911	3651921	6978821031	43·7149860	12·4094794	·0005232862
1912	3655744	6989782528	43·7264222	12·4116436	·0005230126
1913	3659569	7000755497	43·7378554	12·4138070	·0005227392
1914	3663396	7011789944	43·7492857	12·4159697	·0005224660
1915	3667225	7022735875	43·7607129	12·4181316	·0005221932
1916	3671056	7033743296	43·7721373	12·4202928	·0005219207
1917	3674889	7044762213	43·7835585	12·4224533	·0005216484
1918	3678724	7055792632	43·7949768	12·4246129	·0005213764
1919	3682561	7066834559	43·8063922	12·4267719	·0005211047
1920	3686400	7077888000	43·8178046	12·4289300	·0005208333
1921	3690241	7088952961	43·8292140	12·4310875	·0005205622
1922	3694084	7100029448	43·8406204	12·4332441	·0005202914
1923	3697929	7111117467	43·8520239	12·4354001	·0005200208
1924	3701776	7122217024	43·8634244	12·4375552	·0005197505
1925	3705625	7133328125	43·8748219	12·4397097	·0005194805
1926	3709476	7144450776	43·8862165	12·4418634	·0005192108
1927	3713329	7155584983	43·8976081	12·4440163	·0005189414
1928	3717184	7166730752	43·9089968	12·4461685	·0005186722
1929	3721041	7177888089	43·9203725	12·4483200	·0005184033
1930	3724900	7189057000	43·9317652	12·4504707	·0005181347
1931	3728761	7200237491	43·9431451	12·4526206	·0005178664
1932	3732624	7211429568	43·9545220	12·4547699	·0005175983
1933	3736489	7222633237	43·9658959	12·4569184	·0005173306
1934	3740356	7233848504	43·9772668	12·4590661	·0005170631
1935	3744225	7245075375	43·9886349	12·4612131	·0005167959
1936	3748096	7256313856	44·0000000	12·4633594	·0005165289
1937	3751969	7267563953	44·0113622	12·4655049	·0005162623
1938	3755844	7278825672	44·0227214	12·4676497	·0005159959
1939	3759721	7290099019	44·0340777	12·4697937	·0005157298
1940	3763600	7301384000	44·0454311	12·4719370	·0005154639
1941	3767481	7312680621	44·0567815	12·4740796	·0005151984
1942	3771364	7323988888	44·0681291	12·4762214	·0005149331
1943	3775249	7335308807	44·0794737	12·4783625	·0005146680
1944	3779136	7346640384	44·0908154	12·4805029	·0005144033
1945	3783025	7357983625	44·1021541	12·4826426	·0005141388
1946	3786916	7369338536	44·1134900	12·4847815	·0005138746
1947	3790809	7380705123	44·1248229	12·4869197	·0005136107
1948	3794704	7392083392	44·1361530	12·4890571	·0005133470
1949	3798601	7403473349	44·1474801	12·4911938	·0005130836
1950	3802500	7414875000	44·1588043	12·4933298	·0005128205
1951	3806401	7426288351	44·1701256	12·4954651	·0005125577
1952	3810304	7437713408	44·1814441	12·4975995	·0005122951
1953	3814209	7449150177	44·1927596	12·4997333	·0005120328
1954	3818116	7460598664	44·2040722	12·5018664	·0005117707
1955	3822025	7472058875	44·2153819	12·5039988	·0005115090
1956	3825936	7483530816	44·2266888	12·5061304	·0005112474

No.	Square	Cube	Square Root	Cube Root	Reciprocal
1957	3829849	7495014493	44·2379927	12·5082612	·0005109862
1958	3833764	7506509912	44·2492988	12·5103914	·0005107252
1959	3837681	7518017079	44·2605919	12·5125208	·0005104645
1960	3841600	7529536000	44·2718872	12·5146495	·0005102041
1961	3845521	7541066681	44·2831797	12·5167775	·0005099489
1962	3849444	7552609128	44·2944692	12·5189047	·0005096840
1963	3853369	7564163347	44·3057558	12·5210313	·0005094244
1964	3857296	7575729344	44·3170396	12·5231571	·0005091650
1965	3861225	7587307125	44·3283205	12·5252822	·0005089059
1966	3865156	7598896696	44·3395985	12·5274065	·0005086470
1967	3869089	7610498063	44·3508737	12·5295802	·0005083884
1968	3873024	7622111232	44·3621460	12·5316531	·0005081301
1969	3876961	7633736209	44·3734155	12·5337753	·0005078720
1970	3880900	7645373000	44·3846820	12·5358968	·0005076142
1971	3884841	7657021611	44·3959457	12·5380176	·0005073567
1972	3888784	7668682048	44·4072066	12·5401377	·0005070994
1973	3892729	7680354317	44·4184646	12·5422570	·0005068424
1974	3896676	7692038424	44·4297198	12·5443757	·0005065856
1975	3900625	7703734375	44·4409720	12·5464936	·0005063291
1976	3904576	7715442176	44·4522215	12·5486107	·0005060729
1977	3908529	7727161833	44·4634681	12·5507272	·0005058169
1978	3912484	7738893352	44·4747119	12·5528430	·0005055612
1979	3916441	7750636739	44·4859528	12·5549580	·0005053057
1980	3920400	7762392000	44·4971909	12·5570723	·0005050505
1981	3924361	7774159141	44·5084262	12·5591860	·0005047956
1982	3928324	7785938168	44·5196586	12·5612989	·0005045409
1983	3932289	7797729087	44·5308881	12·5634111	·0005042864
1984	3936256	7809531904	44·5421149	12·5655226	·0005040323
1985	3940225	7821346625	44·5533388	12·5676334	·0005037783
1986	3944196	7833173256	44·5645599	12·5697435	·0005035247
1987	3948169	7845011803	44·5757781	12·5718529	·0005032713
1988	3952144	7856862272	44·5869936	12·5739615	·0005030181
1989	3956121	7868724669	44·5982062	12·5760695	·0005027652
1990	3960100	7880599000	44·6094160	12·5781767	·0005025126
1991	3964081	7892485271	44·6206230	12·5802832	·0005022602
1992	3968064	7904383488	44·6318272	12·5823891	·0005020080
1993	3972049	7916293657	44·6430286	12·5844942	·0005017561
1994	3976036	7928215784	44·6542271	12·5865987	·0005015045
1995	3980025	7940149875	44·6654228	12·5887024	·0005012531
1996	3984016	7952095936	44·6766158	12·5908054	·0005010020
1997	3988009	7964053973	44·6878059	12·5929078	·0005007511
1998	3992004	7976023992	44·6989933	12·5950094	·0005005005
1999	3996001	7988005999	44·7101778	12·5971103	·0005002501
2000	4000000	8000000000	44·7213596	12·5992105	·0005000000
2001	4004001	8012006001	44·7325385	12·6013101	·0004997501
2002	4008004	8024024008	44·7437146	12·6034089	·0004995005
2003	4012009	8036054027	44·7548880	12·6055070	·0004992511
2004	4016016	8048096064	44·7660586	12·6076044	·0004990020
2005	4020025	8060150125	44·7772264	12·6097011	·0004987531

No.	Square	Cube	Square Root	Cube Root	Reciprocal
2006	4024036	8072216216	44·7883913	12·6117971	·0004985045
2007	4028049	8084294343	44·7995585	12·6138924	·0004982561
2008	4032064	8096384512	44·8107130	12·6159870	·0004980080
2009	4036081	8108486729	44·8218697	12·6180810	·0004977601
2010	4040100	8120601000	44·8330235	12·6201743	·0004975124
2011	4044121	8132727331	44·8441746	12·6222669	·0004972650
2012	4048144	8144865728	44·8553280	12·6243587	·0004970179
2013	4052169	8157016197	44·8664685	12·6264499	·0004967710
2014	4056196	8169178744	44·8776113	12·6285404	·0004965243
2015	4060225	8181353375	44·8887514	12·6306301	·0004962779
2016	4064256	8193540096	44·8998886	12·6327192	·0004960317
2017	4068289	8205738913	44·9110231	12·6348076	·0004957858
2018	4072324	8217949832	44·9221549	12·6368953	·0004955401
2019	4076361	8230172859	44·9332839	12·6389823	·0004952947
2020	4080400	8242408000	44·9444101	12·6410687	·0004950495
2021	4084441	8254655261	44·9555336	12·6431543	·0004948046
2022	4088484	8266914648	44·9666543	12·6452393	·0004945598
2023	4092529	8279186167	44·9777723	12·6473235	·0004943154
2024	4096576	8291469824	44·9888875	12·6494071	·0004940711
2025	4100625	8303765625	45·0000000	12·6514900	·0004938272
2026	4104676	8316073576	45·0111097	12·6535722	·0004935834
2027	4108729	8328393683	45·0222167	12·6556538	·0004933399
2028	4112784	8340725952	45·0333210	12·6577346	·0004930966
2029	4116841	8353070389	45·0444225	12·6598148	·0004928536
2030	4120900	8365427000	45·0555213	12·6618943	·0004926108
2031	4124961	8377795791	45·0666173	12·6639731	·0004923683
2032	4129024	8390176768	45·0777107	12·6660512	·0004921260
2033	4133089	8402569937	45·0888013	12·6681286	·0004918839
2034	4137156	8414975304	45·0998891	12·6702053	·0004916421
2035	4141225	8427392875	45·1109743	12·6722814	·0004914005
2036	4145296	8439822656	45·1220567	12·6743567	·0004911591
2037	4149369	8452264653	45·1331364	12·6764314	·0004909180
2038	4153444	8464718872	45·1442134	12·6785054	·0004906771
2039	4157521	8477185819	45·1552876	12·6805788	·0004904365
2040	4161600	8489664000	45·1663592	12·6826514	·0004901961
2041	4165681	8502154921	45·1774280	12·6847234	·0004899559
2042	4169764	8514658088	45·1884941	12·6867947	·0004897160
2043	4173849	8527173507	45·1995575	12·6888654	·0004894762
2044	4177936	8539701184	45·2106182	12·6909354	·0004892368
2045	4182025	8552241125	45·2216762	12·6930047	·0004889976
2046	4186116	8564793336	45·2327315	12·6950733	·0004887586
2047	4190209	8577357823	45·2437841	12·6971412	·0004885198
2048	4194304	8589934592	45·2548340	12·6992084	·0004882813
2049	4198401	8602523649	45·2658812	12·7012750	·0004880429
2050	4202500	8615125000	45·2769257	12·7033409	·0004878049
2051	4206601	8627738651	45·2879675	12·7054061	·0004875670
2052	4210704	8640364608	45·2990066	12·7074707	·0004873294
2053	4214809	8653002877	45·3100430	12·7095346	·0004870921
2054	4218916	8665653464	45·3210768	12·7115978	·0004868549

No.	Square	Cube	Square Root	Cube Root	Reciprocal
2055	4223025	8678316875	45·3321078	12·7136603	·0004866180
2056	4227136	8690991616	45·3431362	12·7157222	·0004863813
2057	4231249	8703679193	45·3541619	12·7177835	·0004861449
2058	4235364	8716379112	45·3651849	12·7198441	·0004859086
2059	4239481	8729091879	45·3762052	12·7219040	·0004856727
2060	4243600	8741816000	45·3872229	12·7239632	·0004854369
2061	4247721	8754552981	45·3982378	12·7260218	·0004852014
2062	4251844	8767302328	45·4092501	12·7280797	·0004849661
2063	4255969	8780064047	45·4202598	12·7301370	·0004847310
2064	4260096	8792838144	45·4312668	12·7321935	·0004844961
2065	4264225	8805624625	45·4422711	12·7342494	·0004842615
2066	4268356	8818423496	45·4532727	12·7363046	·0004840271
2067	4272489	8831234763	45·4642717	12·7383592	·0004837929
2068	4276624	8844058432	45·4752680	12·7404131	·0004835590
2069	4280761	8856894509	45·4862616	12·7424664	·0004833253
2070	4284900	8869743000	45·4972526	12·7445189	·0004880918
2071	4239041	8882603911	45·5082410	12·7465709	·0004828585
2072	4293184	8895477248	45·5192267	12·7486222	·0004826255
2073	4297329	8908363017	45·5802097	12·7506728	·0004823927
2074	4301476	8921261224	45·5411901	12·7527227	·0004821601
2075	4305625	8934171875	45·5521679	12·7547721	·0004819277
2076	4309776	8947094976	45·5631430	12·7568207	·0004816956
2077	4313929	8960080533	45·5741155	12·7588687	·0004814636
2078	4318084	8972978552	45·5850853	12·7609160	·0004812320
2079	4322241	8985939039	45·5960525	12·7629627	·0004810005
2080	4326400	8998912000	45·6070170	12·7650087	·0004807692
2081	4330561	9011897441	45·6179789	12·7670540	·0004805382
2082	4334724	9024895368	45·6289382	12·7690987	·0004803074
2083	4338889	9037905787	45·6398948	12·7711427	·0004800768
2084	4343056	9050928704	45·6508488	12·7731861	·0004798464
2085	4347225	9063964125	45·6618002	12·7752288	·0004796163
2086	4351396	9077012056	45·6727490	12·7772709	·0004793864
2087	4355569	9090072503	45·6836951	12·7793123	·0004791567
2088	4359744	9103145472	45·6946386	12·7813531	·0004789272
2089	4363921	9116230969	45·7055795	12·7833932	·0004786979
2090	4368100	9129329000	45·7165178	12·7854326	·0004784689
2091	4372281	9142439571	45·7274534	12·7874714	·0004782401
2092	4376464	9155562688	45·7383865	12·7895096	·0004780115
2093	4380649	9168698357	45·7493169	12·7915471	·0004777831
2094	4384836	9181846584	45·7602447	12·7935840	·0004775549
2095	4389025	9195007375	45·7711699	12·7956202	·0004773270
2096	4393216	9208180736	45·7820926	12·7976558	·0004770992
2097	4397409	9221366673	45·7930126	12·7996907	·0004768717
2098	4401604	9234565192	45·8039299	12·8017250	·0004766444
2099	4405801	9247776299	45·8148447	12·8037586	·0004764173
2100	4410000	9261000000	45·8257569	12·8057916	·0004761905
2101	4414201	9274236301	45·8366665	12·8078239	·0004759638
2102	4418404	9287485208	45·8475735	12·8098556	·0004757374
2103	4422609	9300746727	45·8584779	12·8118866	·0004755112

No.	Square	Cube	Square Root	Cube Root	Reciprocal
2104	4426816	9314020804	45·8698798	12·8139170	·0004752352
2105	4431025	9327307625	45·8802790	12·8159468	·0004750594
2106	4435236	9340607016	45·8911756	12·8179759	·0004748338
2107	4439449	9353919048	45·9020696	12·8200044	·0004746084
2108	4443664	9367243712	45·9129611	12·8220323	·0004743833
2109	4447881	9380581029	45·9238500	12·8240595	·0004741584
2110	4452100	9393931000	45·9347363	12·8260861	·0004739636
2111	4456321	9407293631	45·9456200	12·8281120	·0004737091
2112	4460544	9420668928	45·9565012	12·8301373	·0004734848
2113	4464769	9434056897	45·9673798	12·8321620	·0004732608
2114	4468996	9447457544	45·9782557	12·8341860	·0004730369
2115	4473225	9460870875	45·9891291	12·8362094	·0004728132
2116	4477456	9474296896	46·0000000	12·8382321	·0004725898
2117	4481689	9487735613	46·0108683	12·8402542	·0004723666
2118	4485924	9501187032	46·0217340	12·8422756	·0004721435
2119	4490161	9514651159	46·0325971	12·8442964	·0004719207
2120	4494400	9528128000	46·0434577	12·8463166	·0004716981
2121	4498641	9541617561	46·0543158	12·8483361	·0004714757
2122	4502884	9555119848	46·0651712	12·8503551	·0004712535
2123	4507129	9568634867	46·0760241	12·8523733	·0004710316
2124	4511376	9582162624	46·0868745	12·8543910	·0004708098
2125	4515625	9595703125	46·0977223	12·8564080	·0004705882
2126	4519876	9609256376	46·1085675	12·8584243	·0004703669
2127	4524129	9622822383	46·1194102	12·8604401	·0004701457
2128	4528384	9636401152	46·1302504	12·8624552	·0004699248
2129	4532641	9649992689	46·1410880	12·8644697	·0004697041
2130	4536900	9663597000	46·1519230	12·8664835	·0004694836
2131	4541161	9677214091	46·1627555	12·8684967	·0004692633
2132	4545424	9690843968	46·1735855	12·8705098	·0004690432
2133	4549689	9704486637	46·1844130	12·8725213	·0004688233
2134	4553956	9718142104	46·1952378	12·8745326	·0004686036
2135	4558225	9731810375	46·2060602	12·8765433	·0004683841
2136	4562496	9745491456	46·2168800	12·8785534	·0004681648
2137	4566769	9759185353	46·2276973	12·8805628	·0004679457
2138	4571044	9772892072	46·2385121	12·8825717	·0004677268
2139	4575321	9786611619	46·2493243	12·8845199	·0004675082
2140	4579600	9800344000	46·2601340	12·8865874	·0004672897
2141	4583881	9814089221	46·2709412	12·8885944	·0004670715
2142	4588164	9827847288	46·2817459	12·8906007	·0004668534
2143	4592449	9841618207	46·2925480	12·8926064	·0004666356
2144	4596736	9855401984	46·3033476	12·8946115	·0004664179
2145	4601025	9869198625	46·3141447	12·8966159	·0004662005
2146	4605316	9883008136	46·3249393	12·8986197	·0004659832
2147	4609609	9896830523	46·3357314	12·9006229	·0004657662
2148	4613904	9910665792	46·3465209	12·9026255	·0004655493
2149	4618201	9924513949	46·3573079	12·9046275	·0004653327
2150	4622500	9938375000	46·3680924	12·9066288	·0004651163
2151	4626801	9952248951	46·3788745	12·9086295	·0004649000
2152	4631104	9966135808	46·3896540	12·9106296	·0004646840

No.	Square	Cube	Square Root	Cube Root	Reciprocal
2153	4635409	9980035577	46·4004310	12·9126291	·0004644682
2154	4639716	9993948264	46·4112055	12·9146279	·0004642526
2155	4644025	10007873875	46·4219775	12·9166262	·0004640371
2156	4648336	10021812416	46·4327471	12·9186288	·0004638219
2157	4652649	10035763893	46·4435141	12·9206208	·0004636069
2158	4656964	10049728312	46·4542786	12·9226172	·0004633920
2159	4661281	10063705679	46·4650406	12·9246129	·0004631774
2160	4665600	10077696000	46·4758002	12·9266081	·0004629630
2161	4669921	10091699281	46·4865572	12·9286027	·0004627487
2162	4674244	10105715528	46·4973118	12·9305966	·0004625347
2163	4678569	10119744747	46·5080638	12·9325899	·0004623209
2164	4682896	10133786944	46·5188134	12·9345827	·0004621072
2165	4687225	10147842125	46·5295605	12·9365747	·0004618938
2166	4691556	10161910296	46·5403051	12·9385662	·0004616805
2167	4695889	10175991463	46·5510472	12·9405570	·0004614675
2168	4700224	10190085632	46·5617869	12·9425472	·0004612546
2169	4704561	10204192809	46·5725241	12·9445369	·0004610420
2170	4708900	10218313000	46·5832588	12·9465259	·0004608295
2171	4713241	10232446211	46·5939910	12·9485143	·0004606172
2172	4717584	10246592448	46·6047208	12·9505021	·0004604052
2173	4721929	10260751717	46·6154481	12·9524893	·0004601933
2174	4726276	10274924024	46·6261729	12·9544759	·0004599816
2175	4730625	10289109375	46·6368953	12·9564618	·0004597701
2176	4734976	10303307776	46·6476152	12·9584472	·0004595588
2177	4739329	10317519233	46·6583326	12·9604319	·0004593477
2178	4743684	10331743752	46·6690476	12·9624161	·0004591368
2179	4748041	10345981339	46·6797601	12·9643996	·0004589261
2180	4752400	10360232000	46·6904701	12·9663826	·0004587156
2181	4756761	10374495741	46·7011777	12·9683649	·0004585053
2182	4761124	10388772568	46·7118829	12·9703466	·0004582951
2183	4765489	10403062487	46·7225855	12·9723277	·0004580852
2184	4769856	10417365504	46·7332858	12·9743082	·0004578755
2185	4774225	10431681625	46·7439836	12·9762881	·0004576659
2186	4778596	10446010856	46·7546789	12·9782674	·0004574565
2187	4782969	10460353208	46·7653718	12·9802461	·0004572474
2188	4787344	10474708672	46·7760623	12·9822242	·0004570384
2189	4791721	10489077269	46·7867503	12·9842017	·0004568296
2190	4796100	10503459000	46·7974358	12·9861786	·0004566210
2191	4800481	10517853871	46·8081189	12·9881549	·0004564126
2192	4804864	10532261888	46·8187996	12·9901306	·0004562044
2193	4809249	10546683057	46·8294779	12·9921057	·0004559964
2194	4813636	10561117384	46·8401537	12·9940802	·0004557885
2195	4818025	10575564875	46·8508271	12·9960540	·0004555809
2196	4822416	10590025536	46·8614981	12·9980273	·0004553734
2197	4826809	10604499373	46·8721666	13·0000000	·0004551661
2198	4831204	10618986392	46·8828327	13·0019721	·0004549591
2199	4835601	10633486599	46·8934963	13·0039436	·0004547522
2200	4840000	10648000000	46·9041576	13·0059145	·0004545455
2201	4844401	10662526601	46·9148164	13·0078848	·0004543389

Table CXXVII.—Fourth Powers of Numbers.

No.	4th Power.	No.	4th Power.	No.	4th Power.	No.	4th Power.
1	1	26	456,976	51	6,765,201	76	33,362,176
2	16	27	531,441	52	7,311,616	77	35,153,041
3	81	28	614,656	53	7,890,481	78	37,015,056
4	256	29	707,281	54	8,503,056	79	38,950,081
5	625	30	810,000	55	9,150,625	80	40,960,000
6	1,296	31	923,521	56	9,834,496	81	43,046,721
7	2,401	32	1,048,576	57	10,556,001	82	45,212,176
8	4,096	33	1,185,921	58	11,316,496	83	47,458,321
9	6,561	34	1,336,336	59	12,117,361	84	49,787,136
10	10,000	35	1,500,625	60	12,960,000	85	52,200,625
11	14,641	36	1,679,616	61	13,845,841	86	54,708,016
12	20,736	37	1,874,161	62	14,776,336	87	57,289,761
13	28,561	38	2,085,136	63	15,752,961	88	59,969,536
14	38,416	39	2,313,441	64	16,777,216	89	62,742,241
15	50,625	40	2,560,000	65	17,850,625	90	65,610,000
16	65,536	41	2,825,761	66	18,974,736	91	68,574,961
17	83,521	42	3,111,696	67	20,151,121	92	71,639,296
18	104,976	43	3,418,801	68	21,381,376	93	74,805,201
19	130,321	44	3,748,096	69	22,667,121	94	78,074,896
20	160,000	45	4,100,625	70	24,010,000	95	81,450,625
21	194,481	46	4,477,456	71	25,411,681	96	84,934,656
22	234,256	47	4,879,681	72	26,873,856	97	88,529,281
23	279,841	48	5,308,416	73	28,398,241	98	92,236,816
24	331,776	49	5,764,801	74	29,986,576	99	96,059,601
25	390,625	50	6,250,000	75	31,640,625	100	100,000,000

APPENDIX A.

1. The rules following are intended to apply to the construction of steel boilers; where boilers are to be made of iron, they will be specially considered by the committee.

2. **Boiler Steel.**—The quality of steel to be used in the construction of boilers must be as follows, viz.:—

Plates intended for the cylindrical shells and butt straps, and bars intended for stays, are to have an ultimate tensile strength of not less than 27 tons, or more than 32 tons per square inch, with at least 20 per cent. extension in 8 inches of prepared section.

Plates which have to be flanged or welded, and bars intended for rivets, are to be from 25 to 30 tons per square inch in tensile strength, with not less than 20 per cent. extension in 8 inches, and plates for furnaces may be in tensile strength from 25 to 29 tons per square inch, with not less than 20 per cent. extension in 8 inches.

3. Sample pieces for tensile and bending tests must be prepared as follows, viz.:—One each for every boiler plate, one each for every ten bars rolled of angle or stay bars, one bending test for every five bars rolled of rivet steel, and tensile tests for rivet steel as follows, viz.:— One to every twenty bars rolled, for bars 1 inch diameter and under; one to every twelve, for bars from 1 inch to $1\frac{1}{2}$ inches diameter, and one to every eight for bars over $1\frac{1}{2}$ inches diameter.

4. The sample pieces for bending are to be heated uniformly to a dull red, then cooled in water of about 82° Fahrenheit, and afterwards bent double without fracture to a curve, the inner radius of which does not exceed one and a-half times the thickness of the piece.

5. When, in the judgment of the surveyor, the stay bars are of so large a diameter as to cause difficulty in bending, the temper-bending test may be carried out on one-half of the broken tensile piece. Steel stay bars which have been worked in the fire must be subsequently annealed.

6. In addition to the tests for rivet steel, specified above, the manufactured rivets are to be tested, hot and cold, by bending, crushing under the steam hammer, or in such other way as may be required, a sufficient number of sample rivets being taken indiscriminately for the purpose. The rivets are not to be used until such testing has been carried out to the satisfaction of the surveyors.

7. All plates that have been welded, or locally heated for furnace work, and plates in which the rivet holes are punched instead of being drilled, must be annealed after this work is carried out. Any plate which, upon being tested, shows a higher tensile strength than the specified limit, and is, therefore, liable to rejection, may, at the option of the manufacturer, be annealed and re-tested. Such annealing should not take place after the plate is sheared, unless both plate and test piece are placed under exactly similar conditions for annealing. In the event of this second test proving unsatisfactory, the plate must be rejected.

8. The stamping of the material, rejection for defects, and the issuing of advice notes must be carried out in the same way as for ship steel.

9. **Strength Calculations.**—The sizes and arrangement of the different parts for a given working pressure ; or the working pressure suitable to a given size and arrangement of material may be found from the following formulæ and rules :—

10. **Cylindrical Shells—**

$$\frac{C \times (T - 1) \times E}{D} = W.$$

Where C = 18·75 when the longitudinal seams are fitted with double butt straps of equal width.

C = 18·125 when the double butt straps are of unequal width —*i.e.*, one strap not covering the outer row of rivets.

C = 17·5 when the longitudinal seams are lap joints.

T = thickness of plate, in sixteenths of an inch.

E = the least percentage of strength of longitudinal joints, found as follows :—

For the plate at the joint $E = \dfrac{p - d}{p} \times 100.$

For the rivets at the joint $E = \dfrac{n \times a}{p \times t} \times 85.$

Where p = pitch of rivets, in inches.

d = diameter of rivet holes, in inches.

n = number of rivets used per pitch.

a = sectional area of rivets, in square inches.

t = thickness of plate, in inches.

Where rivets are in double shear, 1·75a is to be used instead of a.

D = mean diameter of shell, in inches.

W = working pressure in lbs. per square inch.

The inside butt strap to be at least three-quarters the thickness of the plate.

Note.—The constant (C) is to be used with steel of the minimum strength of 27 tons per square inch required by par. 2 ; when a higher minimum strength is guaranteed, the constant may be proⸯtionately increased.

11. Flat Surfaces, supported by Stays—

$$\frac{C \times (T - 1)^2}{S} = W.$$

Where C = 125 for plates fitted with screwed stays having riveted heads.

C = 165 for plates fitted with screwed stays and nuts.

C = 200 for plates fitted with stays and double nuts.

C = 218 for plates with stays, double nuts, and washers outside. The washers to be at least half the thickness of the plate, and in diameter equal to one-third the pitch of stays.

C = 238 for plates fitted with stays, double nuts, and washers outside riveted to the plate. The washers to be at least half the thickness of the plate, and in diameter equal to two-fifths the pitch of the stays.

C = 260 for plates fitted with stays, double nuts, and washers outside riveted to the plate. The washers to be the same thickness as the plate, and in diameter equal to two-thirds the pitch of the stays.

C = 284 for plates fitted with stays, double nuts, and doubling strips outside, riveted to the plates. The strips to be the same thickness as the plate, and in width equal to two-thirds the pitch of the stays.

T = thickness of plate, in sixteenths of an inch.

S = surface in square inches, supported by one stay.

W = working pressure, in lbs. per square inch.

Flat plates, having doublings, at least two-thirds of their thickness, riveted to them,—

$$\frac{C \times (T + \frac{t}{2} - 1)^2}{S} = W.$$

Where t = thickness of doubling plates, in sixteenths of an inch.

C, T, S, and W, as before.

Note.—For front plates in the steam space, which are not protected against the direct action of the flame, the constants given above are to be reduced 20 per cent.

12. Tube Plates—

$$\frac{160 \times (T-1)^2}{P^2} = W.$$

Where T and W are as before.

P = mean pitch of stay tubes from centre to centre, in inches.

When girders are fitted to the tops of combustion chambers, the thickness of the tube plates must be found from the formula—

$$\frac{W \times L \times P}{(P - d) \times 1,600} = T.$$

Where W = working pressure, in lbs. per square inch.
    L = width of combustion chamber over the plates, in inches.
    P = horizontal pitch of tubes, in inches.
    d = inside diameter of plain tube, in inches.
    T = thickness of tube plate, in sixteenths of an inch.

13. Stays supporting Flat Surfaces—

$$\sqrt{\frac{S \times W}{8,000}} + \tfrac{1}{8} = D.$$

Where S = surface, in square inches, supported by the stay.
    W = working pressure, in lbs. per square inch.
    D = Effective diameter of stay, in inches.

14. Circular Furnaces.—Thickness of plain furnaces to resist collapsing—

$$\frac{420 \times (T - 1)^2}{L \times D} = W.$$

Where T = thickness of plate, in sixteenths of an inch.
    L = length of furnace in feet, or length between stiffening rings, if fitted.
    D = outside diameter of furnace, in inches.
    W = Working pressure, in lbs. per square inch.

The thickness, however, must not be less than that given by the following formula—

$$\frac{C \times (T - 2)}{D} = W.$$

Where C = 655 for plain furnaces.
    C = 810 for furnaces fitted with one Adamson ring, pitched not more than 42 inches apart.
    C = 905 for furnaces fitted with two Adamson rings, pitched not more than 28 inches apart.
    C = 1,050 for furnaces fitted with a series of Adamson rings, pitched not more than 23 inches apart.
    C = 1,160 for furnaces fitted with a series of Adamson rings, pitched not more than 20 inches apart; or for corrugated, ribbed, and suspension furnaces (Fox, Purves, and Morison).
    C = 950 for Farnley spiral furnace and Holmes' furnace.

15. Girders for Combustion Chamber Tops—

$$\frac{14,500 \times d^2 \times T}{D \times L^2} \times \frac{(n+1)^2}{n(n+2)} = W \text{ when } n = 2, 4, \text{ or } 6.$$

$$\frac{14,500 \times d^2 \times T}{D \times L^2} = W \text{ when } n = 1, 3, \text{ or } 5.$$

Where d = depth of girder at the centre, in inches.

    T = thickness of girder, in inches.

    D = distance from centre to centre of girders, in inches.

    L = length from tube plate to tube plate, or from tube plate to back of combustion chamber, in inches.

    n = number of stays fitted with each girder.

GENERAL CONSTRUCTION.

1. Each boiler must have at least one glass water gauge, three test cocks, and one steam pressure gauge. Double-ended boilers are to have these fittings at each end.

2. A stop-valve is to be fitted to each boiler, so that any one of a series of boilers may be worked independently if required. The neck of the stop-valve to be as short as possible.

3. Two safety-valves will be required for each main boiler, and they must be tested under steam to the satisfaction of the surveyors, and set to a pressure not more than 3 per cent. in excess of the intended working pressure. The combined area of the valves is to be sufficient to prevent the steam accumulating to more than 10 per cent. of the working pressure during fifteen minutes full firing. If the boilers be supplied with forced draught, the valve area must be increased so that the same conditions may be met.

4. Easing gear is to be provided, and so arranged that the safety-valves on any one boiler may be lifted simultaneously, without interfering with those on any other boiler.

5. Surface and bottom blow-off cocks should be fitted to the boiler, in addition to the cock on the hull of the vessel, and all cocks must have spigots extending through the hull plating, with a plate flange round same on the outside.

6. Upon completion, the boilers are to be tested by hydraulic pressure to twice the intended working pressure, and after being placed in position in the vessel, they must be efficiently secured by brackets and stays to prevent any fore and aft, or athwartship movement.

7. Donkey boilers need not have more than one safety-valve, provided the valve area be not less than $\frac{1}{2}$ square inch for each foot of grate surface. In other respects the requirements for donkey boilers are the same as for main boilers.

APPENDIX B.

BUREAU VERITAS RULES FOR BOILER SHELLS, WORKING PRESSURE, OR THICKNESS OF PLATES, AND FOR SIZES OF STAYS.*

Circular Shells and Steam-holders with Internal Pressure.

A riveted joint may fail through the tearing of the plate or butt-strap between the rivets, the shearing of all the rivets, or by a combination of the two. The following formulæ apply to these several cases. The plate thickness and the diameter of rivets to be applied to have the highest values which each formula would give separately.

I. *Rupture through Plate.*

The formulæ for working pressure and plate thickness are in this case :—

$$P = \frac{2 \, a \, R \, (t - |0 \cdot 04)}{D}$$

and

$$t = \frac{P \, D}{2 \, a \, R} + 0 \cdot 04 \text{ inch} \qquad \cdots \cdots \quad (I.)$$

Where P = allowed working pressure, above atmosphere, in pounds per square inch.

D = greatest inside diameter of boiler shell, or steam-holder in inches.

t = thickness of shell plates in inches. $t - 0 \cdot 04$ inch represents the thickness left after a reduction of $0 \cdot 04$ inch through corrosion.

R = the tensile stress, in pounds per square inch, which will be allowed in the plate. The value of R will be the breaking strength divided by $4 \cdot 4$, the latter figure representing the factor of safety for the plate after it has been corroded away by $0 \cdot 04$ inch.

If the actual breaking strength happens to be known by tests carried out to the Administration's satisfaction, it may be applied for finding R ; but when, as usual, it is not known, the value of R will be :—

For Steel : the $4 \cdot 4$th part of the lower limit of tensile strength

---

* *N.B.*—If a boiler is intended for a vessel belonging to a country where the law prescribes heavier scantlings than those required by the following paragraphs, the builders have, of course, to comply with the legal requirements.

ohosen by the designer, which in such case is to be stated when a boiler design is submitted for approval.

For Iron: 10,700 lbs. per square inch, corresponding with a tensile strength of 21 tons. A table annexed shows the values of 2 R for various tensile strengths.

a = ratio of the resistance of the plate left between the holes, to that of the full plates. It will be determined from the following expression :—

$$a = \frac{p - d}{p}.$$

Where p = pitch of rivets in outer row, in inches.

d = diameter of rivet holes, in inches, either the real diameter or a corrected one, according to the following clauses :—

1. When the rivet holes are drilled, or when, having been punched, they are afterwards drilled or rimered out so that the injured metal around is completely removed, the real diameter may be taken.

2. When the holes are simply punched they will be considered as being $\frac{1}{4}$ inch larger in diameter than as punched.

TABLE SHOWING THE VALUES OF 2 R.

In Formulæ (I.) and (IV.) for various Tensile Strengths of the Material.

Tensile Strength of Plates in tons per sq. inch.	Value of 2 R.	Tensile Strength of Plates in tons per sq. inch.	Value of 2 R.
29	29,500	24	24,400
28	28,500	$23\frac{1}{2}$	23,900
27	27,500	23	23,400
$26\frac{1}{2}$	27,000	$22\frac{1}{2}$	22,900
26	26,500	22	22,400
$25\frac{1}{2}$	26,000	$21\frac{1}{2}$	21,900
25	25,500	21	21,400
$24\frac{1}{2}$	25,000	$20\frac{1}{2}$	20,900

II. *Rupture through Rivets.*

In this case the following are the formulæ for finding the allowed working pressure, or required rivet section :—

$$P = \frac{2 A S}{D l}$$

and

$$A = \frac{P D l}{2 S} \qquad \cdots \qquad \text{(II.)}$$

Where P and D have the same meaning as before, and

l = the length, in inches, of the identical parts into which a riveted joint can be subdivided. In most cases l is the pitch of the rivets in the outer rows (figs. 1 and 2).

In general it depends upon the system of joint adopted.

S = the maximum shearing stress, in pounds per square inch, which will be allowed on the rivets. It will be the 4·4th part of the actual shearing resistance of the material, if known, from tests. If the actual shearing resistance of the rivet bars is not known, it will be assumed to amount to 0·8 of their tensile strength, and the value of s will be :—

For Steel.—5·5th part of the lower tensile limit adopted by the designer.

For Iron.—8,600 lbs. per square inch, corresponding with a tensile strength of about 21 tons per square inch.

A = the total shearing surface, in square inches, of the rivets (that is, twice the area of the rivet hole when a rivet is in double shear), with or without corrections, according to the following rules :—

1. The total area of the rivet hole, without any reduction, may be brought into account when the rivet holes are drilled in place after the plates are bent, and the longitudinal and circumferential seams at least double riveted, the former by machine.

This clause also applies to the case where the holes having been punched, are afterwards drilled out in place, so as to correspond perfectly with each other.

2. $\frac{1}{1}$ of the full area are to be taken when the joint is made, as in the preceding case, with the exception that the riveting is done by hand.

3. $\frac{3}{4}$ of the full area are to be applied when the rivet holes are punched after the plates are bent, and the longitudinal seams at least double riveted.

III.—*Combined Rupture through Plate and Rivets.*

This case is only to be examined when the outer row has a wider pitch than the inner ones.

The formula to be applied in this case is :—

$$P = \frac{2(B \times R + C \times S)}{D \times l} \quad \ldots \quad \text{(III.)}$$

Where P R S D and l have the same meaning as before.

B = the sectional area, in square inches, of the plate on the portion l of the joint along the line of its supposed rupture, assuming that in case the plate is liable to corrosion, its thickness has been reduced by 0·04 inch.

The area of the plate is to be corrected as per Case I., No. 2, when the rivet holes are simply punched.

C = the total area of the rivets which are supposed to shear on the length l, corrected, if required, in the same way as prescribed in Case II., Nos. 2 and 3.

For a rivet in double shear the resistance will be considered as being twice that of one in single shear.

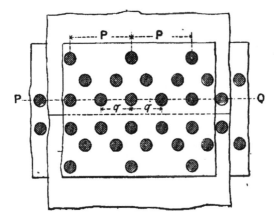

FIG. 1.

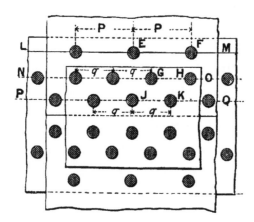

FIG. 2.

IV.—*Rupture through Butt Straps.*

Rupture may take place along one of the inner rows of rivets (see P Q, figs. 1 and 2). The formulæ for this case, based on the same principle as (I.), are :—

$$P = \frac{2\,a\,R\,(t - 0\cdot 04)}{D}$$

and

$$t = \frac{P \times D}{2\,a\,R} + 0\cdot 04 \text{ inch}$$

. . . . . (IV.)

Where P D and R have the same meaning as before.

$t =$ thickness, in inches, of butt strap, or sum of thicknesses, if there are two straps. (The thickness, of course, not to be less than required for caulking).

$$a = \frac{q - d}{q}.$$

Where $q =$ pitch of rivets in the inner row, in inches.

$d =$ diameter in inches of the rivet holes in the inner row; increased by $\frac{1}{4}$ of an inch, if the holes are simply punched.

V.—*Combined Rupture through Butt Straps and Rivets.*

Formula (1II.) applies to this case, B being the section of the butt strap, or straps, along which rupture would take place.

Corrections as for Case III., Nos. 2 and 3.

Remarks.—In zig-zag riveting, the distance between the rows is to be such that no rupture through plate or butt strap is to be feared along the zig-zag line.

When stays are bolted through the shell, they should be so arranged that they do not weaken the shell plates more than the riveted joints. If the resistance at the stay bolts is the smaller of the two, the plate's thickness shall be determined by it. It will be found from a formula the same as (I.), p and d applying to the stay bolts.

For circumferential seams, double riveting will be required, if the thickness of the plates exceeds $\frac{1}{2}$ inch.

In double-ended boilers with six furnaces, treble riveting will be required for the circumferential seams connecting the courses with each other ; it is not required for the end seams.

In double-ended boilers with four furnaces the same arrangement is recommended.

Example showing the application of the above rules to the joint in ꞈ. 2.

I. The value, $p-d$, is the clear distance between the rivet holes, E and F. If the holes are punched,

$$p - (\text{dia. hole } E + \tfrac{1}{4} \text{ inch}),$$

is to be considered as the equivalent length.

II. The length, l, is equal to p. The rivets, which would be sheared on this length, would be :—

> In single shear : one-half of rivet E, one-half of rivet F—total, one rivet E.

> In double shear : one-half of rivets H and J, the whole of G and K—total, three rivets ; each counting double. Consequently,

> A = area of one hole E + 6 × area of one hole G. A is to be multiplied by $\tfrac{11}{8}$ or $\tfrac{7}{8}$, in accordance with Case II., 2 and 3, if the conditions mentioned therein take place.

III. Rupture of plate along the line N O, and shearing of the rivets in the outer row.

No corrosion is likely to take place in the interior of the joint.

Consequently, in formula (III.)—

> B = $(p - 1\tfrac{1}{2} \times$ dia. of rivet hole G) × thickness of plate ; the dia. of the rivet hole G, increased by $\tfrac{1}{4}$ inch, if the holes are punched.

>> The shearing surface of the rivets is one-half of E and one-half of F, therefore,

> C = area of one rivet hole, E, to be multiplied by $\tfrac{11}{8}$ or $\tfrac{7}{8}$, if required, according to clauses Nos. 2 and 3 (Case II.).

IV. Rupture of butt straps through P Q.

The area which resists rupture is, on the length, q : $(q -$ dia. of hole, J) × sum of thickness of straps, the diameter of J to be increased by $\tfrac{1}{4}$ inch, if the holes are simply punched.

V. Rupture of butt strap through L M.

In this case the rivets in the rows, N O and P Q, must shear (double shear), and we have in formula (III.) :—

> B = $(p -$ dia. of hole, E) × thickness of wide strap.
>> The dia., E, being increased by $\tfrac{1}{4}$ inch, if the holes are simply punched.
> C = 3 × area of hole G + 3 × area of hole K.
>> Which value has to be multiplied by $\tfrac{7}{8}$ or $\tfrac{11}{8}$, if the Case II., Nos. 2 and 3, occur.

Shells of Superheaters.—The same mode of determining the working pressure or the thickness of shell plates will be followed for cylindrical superheaters, and the same formulæ may be used, but with the following alterations :—

1. When the plates are exposed to the direct action of the products

of combustion, the values of R and S, as given for Cases I. and II. will have to be multiplied by 0·8 and the addition to the thickness of plates, on account of corrosion, will be increased from 0·04 to $\frac{3}{16}$ of an inch, to compensate for the corrosive action of the gases.

Formulæ (I.) become, therefore :—

$$P = \frac{1·6 \ a \ R \ (t - \frac{3}{16})}{D},$$

and

$$t = \frac{P \ D}{1·6 \ a \ R} + \frac{3}{16} \text{ inch.}$$

2. When the plates are protected from the direct action of the products of combustion, R and S will be multiplied by 0·9, and the additional thickness for burning away will be $\frac{1}{8}$ inch.

Formulæ (I.) become in this case :—

$$P = \frac{1·8 \ a \ R \ (t - \frac{1}{8})}{D},$$

and

$$t = \frac{P \ D}{1·8 \ a \ R} + \frac{1}{8} \text{ inch.}$$

FLAT PLATES.

The allowed working pressure or the thickness of flat plates is to be determined by the following formulæ :—

$$P = \frac{(t-1)^2}{a^2 + b^2} \times \frac{T}{C},$$

and

$$t = 1 + \sqrt{(a^2 + b^2) \frac{P \ C}{T}}.$$

Where P = allowed working pressure above atmosphere, in lbs. per square inch.

t = thickness of plate in sixteenths of an inch.

a = pitch of stays, in inches, in one row.

b = distance, in inches, between two rows of stays.

In case of irregular staying, such as in the annexed sketch (fig. 3),

$$\frac{1}{2} (P_1 + P_2),$$

shall be taken instead of $\sqrt{a^2 + b^2}$.

FIG. 3.

T = tensile strength of the material in tons per sq. inch of the original section.

It is to be determined in the same way as for shell plates, that is :—

For steel it will be equal to the lower limit of the tensile stress, which is to be stated on the drawing.

For iron, 21 tons per square inch.

$C =$ a constant, the value of which depends upon the mode of staying, as follows :—

$C = 0.084$ when the stays are screwed into the plates and riveted over.

$C = 0.065$ when the stays are screwed into the plates and fitted with an outside nut.

$C = 0.062$ when the stays are fitted with inside and outside nuts and washers, provided the diameter of the outside washers be at least 0.4 of the pitch between the rows of stays. The thickness of the washer to be at least $\frac{3}{8}$ that of the plate, and to be increased if the diameter of the washer exceeds $1\frac{1}{2}$ times the diagonal width of the nut.

$C = 0.055$ when the stays are fitted with inside and outside nuts and washers, the outside washer being riveted to the plate and having $\frac{3}{4}$ of the plate's thickness and a diameter equal to 0.6 of the pitch between the rows of stays.

For the values of $\frac{T}{C}$, see Table below.

VALUES OF $\frac{T}{C}$ IN THE FORMULA FOR FLAT PLATES.

Tensile Strength of Plate in tons per square inch.	C=0·084	C=0·065	C=0·062	C=0·055
20	$\frac{T}{C} = 238.0$	$\frac{T}{C} = 307.6$	$\frac{T}{C} = 322.6$	$\frac{T}{C} = 363.6$
21	250·0	323·0	338·6	381·8
22	262·0	338·4	354·8	400·0
23	273·8	353·8	371·0	418·0
24	285·8	369·2	387·0	436·2
25	297·6	384·6	403·2	454·4
26	309·6	400·0	419·4	472·6
27	321·4	415·3	435·4	490·8
28	335·4	430·7	451·6	509·0
29	345·2	446·1	467·8	527·2
30	357·2	461·5	483·8	545·4

When the plates are in contact with steam on one side and flame or hot gases on the other, the thickness is to be increased.

For instance, when in return tube boilers, the top front plates are in no way protected from the hot gases, the working pressure or thickness will, in such a case, be determined by the formulæ :—

$$P = \frac{(t-2)^2}{a^2+b^2} \times \frac{0\cdot9\ T}{C}$$

and
$$= 2 + \sqrt{(a^2+b^2)\frac{P\,C}{0\cdot9\ T}}.$$

When the said front plates are protected by a flame plate, no increase of thickness will be required.

STAYS.

The diameter of stays supporting flat surfaces is to be determined by the following formula :—

$$d = \tfrac{1}{8}\ \text{inch} + \sqrt{\frac{Q}{300\ T}}.$$

Where d = effective diameter in inches (for instance, the diameter at bottom of thread in screw stays).
Q = total load on stay in lbs.
T = tensile strength of the material, in tons per square inch.

For steel this tensile strength will be the lower limit chosen by the boiler designer ; for iron it will be taken at 22 tons. In both cases the actual strength may be applied if it is known from tests.

If the stays are not round, their cross section must be such that the stress per square inch, caused by the load Q, nowhere exceeds one 5·75th part of the tensile strength, after deducting $\frac{1}{16}$ of an inch all round as an allowance for corrosion or wear.

In welded stays the stresses just described will be reduced by 20 per cent. Welding of steel stays is only allowed for very mild qualities.

For high working pressures, such as used in triple-expansion engines, it is recommended to screw all stays into the plates they support, in addition to fitting them with nuts.

This also applies to stay tubes, with the exception that it is recommended that nuts should not be fitted in combustion chambers.

CIRCULAR FURNACES.

Plain Cylindrical Furnaces.

The working pressure and the thickness of the plates may be calculated from the following formulæ :—

$$P = \frac{C\,t^2}{D\,L},$$

and
$$t = \sqrt{\frac{D\,P\,L}{C}}.$$

Where t = the required thickness of plates, in inches.

D = outside diameter of furnace, in inches.

P = working pressure in lbs. per square inch (above atmosphere).

L = length of furnace in feet ; or if made or fitted with efficient rings, the length between the rings.

C = 70,000 when the furnace is truly circular and the longitudinal seams are welded, butted, or lapped and bevelled, and double riveted in the last-named case.

C = 60,000 when the furnace is not truly circular, or when the longitudinal seams are simply lapped.

The above constants apply to iron plates of fair quality.

They may be multiplied by 1·2 when the plates are of mild steel, or of iron of the best quality with regard to behaviour under the action of the fire, having a uniform structure without blisters or other defects, and fulfilling the following conditions, to be ascertained by testing samples :—

	With the Grain.	Across the Grain.
Tensile strength, . . .	23 tons per sq. in.	21½ tons per sq. in.
Elongation in 8 inches, .	16 per cent.	10 per cent.
Cold bending without fracture to an angle of . .	60°	35°
Hot ,, ,, ,,	180°	180°

The thickness should in no case be less than :—

$$t = \frac{P\,D}{8,000} \text{ for iron plates.}$$

$$t = \frac{P\,D}{9,000} \text{ for steel plates.}$$

Corrugated and Ribbed Furnaces.

The plate thickness is to be found by the following formulæ :—

1. For corrugated furnaces :—

$$T = \frac{P\,D}{1,000} + 2.$$

Where P = working pressure in lbs. per square inch.

T = thickness of plate in sixteenths of an inch.

D = outside diameter in inches measured on the top of the corrugations.

The formula applies to corrugations 6 inches long and 1½ inch deep.

2. For ribbed furnaces, when manufactured to the satisfaction of the Administration :—

$$T = \frac{P\,D}{1,160} + 2.$$

28

Where P and T are as above.

D = outside diameter of plain parts between the ribs, in inches. The formula applies to ribs spaced 9 inches and projecting 1⅜-inch, the difference between the greatest and the smallest diameters in any part of the furnace not exceeding $\frac{2}{1000}$.

GENERAL AS TO BOILER ARRANGEMENTS AND FITTINGS.

When two or more boilers are fitted, they should be arranged to work separately or independently of each other, either by stop-valves between the boilers and the common superheater, or by stop-valves between the separate superheaters and the main steam pipes.

Each boiler must be fitted with a separate pressure gauge. Double-ended boilers to have a pressure gauge at each end.

Steam pipes of donkey engines must be independent of the main pipes, so as to keep the steam off the main engines when only winches or other auxiliary engines are working.

The bottom blow-off to be arranged with one cock directly attached to the shell of the boiler, and another directly attached to the skin of the vessel. The surface blow-off must be similarly arranged. The main stop-valve to be so situated and fitted that it can be worked from the starting platform or the stoke-hole floor.

To protect the plating, the cocks on the ship's bottom should be fitted with spigots passing through the plating and through a flange on the outside. If this flange is of iron it must be galvanised.

Steam domes or superheaters, when placed in the uptake and exposed to the direct impact of the flame, will only be allowed as an exception, and must be efficiently protected by flame plates. In all cases it must be possible to examine efficiently the interior and the exterior of the domes or superheaters.

To prevent the boilers shifting in a transverse direction through the rolling of the vessel, or longitudinally in case of collision, they must be properly secured in their seats.

All manholes to be fitted with compensating rings.

At least two safety-valves of an approved design must be fitted to each main boiler.

Their total area must be such that with at least twenty minutes' hand-firing the pressure does not rise more than 10 per cent. of the effective pressure for which the boiler has been approved.

When forced draught is provided for, the area is to be increased in proportion to the increased evaporative power of the boilers.

Suitable arrangements and gear to be fitted in connection with the safety-valves, whereby they may be lifted from the deck as well as from the stoke-hole floor.

If it be practicable to isolate a superheater communicating simultaneously with two or more boilers, it will be necessary to furnish it with a safety-valve of suitable dimensions.

APPENDIX C.

Regulations respecting Steam Boilers.

1. *Plating of Boilers.*—When the smallest dimension of a cylindrical boiler exceeds 25 centimetres, or of a spherical-shaped boiler 30 centimetres, the portions of the boiler exposed to the flame, furnaces, and tubes shall *not* be made of cast iron.

Brass (copper) plates are only to be used for furnaces where their smallest dimension does not exceed 10 centimetres.

2. *Furnaces, Flues, and Combustion Chambers.*—Flues running round or through steam boilers must, at their highest point, be at least 10 centimetres below the minimum water level. This minimum level must, in lake and river vessels, be maintained when the vessel is inclined at an angle of 4°, and for sea-going ships at an angle of 8°. In boilers with a breadth of 1 to 2 metres, the distance to the water level must be at least 15 centimetres, and in boilers of greater breadth at least 25 centimetres.

These conditions do not apply to boilers in which the tubes are less than 10 centimetres diameter, nor to those wherein the portions of the plates in contact with steam are not liable to become red hot. The risk of plates becoming red hot may be assumed to be done away with if the heating surface exposed to the flame, before the flame reaches the point in question is, with natural draught 20 times, and with forced draught 40 times, greater than the area of the grate.

3. *Feed Valves.*—Each boiler must have a feed valve which will close by boiler pressure when the feed is off.

4. Each boiler must be fitted with its independent feed apparatus, each worked separately, and each to be capable, by itself, of keeping the boiler supplied. Several boilers connected together and used for the same purpose may, in this respect, be regarded as one boiler.

5. *Water Gauges.*—Each boiler must be fitted with a gauge glass, and also with a second means of ascertaining the height of the water. Each of these fittings must have a separate connection to the boiler, or if they have a common connection, the stand pipe must be at least 60 square centimetres in area, or 88 millimetres in diameter.

6. Water test cocks to be fitted ; the lowest one must be placed at the level of the regulation water line. All gauge cocks must

so fitted as to have a straight way through for cleaning them from scale or salt.

7. *Water Level.*—The regulation lowest water level is to be distinctly marked on the gauge glass, and also in a prominent place on the boiler shell. The level of the flue, at its highest point, is also to be plainly and permanently marked on the shell of the boiler in way of the ship's beam.

Two water-gauge glasses are to be fitted on the boiler at the standard level of the water when the vessel is inclined as above, one on each side of the boiler, as far apart as possible, and symmetrically placed right and left of centre line of boiler.

When these two gauge glasses are fitted, the additional means named in Rule 5, for ascertaining the water level, may be dispensed with.

8. *Safety-Valves.*—Each boiler must be fitted with, at the least, one reliable safety-valve.

When several boilers have a common steam chest, from which they cannot be separately disconnected, two safety-valves will be sufficient.

Boilers of steamships, locomotives, and portable engines must have at least two safety-valves.

In steamships, except those that are sea-going, one valve is to be placed in such a position that the load on it can easily be ascertained from the deck.

The valves must be so arranged that they can be readily lifted at any time. They are to be loaded so that they will blow off immediately the working pressure is reached.

9. *Pressure Gauge.*—Each boiler must have a reliable pressure gauge, on which the highest regulation working pressure is to be plainly marked.

Boilers of steamships must have two pressure gauges—one clearly visible to the firemen, the other, except when the vessel is sea-going, in a convenient position for being observed from the deck. If a vessel has several boilers connected to a common steam chest, one pressure gauge on deck will be sufficient, in addition to the one on each boiler.

10. *Marking on Boiler.*—(1) Each boiler must have the highest working pressure, the maker's name, the ship number of the boiler, and the date of completion plainly and permanently stamped on it. Steamships, moreover, must also have the standard lowest water level marked on.

(2) Boilers that have already been constructed, at the date of these regulations, are not required to be altered to comply with them.

(3) The regulations for boilers of steamships apply in all cases in which boilers are permanently fixed in, or connected with the vessel.

10a. These particulars must be stamped on a metal plate, which is to be fixed on the boiler by copper rivets, and in such a position that it is easily visible when there are casings on the boiler.

11. *Testing of Boilers.*—Every new boiler must, on completion, be ested by water pressure.

The test pressure for boilers intended for a working pressure of not more than 5 atmospheres to be double the working pressure, and for boilers intended for a working pressure of more than 5 atmospheres, the test pressure is to be 5 atmospheres above the working pressure.

For boilers where the working pressure is under atmospheric pressure, the test pressure is to be 1 kilogramme per square centimetre. The plates of the boiler must withstand the test pressure without showing any permanent change of form, and must also remain tight under the pressure.

The boiler is to be regarded as leaky, if the water at the highest pressure leaks through in other forms than dewy moisture or pearly drops.

11a. When the boilers have been tested and found satisfactory, the copper rivets which fix the brass plate are to be officially stamped, and an impression of this mark is to be inserted on the certificate of test.

12. When boilers have undergone a thorough repair in the boiler shop, or when they have had to be stripped and extensively repaired on board, they must be tested by hydraulic pressure the same as new boilers.

When in boilers with internal furnace, the furnace has been taken out to repair or renew, and when in locomotives the fire-box has been taken out to repair or renew, or when in cylindrical boilers one or more plates have been renewed, the boilers must, on completion of such repairs, be tested by hydraulic pressure as above.

13. *Test Gauge.*—The test pressure is to be registered by an open mercurial gauge, or by the standard pressure gauge provided by the inspecting official.

Each boiler to be tested must be fitted with a connection to suit the standard gauge, and by which the gauge may be applied by the inspecting officer.

Rules for Strength of Boilers.

HAMBURG, *20th Nov.* 1888.

1 kilo. per sq. cm. = 1 atm. = 14·223 lbs., English.

TABLE FOR CALCULATING THICKNESS OF PLATES IN SHIP BOILERS.

Section I.—Shell Plates.

The strength of cylindrical plates is to be calculated according to the following formula :—

$$p = \frac{2\,S\,a\,\delta}{D\,\chi},$$

and therefore

$$\delta = \frac{D\,p\,\chi}{2\,S\,a}.$$

Where p = highest steam pressure in kilos. per sq. cm.

δ = thickness of plates in millimetres.

D = greatest inside dia. of boiler in millimetres.

χ = coefficient of safety = 5, but can be reduced to 4·7 when the longitudinal seams are double riveted, and the rivet holes drilled.

a = coefficient of strength of riveting according to following formula :—

a = for strength of plates between holes $\dfrac{b-c}{b}$.

a = for strength of rivets $\dfrac{q}{b}\dfrac{Z}{\delta}\chi$.

In this formula :—

b = spacing of rivets.

c = dia. of rivet holes.

q = area ,,

δ = thickness of plates.

Z = number of rows of rivets with equal pitch.

χ = 1 for single riveted lap joints.

χ = 1·75 for double riveted lap joints.

The least value of a is to be used in formula L.

In case iron rivets are used for steel plates, the calculation for coefficient a, as also the thickness, must be reckoned for iron plates in the formula.

For welded iron plates, a may be taken as 70 per cent.

S = absolute strength of the material. For wrought iron, 33 kilos. per sq. mm.; for steel, 42 kilos. per sq. mm.

When experience shows that the steel shell plates employed have a greater strength than 42 kilos. per sq. mm., the nett strength may be used in the calculation, nevertheless the steel must not stand more than 48 kilos. per sq. mm., and must have an extension of at least 20 per cent. in 200 mm.

For superheaters S = 28 for iron.

,, ,, S = 35 ,, steel.

Section 2.—Cylindrical Furnaces.

The thickness of plates for cylindrical furnaces must be calculated according to the following formula :—

$$d = R\sqrt{p \times l \times D} + 1 \text{ mm.}$$

p = steam pressure in kilos. per sq. centimetre.

d = thickness of plate in millimetres.

D = outside dia. of furnace in centimetres.

l = length of furnace in metres between stiffeners.

R = 0·33 when the furnace is perfectly circular, the longitudinal seam welded, or with double butt straps and double riveted.

R = 0·35 when the furnace is not quite circular, or when the longitudinal seam is only lapped.

Corrugated Furnaces or Brown's Furnaces.

$$d = \frac{p \times D}{100} + 1 \text{ mm.}$$

In this formula :—

d = thickness of plate in millimetres.

D = mean dia. of furnace in centimetres.

p = steam pressure in kilos. per sq. centimetre.

For furnaces of sea-going ships in both cases, add 2 mm. instead of 1 mm.

Section 3.—*Screw and Through Stays.*

The strength of screw and through stays must not be taken higher than :—

500 kilos. per sq. centimetre for iron.
600 ,, ,, ,, steel.

Steel stays must not be welded.

In welded iron stays the section must be increased 20 per cent.

Section 4.—*Flat Surfaces.*

The thicknesses of stayed flat surfaces are to be calculated according to formula :—

$$d = 1 \cdot 5 + e \sqrt{\frac{p}{R}}.$$

In this formula :—

d = thickness of plate in millimetres.

e = greatest spacing of stays in centimetres.

p = steam pressure in kilos. per sq. centimetre.

R = coefficient, values of which are as follows :—

R = 22 for stays screwed into the plates and riveted.

R = 28 for stays screwed into the plates and fitted with nuts.

R = 30·6 for stays provided with washers inside and outside, the outside washer being three times the dia. of the stay, and ·511 times the thickness of the plate.

R = 34·2 for stays provided with nuts and washers inside and outside, the outside washer being riveted to the plate, the thickness being ·66, the thickness of the plate and the dia. being ·6 of the spacing of the stays.

The thickness of plate can be reduced 12½ per cent., if of steel.

Section 5.—Dog Stays on tops of Combustion Chambers.

The strength of these stays is calculated by the following formula :—

$$d = \frac{p\,(a-e)\,l\,a}{R\,h^2}.$$

$d =$ thickness of dog in centimetres.
$p =$ pressure in kilos. per sq. centimetre.
$a =$ length of dog in centimetres.
$e =$ pitch of stays in ,,
$l =$ pitch of dogs in ,,
$h =$ depth ,, ,,
$R = 420$ for one stay bolt.
$R = 630$ for two or three stay bolts.
$R = 720$ for four stay bolts.

INDEX.

2

31

NEILL AND COMPANY, PRINTERS, EDINBURGH.